Wildlife Stewardship on Tribal Lands

Wildlife Stewardship on Tribal Lands

Our Place Is in Our Soul

Edited by **SERRA J. HOAGLAND** *and* **STEVEN ALBERT**

JOHNS HOPKINS UNIVERSITY PRESS
Baltimore

© 2023 Serra J. Hoagland and Steven Albert
All rights reserved. Published 2023
Printed in Canada on acid-free paper
2 4 6 8 9 7 5 3 1

Johns Hopkins University Press
2715 North Charles Street
Baltimore, Maryland 21218
www.press.jhu.edu

Library of Congress Cataloging-in-Publication Data
Names: Hoagland, Serra J., 1985– editor. | Albert, Steven K., 1960– editor.
Title: Wildlife stewardship on tribal lands : our place is in our soul / edited by Serra J. Hoagland and Steven Albert.
Description: Baltimore, Maryland : Johns Hopkins University Press, 2023. | Includes bibliographical references and index.
Identifiers: LCCN 2022030940 | ISBN 9781421446578 (hardcover ; acid-free paper) | ISBN 9781421446585 (ebook)
Subjects: LCSH: Native American Fish & Wildlife Society. | Wildlife conservation—North America. | Indian reservations—North America. | Indians of North America—Fishing. | Indians of North America—Hunting.
Classification: LCC QL84 .W55 2023 | DDC 333.95/40899707—dc23/eng/20220721
LC record available at https://lccn.loc.gov/2022030940

A catalog record for this book is available from the British Library.

Plate illustrations are used with the permission of the artists: **p. 2**, Kraig Holmes, *Tutuvosift (Animals)*, 2019; **p. 50**, Kraig Holmes, *Tutskwa (The Land)*, 2019; **p. 134**, T. Kyle Secakuku, *Communication*, 2020; **p. 166**, Kraig Holmes, *Tsiro (Birds)*, 2019; **p. 228**, Frank Cerno Jr., *Butterfly and Flower*, 2021; **p. 288**, Frank Cerno Jr., *Mallard*, 2021; **p. 306**, Kraig Holmes, *Tusaqa (Plants)*, 2019; **p. 352**, Jessica Wiarda, *Birds of the Four Directions*, 2018.

Pictographs are used with the permission of the artist, Kimberly Blaeser: **p. 164**, *Dream of Birch-Winged Eagles*, 2019; **p. 266**, *Sandhill Sky*, 2021.

Special discounts are available for bulk purchases of this book.
For more information, please contact Special Sales at specialsales@jh.edu.

Contents

- ix   Contributors
- xxv   Preface
- xxviii   Map

Part I. Background and Policy Issues   3

5   Introduction
Steven Albert and Serra J. Hoagland (Pueblo of Laguna)

12   Chapter 1. Diversity and Complexity of Tribal Fish and Wildlife Programs
Julie Thorstenson (Lakota–Cheyenne River Sioux Tribe)

22   Chapter 2. A Vision of Unity and Equity: Conversations with the Founders of the Native American Fish and Wildlife Society and a Look toward the Future with Native American Youth
Nathan Jim (Confederated Tribes of Warm Springs), John Antonio (Pueblo of Laguna), Gerald Cobell (Blackfeet Nation), Doug Dompier, Ron Skates (Cheyenne River Sioux Tribe), Arthur Blazer (Mescalero Apache Tribe), Ken Poynter (Passamaquoddy Tribe), Sally Carufel-Williams (Flandreau Santee Sioux Tribe), Tamra Jones (Native Village of Tazlina), Mathis Quintana (Jicarilla Apache Nation), Samuel Chischilly (Navajo Nation), Jovon Jojola (Pueblo of Isleta), Ashley Carlisle (Navajo Nation), and Julie Thorstenson (Lakota–Cheyenne River Sioux Tribe)

39   Chapter 3. Connecting People, Science, and Culture
An Interview with Scott Aikin (Prairie Band of the Potawatomi Nation)

45   Chapter 4. Who Stands for the River?
Winona LaDuke (Mississippi Band Anishinaabeg)

Part II. Legal Issues   51

53   Chapter 5. The Importance of Meaningful Federal-Tribal Consultation in Land and Natural Resource Management
Michael C. Blumm and Lizzy Pennock

73   Chapter 6. An Introduction to Indian Reserved Water Rights
Vanessa L. Ray-Hodge (Pueblo of Acoma) and Dylan R. Hedden-Nicely (Cherokee Nation)

99   **Chapter 7. The Promise of Intertribal Wildlife Management**
Bethany R. Berger

120 **Chapter 8. State Regulation and Enforcing Usufructuary Treaty Rights**
Guy C. Charlton

135 **Chapter 9. Tribal Perspectives on the Endangered Species Act**
Steven Albert and Riley Plumer (Red Lake Band of Chippewa Indians)

145 **Chapter 10. Sinixt Hunting: A Test of Tribal Sovereignty**
E. Richard Hart and Cody Desautel (Confederated Tribes of the Colville Reservation)

165 **Chapter 11. "we always knew," "wetlands"**
Poems by nila northSun (Shoshone/Chippewa)

Part III. Resource Use, Protection, and Management   167

169 **Chapter 12. The Indigenous Sentinels Network: Community-Based Monitoring to Enhance Food Security**
Lauren M. Divine, Bruce W. Robson, Christopher C. Tran, Paul I. Melovidov (Aleut Community of St. Paul Island), and Aaron P. Lestenkof (Aleut Community of St. Paul Island)

181 **Chapter 13. The Indigenous Guardians Network for Southeast Alaska**
Michael I. Goldstein, Aaron J. Poe, Raymond E. Paddock III (Central Council of Tlingit and Haida Indian Tribes of Alaska), and Bob Christensen

190 **Chapter 14. Glyph**
A Poem by Kimberly Blaeser (Anishinaabe)

193 **Chapter 15. Case Studies of Species Recovery and Management of Trumpeter Swan and Leopard Frog on the Flathead Indian Reservation**
Kari L. Eneas, Arthur M. Soukkala (Confederated Salish and Kootenai Tribes), and Dale M. Becker (Confederated Salish and Kootenai Tribes)

206	**Chapter 16. Co-Management in Alaska: A Partnership among Indigenous, State, and Federal Entities for the Subsistence Harvest of Migratory Birds**
	Patricia K. Schwalenberg (Lac du Flambeau Band of Lake Superior Chippewa Indians), Liliana C. Naves, Lara F. Mengak, James A. Fall, Thomas C. Rothe, Todd L. Sformo, Julian B. Fischer, and David E. Safine
229	**Chapter 17. Research with Tribes: A Suggested Framework for the Co-Production of Knowledge**
	Caleb R. Hickman (Cherokee Nation), Julie Thorstenson (Lakota–Cheyenne River Sioux Tribe), Ashley Carlisle (Diné), Serra J. Hoagland (Pueblo of Laguna), and Steven Albert
248	**Chapter 18. Thoughts of an Anishinaabe Poet on Wildlife Biology**
	Marcie Rendon (White Earth Nation)
253	**Chapter 19. Protecting What We've Been Blessed With: Big Game and Other Wildlife Programs of the Navajo Nation**
	An Interview with Gloria Tom (Navajo Nation)
256	**Chapter 20. *Shash***
	A Story by Ramona Emerson (Diné)
261	**Chapter 21. A Model for Stewardship: The Lower Brule Sioux Tribal Wildlife Department**
	An Interview with Shaun Grassel (Lower Brule Sioux Tribe)
267	**Chapter 22. Reclaiming Ancestral Lands and Relationships**
	An Interview with Chief James R. Floyd (Muscogee [Creek] Nation) by Janisse Ray
275	**Chapter 23. We Feel Our Place in Our Soul: Perspectives from a Fond du Lac Elder**
	An Interview with Vern Northrup (Fond du Lac Band of Lake Superior Chippewa)
279	**Chapter 24. Partnerships Are the Key to Conservation**
	An Interview with Mitzi Reed (Mississippi Band of Choctaw Indians)
282	**Chapter 25. Burmese Python Impacts and Management on the Miccosukee Reservation, Florida**
	Craig van der Heiden and William Osceola (Miccosukee)
289	**Chapter 26. So Many Things That Humble Me**
	An Interview with John Sewell
295	**Chapter 27. Swamp Boy's Pet and Field Guide**
	A Memoir by Chip Livingston (Creek)

Part IV. Traditional Ecological Knowledge   307

309   Chapter 28. *Talutsa*: Weaving a Cherokee Future
Andrea L. Rogers (Cherokee Nation)

313   Chapter 29. A Traditional Strategy to Promote Ecosystem Balance and Cultural Well-Being Utilizing the Values, Philosophies, and Knowledge Systems of Indigenous Peoples
Richard T. Sherman (Lakota–Oglala Sioux Tribe) and Michael Brydge

332   Chapter 30. The Making and Unmaking of an Indigenous Desert Oasis and Its Avifauna: Historic Declines in Quitobaquito Birds as a Result in Shifts from O'odham Stewardship to Federal Agency Management
Gary Paul Nabhan, Lorraine Marquez Eiler (Tohono O'odham), R. Roy Johnson, Amadeo Rea, Eric Mellink, and Lawrence Stevens

353   Chapter 31. How Traditional Ecological Knowledge Informs the Field of Conservation Biology
Sarah E. Rinkevich and Crystal (Ciisquq) Leonetti (Yup'ik)

366   Chapter 32. Yurok Traditional Ecological Knowledge as Related to Elk Management and Conservation
Juliana Suzukawa (Oglala Lakota), Seafha C. Ramos (Yurok/Karuk), and Tiana M. Williams-Claussen (Yurok Tribe)

379   Chapter 33. *Kue Meyweehl 'esee kue 'Oohl Megetohlkwopew*: Elk and the Yurok People Take Care of Each Other
Seafha C. Ramos (Yurok/Karuk) and James Gensaw (Yurok/Tolowa/Chetco)

384   Chapter 34. Power Parade in Pablo, Montana
Serra J. Hoagland (Pueblo of Laguna)

387   Index

Contributors

EDITORS

Serra J. Hoagland (Pueblo of Laguna, village of Paguate) currently lives in Montana and works for the US Forest Service's Rocky Mountain Research Station at the Missoula Fire Sciences Lab. She earned her PhD in forestry from Northern Arizona University, where she assessed Mexican spotted owl habitat on tribal and nontribal lands. She is an active member of the Native American Fish and Wildlife Society, the American Indian Science and Engineering Society, The Wildlife Society, and the Intertribal Timber Council.

Steven Albert is an assistant director at the Institute for Bird Populations, a conservation science nonprofit, where he works on avian ecology initiatives in North America and Latin America. For many years, he was the Fish and Wildlife Department director and chief conservation law enforcement officer at the Pueblo of Zuni, where he helped develop the nation's first tribal eagle aviary and a program of sacred spring and wetland restoration. He lives in Ramah, New Mexico.

AUTHORS

Scott Aikin is an enrolled member of the Prairie Band of the Potawatomi Nation of northeastern Kansas. He currently serves as the US Fish and Wildlife Service's national Native American coordinator within the Office of the Director. He works with Department of the Interior leadership to ensure that the agency's actions and authorities are implemented in ways that respect and acknowledge tribal sovereignty and the importance of tribal wildlife stewardship.

John Antonio has served as Governor of the Pueblo of Laguna for several terms and was previously on the Laguna Tribal Council. He has held positions with the US Fish and Wildlife Service and the Navajo Fish and Wildlife Department, and he was superintendent of the Bureau of Indian Affairs, Southern Pueblos Agency. He has received numerous awards, including two Distinguished Alumnus Awards from New Mexico State University and the Chief Sealth Award from the Native American Fish and Wildlife Society.

Dale M. Becker (Confederated Salish and Kootenai Tribes) received his BS and MS in wildlife biology from the University of Montana. He retired after 36 years as the Confederated Salish and Kootenai Tribes' wildlife program manager at the Flathead Indian Reservation. His expertise includes raptor ecology and management, mitigation planning, and species reintroductions. He received certification from The Wildlife Society in 2000, was named Montana's TWS

Wildlife Biologist of the Year in 2004, and served as a board member and president of the Trumpeter Swan Society.

Bethany R. Berger, JD, is the Wallace Stevens Professor at the University of Connecticut School of Law. She is a coauthor of *Cohen's Handbook of Federal Indian Law*, leading casebooks in property law and American Indian law, and amicus briefs in five US Supreme Court cases affirming tribal sovereignty. Before entering academia, she worked as a lawyer on the Navajo, Hopi, and Cheyenne River Sioux Reservations.

Kimberly Blaeser, a former Wisconsin Poet Laureate, is the author of five poetry collections, including *Copper Yearning*, *Apprenticed to Justice*, and *Résister en dansant/Ikwe-niimi: Dancing Resistance*. An Anishinaabe activist and environmentalist from White Earth Reservation, Blaeser is a professor emerita at the University of Wisconsin–Milwaukee, an MFA faculty member at the Institute of American Indian Arts, and the founding director of In-Na-Po (Indigenous Nations Poets). Blaeser splits her time between Lyons Township, Wisconsin, and a water-access cabin near the Boundary Waters Canoe Area Wilderness in Minnesota.

Arthur "Butch" Blazer is a member and former President of the Mescalero Apache Tribe. He is a 1975 graduate of New Mexico State University. In 1983, he helped cofound the Native American Fish and Wildlife Society and has served as its national president. Blazer currently serves on the boards of directors for the National Wildlife Federation and Conservation Legacy.

Michael C. Blumm, JD, is the Jeffrey Bain Faculty Scholar and a professor of law at Lewis and Clark Law School in Portland, Oregon. He writes widely on natural resources and federal Indian law issues, including coauthoring the casebooks *Native American Natural Resources Law* (4th ed.) and *Public Trust Doctrine in Environmental and Natural Resources Law* (3rd ed.). His *Pacific Salmon Law and the Environment* was published in 2022 by the Environmental Law Institute.

Michael Brydge co-owns Sweet Grass Consulting, a company engaging in efforts centered on research, impact, and strategy with economically burdened communities. Sweet Grass has worked with 26 Native nations in 18 states. Brydge's research focuses on community-led strategies to enhance affordable, locally designed housing and food sovereignty, and to create integrated workforces with communities such as the Lakȟóta. Brydge was born and raised in the Blue Ridge Mountains of Virginia and has twin 18-year-old sons.

Ashley Carlisle (Diné), the education coordinator for the Native American Fish and Wildlife Society, was raised on the Navajo Nation in Tohatchi, New Mexico. She earned her bachelor of science degree in fish, wildlife, and conservation biology and her master of science in conservation leadership from Colorado State University. Her master's thesis project included working with the Ya'axché

Conservation Trust in Belize to complete an assessment of its jaguar-human conflict program.

Sally Carufel-Williams is a member of the Flandreau Santee Sioux Tribe and is affiliated with the Lac du Flambeau Band of Lake Superior Chippewa Indians. Her career has been one of service to Native American organizations, including the Native American Rights Fund, several public school districts and universities with Native American populations, and Haskell Indian Nations University. In the Native American Fish and Wildlife Society, she has served as executive secretary, board liaison, education-training coordinator, and youth practicum coordinator.

Guy C. Charlton is an associate professor of law at the University of New England and an adjunct professor at Auckland University of Technology and Curtin University. He is from Milwaukee, Wisconsin, where he practiced law, and he has lived and worked in a variety of Commonwealth countries. He holds degrees from the University of Wisconsin–Madison (BA, JD), the University of Toronto (MA), and Auckland University (PhD).

Samuel Chischilly grew up in Crownpoint, New Mexico, and is a member of the Navajo Nation. He is currently in his third year at Fort Lewis College, majoring in biology, and hopes to have a career where he is able to help his community. His immediate plans include furthering his education in graduate school.

Bob Christensen works with the Sustainable Southeast Partnership on community-based natural resource projects that integrate social, ecological, and economic values because he believes that the vitality and resilience of nature and its peoples go hand in hand. Christensen has a BS in sustainable community design from Huxley College and has been working as an environmental consultant in Southeast Alaska for 25 years. Christensen lives in Huna Territory in Lingít Aaní, near Glacier Bay.

Gerald "Buzz" Cobell is an enrolled member of the Blackfeet Nation. His Indian name is Peta Boka (Eagle Child). He received his MS in fisheries and wildlife from Utah State University and has worked for over 40 years in the natural resources profession. He helped found the Native American Fish and Wildlife Society and was its first president. Cobell has also served on the board of directors of the American Indian Science and Engineering Society.

Cody Desautel has served as the natural resource director for the Confederated Tribes of the Colville Reservation, where he is an enrolled member, since 2014. He has also worked for the Bureau of Indian Affairs and the Tribe as a forester and natural resource officer. Desautel has a BS in environmental science and an MS in Indian law. He is president of the Intertribal Timber Council and a member of Washington state's Wildland Fire and Forest Health Advisory Committees.

Lauren M. Divine, PhD, is the director of the Ecosystem Conservation Office for the Aleut Community of St. Paul Island, Alaska, where she has the opportunity to span the boundaries between Western sciences; local and traditional knowledges; tribal, federal, and state management; and stakeholder engagement through community-based monitoring and research. Divine seeks to strengthen relationships across these boundaries to better serve the community, wildlife, and ecosystems of the Bering Sea and the Arctic.

Doug "Miserable Mutt" Dompier worked on Columbia River salmon management issues from 1972 to 2005. In 1979, the Columbia River Inter-Tribal Fish Commission hired him to develop the Fisheries Technical Services Division. He supervised programs for harvest management, fish passage at mainstem dams, fish production (both natural and artificial), and fish habitat protection and restoration. He is the author of *The Fight of the Salmon People: Blending Tribal Tradition with Modern Science to Save Sacred Fish*.

Lorraine Marquez Eiler is the founder of the Hia c-ed O'odham Alliance and a member of the Tohono O'odham Legislative Council. She has served as a board member with the International Sonoran Desert Alliance, the Sonoran Institute, the Arizona-Sonora Health Commission, and interagency advisory councils along the US-Mexico border, and has coauthored scholarly articles on the archaeology and history of the O'odham in Mexico and the United States. Eiler is a descendant of O'odham dwellers at the Quitobaquito Oasis.

Ramona Emerson is a Diné writer and filmmaker originally from Tohatchi, New Mexico. She received a degree in media arts from the University of New Mexico and an MFA in creative writing (fiction) from the Institute of American Indian Arts. She is an Emmy nominee, a Sundance Native Lab Fellow, a Time-Warner Storyteller Fellow, a Tribeca All-Access grantee, and a WGBH Producer Fellow. Emerson's first novel, *Shutter*, was published in 2022.

Kari L. Eneas is a certified wildlife biologist for the Confederated Salish and Kootenai Tribes. She received a BS and an MS from the University of Montana and has worked for the Tribe's Wildlife Management Program since 2008 on various wildlife projects pertaining to upland game and grassland birds, carnivores, and wildlife conflict reduction. She currently focuses on grizzly bear and other carnivore management, including reducing carnivore conflicts with landowners and the public.

James A. Fall, PhD, has worked since the 1970s as a cultural anthropologist researching subsistence harvests and uses of wild resources throughout Alaska. He is also a dedicated scholar of the Dena'ina culture. Fall's work at the Division of Subsistence of the Alaska Department of Fish and Game (1981–2020) was instrumental to translating the importance of fish and wildlife resource uses in Indigenous cultures into harvest management and regulatory processes.

Julian B. Fischer is a wildlife biologist for the US Fish and Wildlife Service, Waterfowl Section of Migratory Bird Management, in Anchorage, Alaska. He supervises waterfowl biologists and oversees data collection for aerial and ground-based surveys that support waterfowl harvest management, habitat conservation, and recovery actions. Fischer has participated in various Alaska Migratory Bird Co-Management Council subcommittees and has provided technical expertise to the council since 2003.

James R. Floyd was Principal Chief of the Muscogee (Creek) Nation from 2016 to 2019. In 1978, he established the first tribally owned health-care system in the United States. He was a health administrator with the Veterans Administration for 30 years. He has received the VA's Diversity Award, the Medal for Meritorious Service, and the Ray E. Brown Award for his advocacy in federal health-care leadership. He currently serves on the board of trustees of the National Parks Conservation Association.

James Gensaw (Yurok/Tolowa/Chetco) is a California-credentialed Yurok language teacher. He currently teaches at Eureka and McKinleyville High Schools. He has been dedicated to revitalizing the Yurok culture and language for decades.

Michael I. Goldstein, PhD, is a planner for the US Forest Service in Juneau, Alaska. He has worked on applied management issues across terrestrial and aquatic systems, addressing pesticides, dispersed recreation, timber harvest, and other forms of resource extraction. Other projects include serving as co-editor in chief of the *Encyclopedia of the Anthropocene*, *Encyclopedia of the World's Biomes*, and *Imperiled: The Encyclopedia of Conservation*. Goldstein helps lead the Guardians collaboration for the US Forest Service in Southeast Alaska.

Shaun Grassel is an enrolled member of the Lower Brule Sioux Tribe, where he has worked as a wildlife biologist for nearly 25 years. Grassel and his team are self-supported, raising all required funding from external sources. Grassel has a BS and an MS in wildlife and fisheries sciences from South Dakota State University and a PhD in natural resources from the University of Idaho. He spends his free time managing his cattle ranch and raising native grasses for seed production.

E. Richard Hart provides historical, ethnohistorical, and environmental historical services and expert testimony for North American tribes. He has authored or edited 11 books and published more than 50 articles. He has testified before the US Congress on numerous occasions and served as an expert witness for tribes in Canada and the United States and for other entities. He has received numerous awards, including the 2012 Washington state Peace and Friendship Award. His papers form a special collection at the University of Utah's Marriott Library.

Dylan R. Hedden-Nicely is a citizen of the Cherokee Nation, an associate professor of law at the University of Idaho College of Law, and the director of the University of Idaho Native American Law Program. He also consults with tribes on water rights and natural resource issues. Hedden-Nicely graduated from the University of Idaho College of Law, magna cum laude, with an emphasis in Native American law, natural resources, and environmental law. Concurrently, he earned a master's degree in water resources.

Caleb R. Hickman, PhD, is the supervisory biologist for the Eastern Band of Cherokee Indians and is a citizen of the Cherokee Nation in Oklahoma, where he grew up. With a small team, he works to study and conserve a variety of game and nongame fish and wildlife species and their habitats. He combines science-based management with a socioecological perspective that includes an eye toward preserving traditional knowledge for the benefit of future generations.

Nathan "Eight Ball" Jim was a member and lifetime resident of the Confederated Tribes of Warm Springs. He overcame his BIA boarding school experience, returning home to work as a firefighter and printer, and helped to establish his Tribe's cultural resource center, where he recorded tribal history and traditions from elders. He was a founding member of the Native American Fish and Wildlife Society and traveled across the country as a renowned powwow MC, sharing his humor, heritage, and enduring spirit. Jim passed away in 1999.

R. Roy Johnson has studied plants and animals of the southwestern US and northwestern Mexico for more than 65 years. Although his PhD is in systematic botany, more than 50 of his more than 200 publications have been on birds. Ten of his publications have been on Organ Pipe Cactus National Monument, and three of those were on the birds of Quitobaquito Springs, Arizona. His desert home is east of Tucson where, with his wife, Lois Haight, he continues biological studies.

Jovon Jojola is from the Pueblo of Isleta, New Mexico. She is enrolled in the natural resources management curriculum at Southwestern Indian Polytechnic Institute in Albuquerque. In her spare time, she gardens and spends time with her family.

Tamra Jones, an Alaskan Native and part of the Raven Clan, is an education intern at the Native American Fish and Wildlife Society. From 2017 to 2021, she studied natural resources and the environment with a minor in business economics at the University of Alaska, Fairbanks. To put herself through school, during the summers she worked as a sub-journey equipment operator, which is, she says, "a fancy way of saying I was a flagger for the Alaska Department of Transportation."

Winona LaDuke is an activist, economist, and author, and has devoted her life to advocating for Indigenous peoples' control of their homelands, natural resources, and cultural practices. She is an Anishinaabekwe (Ojibwe) enrolled member of the Mississippi Band Anishinaabeg. She combines economic and environmental approaches in her efforts to create a thriving and sustainable community for her own White Earth Reservation and for Indigenous populations across the country.

Crystal (Ciisquq) Leonetti is Yup'ik and a citizen of Curyung Tribal Council, born and raised in Alaska. As a Native affairs specialist for the US Fish and Wildlife Service, she represents the regional director in regard to Alaska Native interests, including government-to-government relations with Alaska's 229 tribes. Leonetti has written policies on agency interaction with tribes for several federal agencies. Conservation is a passion handed down to her from her ancestors. She and her husband, Ed, have two daughters, Audrey and Gigi.

Aaron P. Lestenkof is a tribal member and Island Sentinel for the Ecosystem Conservation Office of the Aleut Community of St. Paul Island. Lestenkof is a remote pilot and conducts many of the island's environmental and subsistence resource surveys, including surveys of marine debris, seabirds, marine mammals, water quality, invasive species, and coastal erosion. He also represents the tribal government on the Subsistence Co-Management Council, the co-management body for marine mammals on St. Paul.

Chip Livingston is the mixed-blood Creek author of four books of poetry, fiction, and nonfiction. His essays have appeared in *Cincinnati Review*, *Ploughshares*, *Los Angeles Review*, *Carve*, *Shapes of Native Nonfiction*, and elsewhere. He teaches in the low-rez MFA program at the Institute of American Indian Arts in Santa Fe, New Mexico. He lives in Montevideo, Uruguay.

Eric Mellink is *investigador titular* at the Center for Scientific Research and Higher Education in Ensenada, Baja California. He has published more than 130 technical papers on wildlife ecology and conservation in deserts and on islands off the Pacific coast of Baja California and in the Sea of Cortez. He published his field research on the vertebrates of Quitobaquito Oasis while carrying out his PhD research in arid lands natural resources at the University of Arizona.

Paul I. Melovidov is a tribal member and Island Sentinel coordinator for the Ecosystem Conservation Office of the Aleut Community of St. Paul Island, where he leads environmental and subsistence resource management programs. Melovidov previously worked the commercial harvests of northern fur seals operated under the federal government until the 1980s, when it was discontinued. He is an avid marine mammal and bird hunter, as well as a skilled nature and landscape photographer.

Lara F. Mengak is a conservation social scientist with an interest and background in wildlife ecology and management, particularly focusing on birds. Mengak currently works for the Division of Subsistence of the Alaska Department of Fish and Game, where she works with managers and subsistence users on bird harvest management and emerging bird conservation issues.

Gary Paul Nabhan, PhD, holds the Kellogg Endowed Chair in Southwest Borderlands Food and Water Security at the University of Arizona. He is the author of 33 books on ethnobiology, wildlife, and plant conservation, and has collaborated with a dozen First Nations regarding the protection and use of native plants and wildlife. His collaborative work has been honored with a MacArthur fellowship and lifetime achievement awards from several professional societies. He lives and tends his garden on the US-Mexico border.

Liliana C. Naves, PhD, is an oceanographer with a background in avian ecology research. Since 2007, she has worked for the Division of Subsistence of the Alaska Department of Fish and Game in Anchorage. She leads harvest research and coordinates the Harvest Assessment Program on behalf of the Alaska Migratory Bird Co-Management Council. Her work integrates biological and social sciences to support sustainable bird populations, harvest opportunities, and subsistence ways of life in Alaska.

Vern Northrup (Fond du Lac Band of Lake Superior Chippewa) is an artist and retired wildlands firefighter. His photographs and teachings offer Anishinaabe perspectives on the place of humans in the world, the seasons, and more. He uses photography as a tool to educate both himself and the viewer about the rhythms of nature, the preservation of tradition, and the relationship between resilience and sustainability.

nila northSun (Shoshone/Chippewa) has been writing poetry for over three decades. Her books include *diet Pepsi & nacho cheese*; *coffee*; *small bones, little eyes* (with Jim Sagel); *a snake in her mouth*; *love at gunpoint*; and *whipped cream & sushi*. She has received the Silver Pen Award from the University of Nevada, Reno; the Indigenous Heritage Award from Atayal; and the Sierra Arts Foundation Literary Award. She is now retired and living on her Reservation in Fallon, Nevada.

William Osceola is a tribal member and member of the Miccosukee Business Council. He taught language, culture, and digital arts at the Miccosukee Indian School for nearly a decade before being elected secretary of the Miccosukee Business Council in 2021. Although born in Florida and raised in different camps inside the Reservation, Osceola lived in Connecticut for seven years, which increased his appreciation for his tribal community.

Raymond E. Paddock III works for the Central Council of Tlingit and Haida Indian Tribes (Tlingit & Haida) of Alaska as their environmental coordinator, providing training activities, educational assistance, and coordination regionally on environmental and natural resource concerns in Southeast Alaska. Tlingit & Haida contributes to the capacity growth of Southeast Alaska tribes and provides a wide variety of services to assist tribes as they address local and regional environmental issues.

Lizzy Pennock is the carnivore coexistence advocate at WildEarth Guardians. She earned a JD, magna cum laude, from Lewis and Clark Law School; received the 2021 Environmental, Natural Resources, and Energy Law Leadership Award; and was elected to the Cornelius Honor Society. She earned a bachelor of science, summa cum laude, in marine science from the University of Hawai'i at Hilo.

Riley Plumer is an attorney and an enrolled member of the Red Lake Band of Chippewa Indians. He is dedicated to working on identifying innovative solutions to the complex legal issues confronting tribes in the exercise of their sovereignty. Plumer has extensive experience representing tribes and tribal organizations on a broad spectrum of legal issues, including treaty rights, tribal land status, gaming, housing, labor and employment, and government contracting.

Aaron J. Poe has worked in Alaska for 23 years, building equitable partnerships between agency managers, tribes, researchers, industries, and communities to address large-scale issues, such as climate change. He is the network program officer for the Alaska Conservation Foundation and the coordinator for the Aleutian Bering Sea Initiative. He has bachelor's degrees in fisheries and wildlife management and in geography from Utah State University and a master's degree in natural resource management from the University of Arizona.

Ken Poynter is a member of the Passamaquoddy Tribe of Maine, where he served as tribal biologist. Poynter has an associate of arts degree and a BS in biology from the University of Maine and an MS from New Hampshire College. He joined the Native American Fish and Wildlife Society Board of Directors in 1989, representing the Northeastern Region. Poynter served as secretary-treasurer from 1988 to 1992 and as executive director from 1993 to 2001.

Mathis Quintana, an intern at the Native American Fish and Wildlife Society, grew up in Dulce, New Mexico, and is a member of the Jicarilla Apache Nation. He is currently a senior at Colorado State University, majoring in natural resource management. He attended Navajo Preparatory High School in Farmington, New Mexico, and is grateful to have opportunities through the Native American Fish and Wildlife Society.

Seafha C. Ramos, PhD (Yurok/Karuk), does research on the integration of Western and Indigenous sciences in wildlife research. With an emphasis on Traditional Ecological Knowledge (TEK) through an Indigenous lens, she strives to conduct culturally sensitive research with Indigenous communities. She has conducted studies in molecular scatology in partnership with the Yurok Tribe Wildlife Program. She practices TEK through gathering and processing traditional foods, learning the Yurok language, and participating in ceremony.

Janisse Ray is the author of five books of literary nonfiction and two volumes of eco-poetry. Her first book, *Ecology of a Cracker Childhood*, a memoir about growing up on a junkyard, was a New York Times Notable Book and was chosen as the Book All Georgians Should Read. She has won a Pushcart Prize, a Lifetime Achievement Award from the Georgia Writers Association, and the Southern Environmental Law Center Award in journalism. She is an inductee in the Georgia Writers Hall of Fame.

Vanessa L. Ray-Hodge, JD, is a member of the Pueblo of Acoma and a partner at Sonosky, Chambers, Sachse, Endreson & Perry, LLP. She works in all areas of the firm's practice with a special focus on Indian water rights. Prior to rejoining Sonosky, Ray-Hodge served as the senior counselor to the first Native American solicitor, Hilary Tompkins, at the Department of the Interior, advising her on tribal issues across the nation. Ray-Hodge graduated from Columbia Law School with honors.

Amadeo Rea, PhD, is a retired ornithologist and ethnobiologist who has authored five books on O'odham and Pima knowledge of plants and animals and the historic shifts in habitats on Indian reservations in Arizona. He has been affiliated with the San Diego Natural History Museum and three universities in Arizona and California as a researcher and curator. He was among the founders and is a past president of the Society of Ethnobiology. Rea lives in San Diego, California.

Mitzi Reed is one of the Southeast Region's directors for the Native American Fish and Wildlife Society. She serves as the director and biologist of the Choctaw Wildlife and Parks Department for the Mississippi Band of Choctaw Indians. Reed received a BS and an MS in biology from the University of West Alabama. She is passionate about natural resources conservation, and she works with the Choctaw Youth Conservation Corps to promote appreciation and understanding of the importance of protecting our natural resources.

Marcie Rendon is a citizen of the White Earth Nation and is on *Oprah Magazine*'s list of 31 Native American Authors to Read. She has received numerous awards, including a McKnight Distinguished Artist Award, a 50over50 Minnesota AARP Award, and a Pollen Award. Along with Diego Vazquez, she received the Loft's 2017 Spoken Word Immersion Fellowship for her work with incarcerated women. Rendon's crime novels include *Girl Gone Missing* and *Murder on the Red River*.

Sarah E. Rinkevich received her PhD in wildlife conservation from the University of Arizona, where she is currently an assistant research scientist. Her early work focused on population estimates of Mexican wolves on the Fort Apache Indian Reservation. She has worked for the US Fish and Wildlife Service in the Endangered Species Conservation Program since 1993 and works extensively with tribes in the Southwest on the management and conservation of endangered species.

Bruce W. Robson is a codirector of Community and Ecology Resources, LLC, and the technical director of the Indigenous Sentinels Network for the Aleut Community of St. Paul Island, Alaska. He has worked in the Pribilof Islands since 1989 as a marine wildlife and fisheries biologist. From 1990 to 2004 he was a research biologist at the National Oceanic and Atmospheric Administration Marine Mammal Laboratory, conducting research on northern seals and Steller sea lions in the Bering Sea and Aleutian Islands.

Andrea L. Rogers (Cherokee Nation) grew up in Tulsa, Oklahoma. She received an MFA from the Institute of American Indian Arts. She has published a book, *Mary and the Trail of Tears*, and her essays have appeared in *You Too?* and *Allies*. Her short story "The Ballad of Maggie Wilson" is included in *Ancestor Approved: Intertribal Stories for Kids*. Her picture book, *When We Gather*, a story about wild onion dinners, is forthcoming.

Thomas C. Rothe managed the Alaska Department of Fish and Game's Waterfowl Program from 1983 to 2009, leading diverse research, management, and public education projects. He served on the Pacific Flyway Study Committee and the Sea Duck Joint Venture, and he assisted in the Yukon-Kuskokwim Delta Goose Management Plan, the 1997 amendments to the US migratory bird treaties with Canada and Mexico, and the formation of the Alaska Migratory Bird Co-Management Council. Rothe has dedicated his career to collaborative waterfowl harvest management and conservation along the Pacific Flyway.

David E. Safine is a waterfowl biologist for the US Fish and Wildlife Service's Alaska Region in the Division of Migratory Bird Management. Safine serves as a member of several subcommittees of the Alaska Migratory Bird Co-Management Council and as an Alaska Region liaison for the Pacific Flyway Study Committee. His professional interests are focused on harvest management for migratory birds and the demographics of breeding waterfowl, particularly sea ducks.

Patricia K. Schwalenberg is Anishinaabe, a member of the Lac du Flambeau Band of Lake Superior Chippewa Indians in Wisconsin. For 35 years she has worked with tribes across the country on natural resource program development and management. With a background in cultural anthropology, she works across cultures in partnerships including tribal, state, and federal governments and nongovernmental organizations that utilize and honor all ways of knowing. She serves as the executive director of the Alaska Migratory Bird Co-Management Council.

John Sewell is the wildlife biologist for the Passamaquoddy Tribe at Indian Township, Maine, where he has worked since 2002, striving to conserve tribally important fish and wildlife. Sewell has also worked across the country in various wildlife management positions and with the Maine Department of Inland Fisheries and Wildlife. He received his BS from the University of Maine, Orono, in wildlife ecology.

Todd L. Sformo is a wildlife biologist for the North Slope Borough of the Department of Wildlife Management in Utqiaġvik, Alaska, and a research scientist with the Institute of Arctic Biology at the University of Alaska, Fairbanks. Sformo is a technical advisor to the Alaska Migratory Bird Co-Management Council Native Caucus and chair of the AMBCC Handicraft Committee. He also works on issues related to bowhead whales, fish, freshwater mold, ice-binding proteins, and low-temperature-associated life science.

Richard T. Sherman was born and raised on the Pine Ridge Indian Reservation in South Dakota and is a member of the Lakota–Oglala Sioux Tribe. He has worked most of his life on issues of fisheries, wildlife, buffalo management, ethnobotany, and Indigenous stewardship, and he has served the Tribe as a wildlife biologist, executive director, and board member of Oglala Sioux Parks and Recreation. He managed the tribal buffalo herd for more than 30 years. Sherman has a master's degree in regional planning.

Ron Skates is a member of the Cheyenne River Sioux Tribe. He has served as a game warden for his Tribe and for more than 25 years worked for the US Fish and Wildlife Service in the Northern Plains Region. Skates was active in the Native American Fish and Wildlife Society from its inception and for decades afterward; he was elected vice president of the society in 1990 and 1991 and president in 1992 and 1993.

Arthur M. Soukkala has been a wildlife biologist for the Confederated Salish and Kootenai Tribes for over 30 years, working on carnivores, wildlife conflicts, and wildlife habitat mitigation planning. His current work centers on restoring wildlife species and habitats on the Flathead Indian Reservation. He has wildlife management degrees from the University of Minnesota and the University of Maine. He was recognized as the Wildlife Biologist of the Year by the Montana Chapter of The Wildlife Society in 2006.

Lawrence Stevens has studied southwestern ecology for 50 years. He received his PhD in zoology from Northern Arizona University in 1989 and is presently the director of the Springs Stewardship Institute and the curator of biology at the Museum of Northern Arizona in Flagstaff. His more than 150 peer-reviewed and popular articles and books focus on taxonomy, biogeography, springs and river ecology, and water management. In recent publications, he has compared the ecology of Quitobaquito Springs with that of other southwestern springs.

Juliana Suzukawa (Oglala Lakota) is an undergraduate student pursuing a bachelor of science degree in wildlife management and conservation at Humboldt State University. He has participated in a bird diversity project based on bird surveys taken throughout California. The chapter in this book is his first publication. Suzukawa lives with his family in Los Angeles.

Julie Thorstenson (Lakota–Cheyenne River Sioux Tribe) is the executive director of the Native American Fish and Wildlife Society. She grew up on a cattle ranch on the Cheyenne River Sioux Reservation in South Dakota, where a love for the land and the environment was instilled in her. She earned a BS, an MS, and a PhD from South Dakota State University. Thorstenson has worked in various positions for her Tribe, including wildlife habitat biologist and health department CEO.

Gloria Tom has been the director of the Navajo Nation Fish and Wildlife Department since 1998, responsible for managing and conserving the Navajo Nation's fish and wildlife resources on behalf of the Navajo people. She also has worked as the director of the San Carlos Apache Tribe's Recreation and Wildlife Department and as a district supervisor for USDA Wildlife Services in New Mexico. In 2020, Tom was inducted into the Arizona Outdoor Hall of Fame.

Christopher C. Tran is the science technician for the Pribilof Islands' Aleut Community of St. Paul Island, Alaska. He is an environmental scientist and has worked for state and tribal governments on fish passage and habitat issues in the Pacific Northwest. Tran oversees data management, interpretation, and visualization for the Tribe's environmental monitoring programs and participates in federal fisheries policy and management through the regional fishery management council.

Craig van der Heiden is the Miccosukee Fish and Wildlife director. In 2012, he received his PhD in integrative biology from Florida Atlantic University for work on the distribution dynamics of crayfish populations in the Florida Everglades. His current research is on applied conservation biology with a focus on rare and endangered species distribution, restoration, and management.

Tiana M. Williams-Claussen is a Yurok tribal member and was raised within Yurok ancestral territory. She was instrumental in the formation of the Yurok

Tribe Wildlife Department in 2008 and currently serves as the department director. Her Native upbringing and formal education allow her to bridge the gap between traditional understandings of the world and those rooted in Western science, supporting her work toward a cohesive, well-informed approach to holistic ecosystem management in Yurok country.

ARTISTS

Frank Cerno Jr. was born in 1952 and is a member of the Pueblo of Laguna, where he currently resides. His passion for art began at an early age and continued through college, but it came to an abrupt stop due to continuing educational pursuits, work, and raising a family. After a 47-year hiatus from his artwork, he rekindled his passion for creating new, different, and exciting art, incorporating Native designs.

Lauren Helton is a biologist and scientific illustrator with the Institute for Bird Populations, a conservation science nonprofit organization. When she's not painting birds, bats, insects, and other wildlife, or managing IBP's bird-banding data, she can be found outdoors training bird banders and early career scientists on the fine details that can be read in a bird's feathers. Her passions are birds of prey, tropical ecology, science fiction, and tabletop role-playing games. See more of her artwork at tinylongwing.carbonmade.com.

Kraig Holmes is a member of the Hopi Tribe, born into the Snow Clan and born for the Sun-Forehead Clan. He was raised on the Hopi Reservation and has always been deeply connected to the Hopi way of life. He started drawing as a child, following in the footsteps of his fathers and brothers, who are gifted artists. In addition to painting, he enjoys other forms of art, creating Hopi Kachina figures, jewelry, and graphic art.

Melitta Jackson is a Yurok tribal member and Hupa, Karuk, and Modoc descendant. She graduated from Humboldt State University in 2017 and has received several awards for her artwork. Jackson's art is inspired by scientific illustration, pop surrealism, and the cultural stories and practices of her Native American community. She currently works for Cal Poly Humboldt as a resource-sharing and information specialist.

Nicole Marie Pete is an enrolled member of the Colorado River Indian Tribes. She was born and raised in California and now lives in southern California. Her maternal clan is Bitter Water, and she was born for the Big Water Clan. Pete's artistic aspirations stem from the positive influences of her maternal grandfather, John Scott Sr. In addition to creating art, she loves playing video games, drawing anime, creating websites, and designing logos and graphics.

Kyle Secakuku (Hopi name Taqaomauw) is of the Bearstrap/Spider Clan from the Hopi Village of Shungopavi. His mother is Andrea Joshevama, and his father is Charles Secakuku of the Bamboo Clan from the village of Baqavi. Secakuku is currently attending Southern Utah University, studying criminal justice, and he plans to work in law enforcement. Secakuku says, "My art captures my interpretation of Hopi philosophy and experience from farming, ceremonies, teachings, and exposure from outside of Hopi realm."

Jessica Wiarda, a designer and muralist, grew up in Logan, Utah, raised by her Hopi mother and white father in a blend of cultural traditions. Inspired by the designs of Hopi art, she developed an affinity for painting and drawing at an early age. She is a registered matrilineal member of the Hopi Tribe from the Deer Clan in Polacca. Wiarda's work focuses on blending traditional Hopi design with bright color palettes and contemporary designs. She created the fashion brand HONOVI.

Preface

A confluence of several circumstances makes the timing of this book particularly relevant. Twenty-first-century expansions in tribal-federal co-management and co-production have showcased how successful these initiatives can be. The confirmation of the first Native American US secretary of the Interior, Deb Haaland, is a fresh and long-awaited development and seems likely to lead to an increased profile for tribal natural land and wildlife management.

After joining Salish Kootenai College in Montana, I (Serra Hoagland), along with several other Native scholars, had contemplated the idea of combining several works throughout Indian Country to serve as a resource for courses on tribal wildlife and natural resource management. As someone who had been asked to create lectures and curricula on this topic, I knew there were great ideas and projects to serve as examples, but few were specific to contemporary Native American issues. I saw an immense gap in the extant literature, and I knew a volume on tribal wildlife management could help educate the next generation of wildlife students.

At about the same time, I (Steve Albert) came across a remarkable book, *Footprints of Hopi History: Hopihiniwtiput Kukveni'at*, a collection of essays about how Hopi elders, the Hopi Cultural Preservation Office (HCPO), and non-Indian scholars are co-producing important research aimed at understanding Hopi cultural and natural heritage. The crux of these efforts, which are led by former HCPO Director Leigh Kuwanwisiwma and current Director Stewart Koyiyumptewa, is that Hopi traditional knowledge and Western science can work together to develop new insights into the relationship between the Tribe and Hopitutkswa, the vast traditional Hopi homeland. Their work inspired me to try to tell a story that could only be conveyed with collaborators whose work I had seen, admired, and on a few lucky occasions been a part of. I made a presentation of the idea for this book at the Native American Fish and Wildlife Society (NAFWS) conference at the Gila River Indian Community in 2019 and received a positive response. Serra and I met there, and we have not let each other's email inboxes rest ever since!

The third strand in this collaboration was the support of the NAFWS, especially the executive director, Dr. Julie Thorstenson. The NAFWS is a national nonprofit organization that fosters communication among tribes on information related to fish and wildlife management. When we approached Julie with the idea for the book, she was immediately supportive and offered resources and the extra ounce of encouragement we needed. We also spoke to several former and current prominent leaders in tribal wildlife management to get their blessing and approval. Their suggestions and direction helped guide the development of the book.

The final component to the origin story of this book was the support of our publisher, Johns Hopkins University Press. Johns Hopkins is an avid ally and a renowned publisher for wildlife management books. Working with Tiffany

Gasbarrini and her staff to turn our idea into a fully fledged resource for wildlife and natural resource professionals has been a blessing.

Because we know that academic books of this type are often relatively expensive, potentially beyond the reach of Native American elders, students, tribal wildlife departments, and others who might enjoy it, we opened a dialogue with the early contributors to the book about equity, intellectual property, and the long- and short-term benefits to the NAFWS. As a result of those discussions and with the full support of Johns Hopkins University Press, we agreed that the profits from hard-copy sales of the book would be donated to the NAFWS in perpetuity, and after three years the material in the book would become open source.

We would like to thank Julie Thorstenson and the entire staff of the Native American Fish and Wildlife Society, past and present, for their enthusiastic and unwavering support during the development of this book. We are extremely grateful to Hilary Hart, Tania Morrison, Sohrob Nabatian, and Teresa Marmorell of the Kalliopeia Foundation for financial support, which was used for honoraria for tribal elders, writers, and artists who contributed to this project and to support the publication in full color of much of the artwork. We thank the staff at Johns Hopkins University Press for their immediate and enthusiastic support of the idea for the book, especially Tiffany Gasbarrini, who worked with these two book rookies and their endless naive questions. Thanks also to Emily Sylvan Kim, who volunteered her time to edit the creative nonfiction in the book.

Serra Hoagland would like to say *da'wa'eh* (thanks) to my immediate and extended family for their love, support, and guidance. From the day I could not wait to get on the school bus, to the years I spent away from home while in graduate school, you all have prayed for me and kept me motivated, sane, and giggling. Without a doubt, my love for science and protecting creatures comes from my family and my upbringing. I only hope that this book helps honor and serve our ancestors and relatives. And finally, a big thank you to my husband, Brian McGowan, and pup, Bella, who have assisted with moving countless boxes of my books over the years so that I can continue to accrue more.

Steve Albert is extremely grateful to the many tribal elders, biologists, and others who made me feel so welcome in this big, crazy family of tribal wildlife professionals. I would especially like to thank John Antonio, Albert Chopito, Nelson Luna, Melissa Peyketewa, Clybert Peyketewa, Edward Wemytewa, Merlin Simplicio, Leigh Kuwanwisiwma, Stewart Koyiyumptewa, Harry Chimoni, Perry Tsdadiasi, Kirk Bemis, Butch Blazer, and Barton Martza. Thanks to my family—HBP, MJB, AA, and OPA—for their encouragement and support. And thanks to my pals Eric Dexheimer and Bruce Finke for their thoughtful editing.

Finally, from both of us, a big THANK YOU to all of the contributors to this book. We are beyond grateful for your collective knowledge, dedication, and expertise. It's been an immense honor working with you, and we are blessed to have compiled what could be the key to sustaining our relationship to the land.

Note to the Reader

As coeditors of this volume, we have attempted to include Native languages and vocabulary where applicable and appropriate. We recognize there could be inconsistencies in orthography or slight variations in the spelling of Native words within this book. If readers want to delve into this topic more, we suggest reaching out to the individual tribal communities for further information.

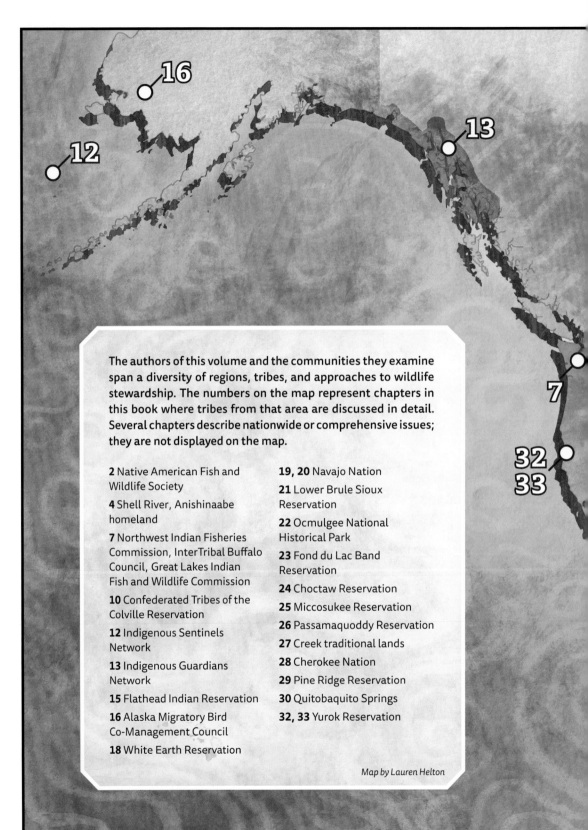

Map by Lauren Helton

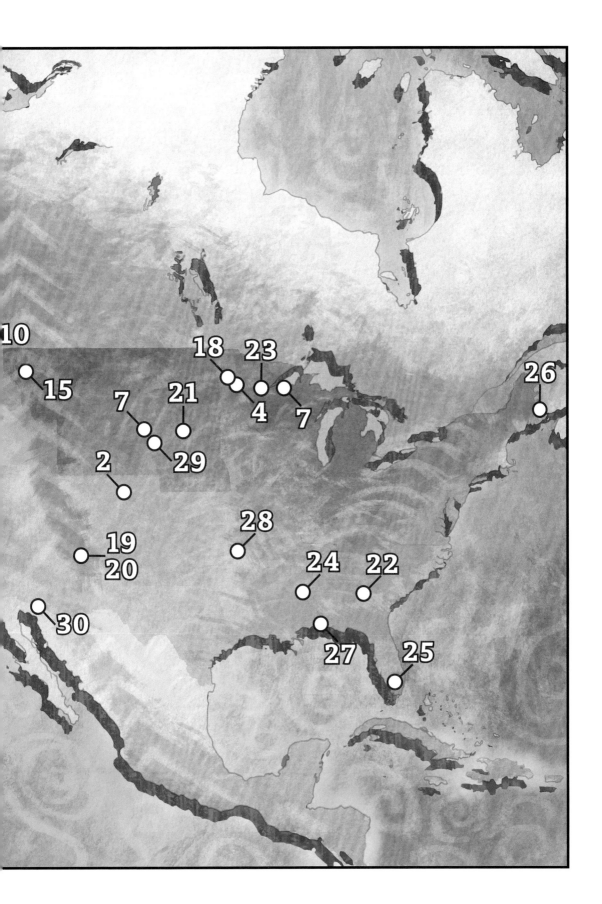

Wildlife Stewardship on Tribal Lands

Part I

Background and Policy Issues

As we describe in the introduction, tribal wisdom has been—and largely still is—passed along orally and visually through countless generations. To represent this facet of tribal relationship with the natural world and to be as inclusive as possible of the wide range of tribal voices, this volume has contributions from tribal elders, storytellers, poets, and artists as well as prominent Native scholars. This first part of the book provides a broad overview of tribal wildlife management, provides context for tribal wildlife management departments, offers a history of the Native American Fish and Wildlife Society, and ends with a look to the future.

Introduction

STEVEN ALBERT *and* SERRA J. HOAGLAND

We feel our place in our soul.
—Vern Northrup, Fond du Lac Band of Lake Superior Chippewa

One day in the fall of 1976, two Louisiana hunters killed a pair of bull elk on the Mescalero Apache Reservation in southern New Mexico. Elk were abundant and well managed on Mescalero lands at that time, almost exclusively due to the efforts of the Tribe. The species had been extirpated from the state in the early 1900s, primarily from overhunting by non-Indians, but in the 1960s, the Tribe had reintroduced a small herd onto the Reservation, where they thrived, expanding their population onto adjacent land and bringing considerable recreational opportunities to nearby communities and revenue to the state.

The hunters had valid hunting permits issued by the Tribe. Nevertheless, on leaving Mescalero lands, they were stopped and searched, and had their elk meat confiscated by federal fish and wildlife agents, who had been alerted by state officials. Even though the hunt occurred on tribal land, at that time the New Mexico Department of Game and Fish required non-Indian hunters to also have state permits. The state brought charges against the men in magistrate court, and both pleaded guilty and paid $600 fines. The Tribe, which wished to press the issue to assert their tribal sovereignty, offered to pay Oliver's and Shaw's legal expenses if they would challenge the verdicts. This set in motion one of the most significant legal battles for tribal sovereignty over wildlife resources in US history, one that eventually worked its way to the US Supreme Court. There, in 1983, the Tribe prevailed by a vote of 9–0, the Court agreeing that the Tribe could issue big game permits to whomever it wished, without state approval or interference.

It may seem surprising to those unfamiliar with the intersection of tribal, state, and federal natural resources law that as late as the 1980s, tribes lacked the full legal authority to manage their own game populations. As Julie Thorstenson describes in chapter 1, in the continental United States, tribes have jurisdiction or management influence over more area than the National Park Service and a long record of sustainable land stewardship. However, to tribes fighting to gain basic jurisdictional authority, the refusal of federal and state governments to acknowledge tribal control is not surprising at all. As described by many of the scholars in part I, including the founders of the Native American Fish and Wildlife Society (NAFWS) in chapter 2, the legal and administrative battles to gain this right, along with accessing adequate resources to carry it out, have been long, arduous, and contentious.[1]

Another message we hope comes through in this volume is the diversity of tribal landscapes, cultures, attitudes, and approaches. Though overall perspectives on sovereignty, funding, and collaboration are common to many tribes, issues such as legal challenges to specific jurisdictional rights, the allocation of water in a time of drought and climate change, co-management agreements affecting tens of millions of acres and hundreds of species, and the co-production of research differ widely across the continent in specific tribal contexts and will play a large role in wildlife management in the United States in the 21st century. The 574+ federally recognized tribes in the United States represent a stunning array of languages, histories, and landscapes. Native peoples on this continent have interacted with the environments they inhabit in radically different ways, from traditional hunting-based societies to highly complex agricultural mega cities (Kraniak 2016). One could reasonably argue that doing adequate service to this topic would require more than 500 separate volumes. We celebrate this diversity and complexity while we also draw attention to several places where tribal interests and attitudes converge, and where tribes stand mostly unified in their vision for sustainable and sovereign management of their lands and wildlife.

Peer-reviewed studies have demonstrated that around the world, Indigenous-managed lands have more intact biodiversity than other lands, even those set aside for conservation (Corrigan et al. 2018, Schuster et al. 2018). In the 21st century, tribes in the United States have integrated cutting-edge technical capabilities, research sophistication, and management expertise, and they are developing some of the most innovative wildlife management programs in the country. This story has largely not been told. In a few cases, academic institutions such as Northern Arizona University are adjusting their curricula to provide students with the opportunity to learn from tribal resource professionals and their work in sustaining Native culture, community, and traditions (Hoagland et al. 2017). However, for many reasons, tribal resource professionals tend not to publish in mainstream scientific journals, and they shy away from overt publicity. Humility is a ubiquitous core value among Native communities. In that sense, this volume is breaking new ground: for several of the book's contributors, this is their first academic publication. We hope these chapters will be of interest to both Indian and non-Indian readers, students, and practitioners in wildlife and other natural sciences.

WHOSE HISTORY?

In popular nonfiction, in undergraduate and graduate wildlife biology textbooks, and in peer-reviewed articles, John Muir, Theodore Roosevelt, Gifford

Pinchot, Aldo Leopold, and Rachel Carson are often lauded as the founders of conservation, and their writings and accomplishments form the basis for the modern "conservation ethic." It is undeniable that these individuals were among the first to write from a European perspective about conservation as a distinct discipline. They promoted and enacted laws and policies that cemented the protection of wildlife as a central American belief. Yet for several thousand years, Native Americans had managed the continent's wildlife quite well—conducting prescribed burns to move and create forage for game, regenerating habitat, and sustainably harvesting wildlife and fish while observing, documenting (usually orally), and responding to environmental and wildlife population changes—the definition of what we now call "adaptive management." But several of the wildlife sciences' so-called founders also were quite harsh in their views of Native people and their land stewardship. In an attempt to convince the American public to suppress all wildfires, Leopold (1920:12–13) wrote that it was "absurd to think that the Indians had any idea of forest conservation in mind." Leopold called the light burning practices of tribes a "fallacy" that caused "destructive effects." His writings insulted and harped on Native land management practices, leading to a century of full fire suppression with immense devastating impacts. Pinchot was a proud member of the American Eugenics Society (Wohlforth 2010), and Roosevelt's comments on the Sand Creek Massacre of hundreds of Cheyenne and Arapahoe men, women, and children paint the nation's 26th president in a highly unflattering light (Schwartz 2020). The Sierra Club has even acknowledged and apologized for Muir's racist views (Melley 2020).

In the 21st century, many Indigenous and other scholars of environmental and wildlife science have advocated for revising or "indigenizing" the North American Model of wildlife management, which has formed the basis for wildlife education and management for nearly a century. Though it is predicated on a sound base of science and adaptive management and serves as the model for almost all wildlife management at the federal, state, and local levels, the model tends to minimize or exclude issues of tribal sovereignty and to dismiss Traditional Ecological Knowledge as incompatible with Western science. Hessami et al. (2021) identified several shortcomings (from an Indigenous perspective) in the North American Model and proposed a revision. Similarly, Eichler and Baumeister (2018) believed that the traditional model does not serve tribal interests and needs to be heavily revised if it is to address them.

As several authors in this volume describe, tribal worldviews on the environment are indistinct from other aspects of Indigenous daily life and spiritual practice. They are what informed the conservation ethic even though few academic volumes on wildlife management contain more than passing references to Indigenous conservation practices or philosophy. This may be partly due to a lack of written records left by the tribal practitioners and to a reticence by tribes to speak and write about their accomplishments.[2] But it also is a result of differing perspectives about what warrants inclusion in the history of the science and whose history gets told.[3] Further, there is an uneasiness by some scientists to cross into what is viewed as other disciplines. Traditional or Indigenous conservation practices are often thoughtfully documented in the literature of the anthropological or sociological sciences while remaining unknown to practitioners or students of wildlife conservation. An interdisciplinary approach and including a wider range of voices in discussions about wildlife conservation may be changing this tendency.

TRIBAL SOVEREIGNTY AND TRUST RESPONSIBILITY

If there is one theme that underlies almost all of the legal and technical chapters in this volume, it is the struggle of tribes to assert their inherent sovereignty. From the earliest contact between foreign nations and the Indigenous people of this continent, the sovereignty of tribes was recognized in law, if not always in practice. As US Supreme Court chief justice John Marshall noted in 1832: "Indian Nations [have] always been considered as distinct, independent political communities, retaining their original natural rights, as the undisputed possessors of the soil. . . . The very term 'Nation' so generally applied to them means 'a people distinct from others.'"[4]

This ostensibly undisputed right of tribes to manage their own affairs has been challenged, ignored, or actively subverted in countless ways. Today, when projects or policies that could affect Indian resources occur on federal or state land or on tribal land with federal or state funding or support, consultation with affected tribes is required by several federal laws, directives, and executive orders. Yet tribal interests are often an afterthought or are grouped with those of the general public—one more stakeholder among hundreds. As one tribal elder put it during a recent (highly flawed) consultation process:

> What does it take for a government agency to involve us at the beginning stages and say, "Hey, we're planning this project; what do you think?" Try to work with Native communities, instead of [having us be] the last check-off because they have to. What is it going to take for agencies to think that way? Why is the tail wagging the dog? I think some of it is just ignorance. Somebody is responsible for this part of it, somebody's responsible for that part, and there is just lack of communication and overall vision. I think there's a path forward to a better way.

Another flawed consultation that made national headlines was the highly contentious effort to run the Keystone XL pipeline through areas sacred to the Sicangu Lakota Oyate, Assiniboine (Nakota), and Gros Ventre (Aaniiih) Tribes, though this process was halted during the writing of this book.

The other core legal concept that greatly affects tribes' ability to manage their affairs is the doctrine of trust responsibility: the requirement that the federal government protect and support tribal self-government and provide tribes with the resources they need so their members can have economically and socially prosperous lives. These rights were guaranteed, of course, in exchange for the hundreds of millions of acres that tribes ceded, upon which the entire social and economic life of the United States is built. The intertwined and counterbalanced concepts of trust responsibility and tribal sovereignty—giving tribes the financial and technical means for their members to carry on their lives on a par with the rest of US society while also allowing them the leeway to do so on their own terms—are frequently referenced in the chapters that follow.

THE UNITY OF ALL LIFE

An element of the tribal conservation ethic that came through clearly in discussions with the elders interviewed for this book is the unity of all life. As one Hopi elder remarked:

> Energy is connected throughout the world, through all of us, and these ancient sites. We are all connected. . . . There's energy there, and that's how we are connected together. . . . [There may be] a forest with a tree on the eastern side of the forest, and another tree way on the west side. It may seem they are far apart, and have no ties together, but actually they do, through the root system. Just like the spider's web, they are all connected together, they are all one. And we are the same way too. We have energy, you have energy, we are all connected together, we're bonded together.

This long-held tribal view of the natural world, often dismissed by Western science, has been confirmed with sometimes surprising discoveries. For example, the aspen trees in a grove are a single individual organism, and other findings have borne out the idea that *entire plant ecosystems*, including diverse and ostensibly competing species, communicate and share resources through their root systems (Guo et al. 2008, Simard 2018).

While Western science has made impressive strides in understanding discrete elements of the natural world, partly by parsing scientific disciplines into finer and finer areas of specialization, it's important to take stock of what has been lost in the process. To take just one example, the damming, dredging, and channelizing of nearly every major river in the West may have enabled the settlement of previously uninhabited areas and the "productive" use of water, but at a cost of near extinction of dozens of species and a loss of ecosystem integrity and great beauty. In this sense, perhaps the very term natural "resources"—things to be taken advantage of for utility or profit—is distracting and discordant. Tribes tend to view plants, animals, and humans as equal parts of an integrated system. This perspective generally stands at odds with the nontribal view that the natural world is something apart from people, made for us to profit from or have dominion over. The sentiment was expressed by Fond du Lac Band of Lake Superior Chippewa Vern Northrup during an interview for this book: "We feel our place in our soul. . . . For us it is part of our body. It is not that we own it, it is part of who we are. Without it we won't be. We have to take care of it and humble ourselves."

The nearly universally held tribal view that all species, including humans, are related has moved beyond metaphor to documented scientific fact. Go back far enough in the family tree and that deer (or mosquito or tree) could be viewed as your cousin.

ABOUT THE ART, POETRY, MEMOIR, AND SPOKEN WORD

As we contacted authors and gathered material for this book, it became clear that the scope of what we wanted to present could not be adequately conveyed by a series of technical and legal chapters, well written and comprehensive as they may be. Some of the diverse voices we wanted to capture were not amenable to that format, and many of our busy colleagues did not have the time to write about their work or much interest in preparing an academic-style manuscript. But more important, tribal wisdom has been, and largely still is, passed along orally and visually. To represent this facet of the tribal relationship to the natural world, we needed to include tribal elders, whose knowledge has been relied on for millennia to guide the use, protection, and management of fish and wildlife; tribal storytellers, whose insights into the natural and human world offer so many lessons; and tribal artists, whose ideas are represented through visual imagery.

We thank these contributors, whose work adds important context to the more technical chapters of this volume. As our Native family tells us, often the beauty, significance, and spiritual connection of wildlife cannot be described in the English language. We hope the inclusion of creative works in this book serves as a symbol for the multitude of dimensions through which wildlife can be perceived, valued, and interpreted.

For tribal fish and wildlife managers and the collaborators working with them, we hope this volume will provide a sense of recognition, camaraderie, and pride. For non-Indian colleagues, students, and others, we hope that developing a deeper understanding of the complexities of working with tribes will provide opportunities for new ideas, innovative collaborations, and most of all, enhanced protection and sound management—make that *co*-management—of the wildlife and lands we all need.

NOTES

1. Under current law, tribes are not eligible to receive funds under the Federal Aid in Wildlife Restoration Act or the Federal Aid in Sport Fish Restoration Act—known by the names of their legislative sponsors: the Pittman-Robertson (16 USC 669–669i; 50 Stat. 917) and Dingell-Johnson (16 USC §§ 777–777l) Acts. These sources, which are supplied by an excise tax on sporting equipment, have distributed over $23 billion to states and territories.

2. The lack of written records does not detract from the reliability of oral histories and transmitted observations, which in many tribes have been not just recounted but

actively taught annually for hundreds or thousands of years. See, for example, Cruikshank 2002, Babcock 2012, Williams and Riley 2020.

3. Reexaminations of the sometimes racist and anti-Indian legacy of the early conservation movement have also shed a more nuanced light on the forebears of the field. See, for example, Johnson 2014, Purdy 2015, Melley 2020, and Schwartz 2020.

4. Worcester v. Georgia, 31 US (6 Pet.) 515, 561 (1832).

LITERATURE CITED

Babcock, H. M. 2012. [This] I know from my grandfather: the battle for admissibility of Indigenous oral history as proof of tribal land claims. American Indian Law Review 37:19–29.

Corrigan, C., H. Bingham, Y. Shia, E. Lewisa, A. Chauvenet, and N. Kingston. 2018. Quantifying the contribution to biodiversity conservation of protected areas governed by Indigenous peoples and local communities. Biological Conservation 227:403–12.

Cruikshank, J. 2002. Oral history, narrative strategies, and Native American historiography: perspectives from the Yukon Territory, Canada. Pages 3–17 in N. Shoemaker, editor. Clearing a path, theorizing a past in Native American studies. Routledge, New York.

Eichler, L., and D. Baumeister. 2018. Hunting for Justice: an Indigenous critique of the North American Model of wildlife conservation. Environment and Society: Advances in Research 9:75–90.

Guo, D., R. J. Mitchell, J. M. Withington, P.-P. Fan, and J. J. Hendricks. 2008. Endogenous and exogenous controls of root life span, mortality and nitrogen flux in a longleaf pine forest: root branch order predominates. Journal of Ecology 96:737–45.

Hessami, M. A., E. Bowles, J. N. Popp, and A. T. Ford. 2021. Indigenizing the North American Model of wildlife conservation. Facets 6(1):1285–1306.

Hoagland, S. J., R. Miller, K. M. Waring, and O. Carroll. 2017. Tribal lands provide forest management laboratory for mainstream university students. Journal of Forestry 115(5):484–90.

Johnson, E. M. 2014. How John Muir's brand of conservation led to the decline of Yosemite. Scientific American, 13 August.

Kraniak, D. 2016. Conserving endangered species in Indian Country: the success and struggles of Joint Secretarial Order 3206 nineteen years on. Colorado Natural Resources, Energy and Environmental Literature Review 26:321–49.

Leopold, A. 1920. "Piute forestry" vs. forest fire prevention. Southwestern Magazine 2 (March):12–13.

Melley, B. 2020. Sierra Club apologizes for founder John Muir's racist views. Washington Post, 22 July.

Purdy, J. 2015. Environmentalism's racist history. New Yorker, 13 August.

Schuster, R., R. R. Germain, J. R. Bennett, N. J. Reo, D. L. Secord, and P. Arcese. 2018. Biodiversity on Indigenous lands equals that in protected areas. BioRxIV:10.1101/321935.

Schwartz, J. 2020. As Teddy Roosevelt's statue falls, let's remember how truly dark his history was. Intercept, 22 June.

Simard, S. W. 2018. Mycorrhizal networks facilitate tree communication, learning, and memory. Pages 191–213 in F. Baluska, M. Gagliano, and G. Witzany, editors. Memory and learning in plants: signaling and communication in plants. Springer Press, New York.

Williams, B., and M. Riley. 2020. The challenge of oral history to environmental history. Environment and History 26:207–31.

Wohlforth, C. 2010. Conservation and eugenics. Orion Magazine, June.

Chapter 1

Diversity and Complexity of Tribal Fish and Wildlife Programs

| JULIE THORSTENSON

INTRODUCTION

Many people are unaware of the number of tribal nations in the United States, how much land they manage, and how diverse they are. There are currently 574 federally recognized tribes in the country (Department of the Interior 2021), and others are still fighting for recognition. Overall, American Indian and Alaska Native tribal nations cumulatively manage or influence a land base area that is larger than that of the National Park Service. In this chapter I focus on tribes in the lower 48 states and offer a brief overview of the factors that contribute to the diversity and complexity of tribal fish and wildlife management programs. I provide an overview of several decades of events that have impacted tribes and shaped their management of wildlife and other natural resources.

THE ERAS OF TRIBAL-FEDERAL RELATIONS

Utter (1993:26) describes the United States' ever-evolving relationship with tribes over two centuries as

> how best to (a) remove the Tribes from the settled parts of the country, (b) conquer them, (c) establish and keep them on reservations and away from American society, (d) take their reservation lands away, (e) extinguish their culture and absorb them within American society, (f) reorganize them for renewed self-government, (g) terminate their self-government, and (h) establish opportunities for Indian Tribes to determine for themselves what their futures should be.

This complex relationship has resulted in heartbreaking losses of cultures, languages, lands, and people. It is estimated that between 50 million and 100 million Native people inhabited the Western Hemisphere at the time of Columbus's arrival (Denevan 1976 in Utter 1993). The American Indian / Alaska Native population in the 2020 US Census was around 6.79 million.

European invasion led to hundreds of years of genocide and encroachment on Indigenous lands, though at times some tribes fought as allies alongside colonial powers. Tribes were recognized by the United States as sovereign entities from the beginning of the establishment of the country, which set the tone for US-Indian relations—in theory, if not always in practice. The first treaty between the fledgling United States and an American Indian tribe was with the Lenape (Delaware) in 1778—before the War of Independence had concluded. This was the first of 370 treaties (Utter 1993) prior to the termination of the practice by the US Congress in 1871 (Legal Information Institute n.d.). The fact that tribes live within the exterior boundary of the country also makes them subject to certain aspects of US law, though the intersection of federal, state, and tribal law and the 370 treaties signed between the United States and tribes makes this area of jurisprudence complex.

As the philosophy of manifest destiny began to take hold, subsequent dealings between the United States and Indian tribes dramatically changed. The *Removal Era* (1820–1850) became official government policy with President Andrew Jackson's signing of the Indian Removal Act in 1830. This act, which sought to move all the tribes in the eastern United States across the Mississippi River, resulted in tens of thousands of Native Americans forcibly removed from their ancestral homelands (Bowes 2016).

The United States adopted a new approach in 1850, referred to as the *Reservation Era*, in which the country sought to create reservations in a tribe's traditional territory. Within these areas it was assumed that tribal culture, law, and government would continue relatively undisturbed by the outside settler community.

However, the confinement of tribes to certain areas did not end federal involvement in Indian affairs. From about 1880 to 1930, the United States, still seeking to solve the "Indian problem," commenced a policy to assimilate Native Americans into the larger society. The *Assimilation and Allotment Period* was an effort to bring Native Americans into mainstream society by changing their customs, dress, occupations, languages, religions, and philosophies (Utter 1993) or, as US Army officer Richard H. Pratt infamously stated, "Kill the Indian and save the man." During this period, in 1887, Congress passed the General Allotment Act (Dawes Act), which began breaking up reservations by granting land allotments to individual Native Americans. By the 1920s the failure of the policy and the poverty it caused had become so evident that it could not be ignored, and the United States moved to restore some self-governance and resources to tribes with the Indian Reorganization Act (IRA) of 1934 (Public Law 73-383). The act promoted tribal self-governing powers. Many tribes organized under the IRA and developed constitutions and tribal governments based on the US system of secular representative government, including elections.

Tribal sovereignty again was threatened during the *Termination Period*. From 1945 to the 1960s, US policy sought to legally eliminate the special relationship between the tribes and the federal government while dismantling tribes' authority over their reservations. The Termination Period also included the Bureau of Indian Affairs (BIA) Relocation Program and the extension of state jurisdiction into Indian Country, which shifted significant areas of federal criminal jurisdiction to the states (Gonzales et al. 2005). Between 1954 and 1964 more than 100 tribes were terminated, and over 2 million acres of land were removed from trust status, the legal means by which the United States stewards tribal lands for tribal benefit.

The *Self-Determination Era* began in 1968 with the passage of the Indian Civil Rights Act, which for the first time granted Native Americans full access to the Bill of Rights, including religious freedom, the right of habeas corpus, and other rights. In 1975, the Indian Self-Determination and Education Assistance Act allowed tribes to access federal funds to undertake services that would otherwise be provided by the federal government. Hundreds of tribes today use this authority to contract for programs of health, education, law enforcement, and management of natural resources.

INDIAN REORGANIZATION: LANDMARK COURT CASES FOR TRIBAL FISH AND WILDLIFE MANAGEMENT

Over the past half century, dozens of specific court cases have impacted the way tribes, states, and the federal government interact with regard to fish and wildlife management. As tribes have exercised more sovereignty and self-governance, conflict has arisen, often with states. In *Menominee Tribe of Indians v. United States* (391 US 404 [1968]), the US Supreme Court held that the hunting and fishing rights granted or preserved by the Wolf River Treaty of 1854 between the United States and the Menominee Tribe survived the Termination Act of 1954. *The White Mountain Apache v. the State of Arizona Department of Game and Fish* and *Confederated Tribes of the Colville Indian Reservation v. State of Washington* also dealt with questions of hunting and fishing rights. These two cases were examples of tribes marketing the opportunity to hunt and fish on their reservations in Arizona and Washington to non-Indians. The rulings in both cases reaffirmed that tribes have jurisdiction over the fish and wildlife on their lands.

In *New Mexico v. Mescalero Apache Tribe* (462 US 324 [1983]), the US Supreme Court affirmed a tribe's ability to sell hunting and fishing tags to non-Indians, stating the "exercise of concurrent jurisdiction by the State would effectively nullify the Tribe's unquestioned authority to regulate the use of its resources by members and nonmembers, would interfere with the comprehensive tribal regulatory scheme and would threaten Congress' overriding objective of encouraging tribal self-government and economic development."

In 2019, the Supreme Court decided another tribal rights case, *Herrera v. Wyoming* (587 US), ruling in favor of off-reservation hunting and fishing treaty rights for the Crow Tribe.

FUNDING

Funding—or rather, the lack of it—has been a central issue of concern for tribal fish and wildlife management. For some tribes, the IRA was the first step in developing their fish and wildlife programs. In contrast, states have had access to funds to manage and implement programs through the Pittman-Robertson Federal Aid in Wildlife Restoration Act (1937) and the Dingell-Johnson Federal Aid in Sport Fish Restoration Act (1950). This system generates revenue from an excise tax on fishing gear, guns, ammunition, and other hunting supplies, and has channeled millions of dollars to the states for wildlife and habitat management. PR/DJ funds, as they are often called, are the backbone of state fish and wildlife funding, providing almost $415 million in 2021 to states and territories. Since the inception of the programs, more than $24 billion has been distributed (USFWS 2021). Tribes are not eligible for any of this money, despite tribal members being required to pay both taxes. Tribes have tried many times to amend PR/DJ to include tribes in the distribution of funds, but so far without success. Tribes must often cobble together funding from various sources, including game license revenues, grants, and contracts with the BIA.

In 2000, the State Wildlife Grant program instituted a funding formula based on land area and population, which provided revenue to states for fish and wildlife management. In its first 20 years, the program awarded $1 billion, an average of $1 million per state per year. By contrast, the Tribal Wildlife Grant (TWG) program, which was established in 2001, is a *competitive* award process for federally recognized tribes with a maximum grant award amount of $200,000. The 574 tribes in the United States have been forced to compete for a small fraction of what is given to the states. From 2002 to 2020, the US Fish and Wildlife Service awarded $94 million through the TWG, providing support for about 506 projects: fewer than one project per tribe over the course of the program, and an average of about $9,000 per tribe per year. (It bears repeating that many tribes manage land areas larger than those of several states. The Navajo Nation, with over 17 million acres located across 3 states, is larger than 10 states.)

Tribal Wildlife Grants have had an undoubtedly positive impact for tribal fish and wildlife programs; however, the inequity in funding is hard to overlook. The TWG program is funded at a significantly lower amount than the State Wildlife Grant program, historically between $4 million and $7 million per fiscal year. The small revenues that tribes use to fund their fish and wildlife programs, such as hunting and fishing license fees, vary greatly based on those resources, accessibility, and marketing. Other types of revenue include land lease fees and casino income, but these also vary widely since not all tribes have large land bases or casinos.

The lack of annual, sustainable funding impacts the ability to recruit and retain professional staff. Many tribal fish and wildlife programs are small and lack capacity, which makes it difficult to engage in and be represented in national or regional conversations about issues such as endangered species and co-management. Tribes are also left out of large projects for which federal and state agencies seek input and invite tribal members to meetings.

THE DIVERSITY OF TRIBAL LANDS AND PROGRAMS

Tribal fish and wildlife programs are as unique and individual as the tribes that run them. Their size and success depend on many factors. The Cheyenne River Sioux Tribe Game, Fish and Parks Department began in 1935 with the Tribe issuing furbearer licenses under its newly formed constitution. Many formal programs were established in the 1970s after tribes were able to contract for the wildlife management that was undertaken by the federal government. The complexity of the programs varies from tribes with one conservation officer or licensing agent to those with more than 200 employees and multiple divisions, such as research branches, cultural preservation offices, fish hatcheries, eagle aviaries, and a zoo.

The 140 million acres of tribal land include more than 730,000 acres of lakes and reservoirs, 10,000 miles of streams and rivers, and 18 million acres of commercial forests. These lands and waters provide habitat for fish and wildlife, including more than 500 threatened or endangered species.

Through various acts of Congress, many tribes have "checkerboarded" reservations—tribal land interspersed with other jurisdictions, including state, private, and federal—a factor that adds another layer of complexity when managing hunting and fishing. Some of this is a result of reservations being opened to homesteaders and lands sold and placed in deeded status instead of remaining under the control of the tribe.

Tribal land also has several status types, including tribal trust lands, individual allotments, deeded lands, and public domain lands. Besides these different types of land tenure across reservations and the public domain, there are several tribes that have interests across international borders. There are two dozen tribes that live along the US-Mexico border and the US-Canadian border, with tribal members on both sides. Numerous territories with international crossings can be found in Alaska on the Bering Sea, the Pacific Ocean, and the Beaufort Sea. And this picture is not static: land status in Indian Country is constantly changing as more lands are taken into trust status and as tribes purchase additional properties.

TRIBAL ORGANIZATIONS

In order to increase cooperation and the strength of their voices, to help manage their collective resources, and to maintain their treaty rights on and off reservations, many tribes have formed commissions that are designed to deal with larger regional issues. One of the first of these was the Northwest Indian Fisheries Commission (NWIFC), which was established in the wake of the

United States v. Washington court case (also known as the Boldt Decision; see chapter 7). NWIFC is made up of 20 treaty tribes in western Washington. This case reaffirmed the tribes' treaty-reserved fishing rights and recognized tribes as natural resource co-managers, along with the state of Washington, with an equal share of the harvestable salmon.

The Columbia River Inter-Tribal Fish Commission (CRITFC) was established in 1977 after several court cases and legislative actions reaffirmed tribal treaty fishing rights.

The creation of the Great Lakes Indian Fish and Wildlife Commission (GLIFWC) was another milestone in the development of tribal fish and wildlife management. Formed in 1984, the GLIFWC represents 11 Ojibwe tribes that have off-reservation reserved hunting, fishing, and gathering rights according to their 1837, 1842, and 1854 treaties with the United States. Additionally, the Great Lakes area includes the Chippewa Ottawa Resource Authority (CORA) and the 1854 Treaty Authority (2021). CORA (2021) serves as an intertribal management body for the five 1836 treaty fishing tribes.

In 1992, the Intertribal Bison Cooperative (now the InterTribal Buffalo Council, ITBC) was established with the assistance of the NAFWS. ITBC comprises 69 federally recognized tribes whose mission is to restore buffalo to Indian Country to preserve their historical, cultural, traditional, and spiritual relationship for future generations.

The most recent of these organizations is the Southwest Tribal Fisheries Commission (SWTFC), formed in 2002 in response to the closure of the Mescalero National Fish Hatchery, which served tribes all over the West, and the subsequent decrease in funding. The SWTFC is a coalition of 18 tribes in the Southwest that seek to assist tribal fisheries programs. This commission is not based on any off-reservation treaty fishing rights.

The Native American Fish and Wildlife Society (NAFWS) is the most accessible and efficient purveyor of information pertaining to tribal wildlife conservation and management. NAFWS is a national 501(c)3 nonprofit organization incorporated and headquartered in the state of Colorado since 1983. NAFWS's mission is to assist Native American and Alaska Native tribes with the conservation, protection, and enhancement of their fish and wildlife resources (see chapter 2 for a history of this organization). The NAFWS is a membership organization and the only national Native organization that provides technical assistance to tribes and to federal, state, and local governments, as well as others working in the area of tribal fish and wildlife resource management. NAFWS has seven regions and 227 supporting member tribes, meaning that 227 different tribes have expressed their support of the work NAFWS does through a tribal resolution. There are also individual members of the NAFWS.

Although each Tribe is unique, these intertribal organizations focused on common regional and national issues help amplify often unnoticed tribal perspectives. There are many other intertribal organizations that serve a similar purpose of uniting Indigenous voices around common concerns.

GROWING INFLUENCE

Tribes are increasingly being recognized as valuable and knowledgeable contributors to the restoration, conservation, and overall management of fish and wildlife species. Tribal management is vital to the recovery and protection of many threatened and endangered species. For example, the black-footed ferret (*Mustela nigripes*), the most endangered mammal in North America, has found success on many tribal lands in South Dakota and Arizona. The White Mountain Apache Tribe has been a crucial partner in the Mexican wolf reintroduction and recovery.

Tribes have also become involved with recovery projects due to species' cultural significance to them. The Eastern Band of Cherokee Indians in North Carolina is working with the endangered species the eastern hellbender salamander (*Cryptobranchus alleganiensis alleganiensis*; ju-wah in Cherokee; Caleb Hickman, pers. comm. 2020), several bat species, and the Carolina northern flying squirrel (*Glaucomys sabrinus coloratus*). The salamander and the squirrel are part of the Cherokee story of the first stickball games, providing important teaching lessons about inclusion and how to respect the underdog (Hickman, pers. comm. 2020). The Columbia River basin tribes are working to ensure the Pacific lamprey (*Entosphenus tridentatus*), a culturally important species as well as a traditional food source, is available for future generations.

Two of the more well-known species restoration projects in Indian Country are bison (*Bison bison*) and salmon (*Salmo* spp.), both considered to be first foods and important traditional animals for many tribes. The return of the Bison Range to the Confederated Salish and Kootenai Tribes in December 2020 through Public Law 116-260 continues tribal efforts to restore bison to their homelands, cultures, and diets, and also recognizes tribal self-governance and capabilities in wildlife management. The Northwest tribal commissions have a strong focus on salmon restoration. To restore salmon spawning migration disrupted by hydropower, efforts by tribes have resulted in dam removal and the construction of fish passages to improve fish habitat.

Tribes are influential in developing innovative ways to coexist with fish and wildlife. In 2010, the Confederated Salish and Kootenai Tribes led and directed efforts to achieve wildlife and wetland mitigation during the reconstruction of the main highway route through the center of their Reservation, resulting in the creation of 43 wildlife underpass crossing structures and one large overpass nicknamed the Animals' Bridge over US Highway 93. This was well before terms like "corridors" and "connectivity" were in mainstream conservation consciousness.

Tribes are also on the forefront of climate change and climate adaptability work. GLIFWC has a climate change program and a vulnerability assessment process that integrates Traditional Ecological Knowledge interviews (GLIFWC 2021). GLIFWC and other tribes are developing seed banks too. The Karuk Tribe of California has a Climate Adaptation Plan (Norgaard et al. 2019), detailing climate impacts and the Tribe's adaptation strategies. Tribes are working throughout the country in cooperation with the USGS Climate Adaptation

Science Centers and research entities like universities and US Forest Service research stations and their scientists.

COLLABORATING WITH TRIBES

Fish and wildlife do not respect political boundaries, so collaboration among all parties is essential for our shared resources to be protected. Tribes, like any other sovereign entity, want to protect their natural resources from threats. These include diseases, such as chronic wasting disease, and the control of invasive species, such as feral hogs by the Mississippi Band of Choctaw Indians and the unwanted expansion of northern pike into Native fisheries. There are many other examples of tribes working collaboratively with their neighbors on difficult multijurisdictional management issues.

Tribes have literally hundreds of issues to deal with and relatively little funding to accomplish their goals and mandates, many of which concern economic prosperity and the health of tribal members. At times, it can be a challenge to elevate fish and wildlife priorities to the top of a tribal government's list. Therefore, when working with tribes, it is important to be patient and respectful of the many challenges they confront on a daily basis. Tribes' lack of immediate response is not a clear indication that they do not care about the issue at hand.

The complexities of state-tribal relations also need to be considered. Some tribes work hand in hand with states and have cooperative management plans for certain species, or even the cross-deputization of conservation law enforcement officers. However, good working relationships are not universal. A state-tribal relationship can be impacted by many things, for example, contradictory wildlife management plans. A Tribe may be managing for the subsistence hunting of big game, ensuring the animal population is large enough to support every tribal member, while at the same time a state may be managing based on depredation complaints and trying to decrease a population. Fire management, too, has the potential for tribes and states to be at cross-purposes. Tribes historically used fire as a management tool, and the model of fire suppression often used in state natural resource management plans is contradictory to what many have been taught.

Another factor to consider is connecting with the right person. Many first attempts at tribal engagement by outside agencies go directly to the President or Chairperson of the Tribe, which is the appropriate avenue. However, it may take time for information to trickle down to the person doing the management or field work, and additional time for it to go through the tribal approval process. Ensuring adequate time—including significant face-to-face time—is essential in collaborative projects.

In summary, partnerships must be mutually beneficial. Many tribes describe being asked for input but not being included in the final outcome. A discussion of several issues related to collaboration with tribes can be found in chapter 17.

WHAT DOES THE FUTURE HOLD?

Tribes share a responsibility and a commitment to planning for the next seven generations. As tribes continue exercising their sovereignty and winning court cases affirming their sovereignty, they will grow more empowered and more capable of managing their own resources. Tribes will grow stronger together. As more Native Americans and Alaska Natives achieve college degrees in resource management, science, and technology and remain committed to their culture and Traditional Ecological Knowledge, tribes will grow stronger together. As tribes unite and assert the need for increased funding and for equity in funding, tribes will grow stronger together. As more fish and wildlife professionals acknowledge the validity and importance of including holistic management, tribes will grow stronger together.

For decades, tribes have had to tell their story as a condition of receiving services their ancestors paid for. Having people at agencies and NGOs that understand how Native Americans view their relationship with the United States is critical. For the first time, a Native American, Deb Haaland, is leading the US Department of the Interior, and tribes no longer need to preface every letter to the secretary of the Interior with an explanation of who they are and where they come from.

Native people and tribal nations are resilient, having faced hardships that the United States has, for the most part, forgotten. Erased from US history books and classrooms, tribes have maintained their sovereignty and culture despite efforts to eradicate them. They have true empathy for their fish and wildlife relatives, which makes them strong stewards of conservation. Tribes are great leaders with their unique perspective, Traditional Ecological Knowledge, and holistic management viewpoint. As more Native people learn their culture and speak their language, no matter if their audience understands them, they fulfill the commitment that their ancestors planned for them to care for Mother Earth and all of our relatives. As one elder stated, "Every time you speak your language, your ancestors smile. We are the ones our ancestors prayed for."

LITERATURE CITED

1854 Treaty Authority. 2021. https://www.1854treatyauthority.org. Accessed 13 July 2021.

Bowes, J. P. 2016. Land too good for Indians: northern Indian removal. University of Oklahoma Press, Norman.

Chippewa Ottawa Resource Authority [CORA]. 2021. Home page. https://www.1836cora.org. Accessed July 2021.

Denevan, W., editor. 1976. The Native population of the Americas in 1492. University of Wisconsin Press, Madison.

Department of the Interior. 2021. Indian entities recognized by and eligible to receive services [from the] United States Bureau of Indian Affairs. Federal Register 86(18).

Gonzales, A. R., R. B. Schofield, and G. R. Schmidt. 2005. Public Law 280 and law enforcement in Indian Country: research priorities. Publication NCJ 209839. National Institute of Justice, Washington, DC.

Great Lakes Indian Fish and Wildlife Commission [GLIFWC]. 2021. Home page. http://glifwc.org. Accessed May 2021.

Legal Information Institute. n.d. 25 US Code § 71—Future treaties with Indian tribes. https://www.law.cornell.edu/uscode/text/25/71. Accessed May 2021.

Norgaard, K. M., W. Tripp, E. Crosby, T. Soto, L. Hillman, S. Fricke, F. Lake, W. Harling, S. Worl, and K. Vinyeta 2019. Karuk climate adaptation plan. Karuk Tribe, California.

US Fish and Wildlife Service [USFWS]. 2021. Wildlife and sport fish restoration program. https://www.fws.gov/wsfrprograms/subpages/grantprograms/FundingIndex.htm. Accessed 25 July 2021.

Utter, J. 1993. American Indians: answers to today's questions. National Woodlands Publishing, Lake Ann, Michigan.

Chapter 2

A Vision of Unity and Equity

Conversations with the Founders of the Native American Fish and Wildlife Society and a Look toward the Future with Native American Youth

> NATHAN JIM, JOHN ANTONIO, GERALD COBELL, DOUG DOMPIER, RON SKATES, ARTHUR BLAZER, KEN POYNTER, SALLY CARUFEL-WILLIAMS, TAMRA JONES, MATHIS QUINTANA, SAMUEL CHISCHILLY, JOVON JOJOLA, ASHLEY CARLISLE, *and* JULIE THORSTENSON

In 1999, the annual conference of the Native American Fish and Wildlife Society was held at the Foxwoods Casino in Connecticut, on the Aboriginal land of the Mashantucket Pequot Tribal Nation. The society was celebrating its 10th year—a good time to look back to its founding. Over the course of the week and, in a few cases, in the ensuing months, Joanna Stancil and Pete Eidenbach conducted interviews with some of the founders of the society. These were transcribed but, until now, never published. These stories present a living history of the founding of one of the seminal tribal natural resources organizations in the United States.

NATHAN "EIGHT BALL" JIM

The Native American Fish and Wildlife Society was based on the vision of our elders, our people before, which was to live a good life, seeking the prophecies that were brought out by songs, dreams, and messages from the Creator. These things have happened from the time the humans were made. We were part of creation, and our feelings and sense of the natural resources are part of that too. We are striving to make an organization where individuals can exercise their own way of thinking, share what they have learned, and teach others for the future. We are teachers, people who have wisdom, a people who have a history, a people who are concerned about our future.

The feeling of spirituality and the belief in creation is similar to all basic Indian ways of respect for the Creator. When the Indian was put here by the Creator, all things were made especially for the use of two-legged people, and we were charged with the responsibility of taking care of them. And one of the things we learned from our legends was there was always a way of protecting the values of our people. We were taught the morals; we were taught a way of life.

Our songs come from the longhouse. Our songs taught us of the creation—that a tear falling from somebody, maybe the Creator on his way down, that life began within that tear. The tear is full of liquid, and life came within it, and from this life inside, it grew and grew. And as it grew, the Earth was formed, and all the universe was like tears. They all grew with life, and they grew and the mountains erupted.

As the water receded, the mountains were given powers and names, and at the time of creation, they could walk and talk and have battles. We have legends of our mountains, and we have stories of how trees became part of it as the water retreated, and then [came] the game animals and the birds and the insects. The water went down to the plateaus, then to the high deserts. The junipers and lowland cedar, sagebrush—all became part of creation as the water went down. The last of the water formed rivers and lakes, and it became evident that there were people in the river—they were the fish people. The seashells, the whales that remained in the water, the fish, some of them were given a different way of life: to go upstream as the water flowed down, as the water came out of the Earth or came through the wonderful ways of creation as rain or snow, bringing water to the inland. We know that the water will be pure and a never-ending source that all life will depend on, will always live forever.

This is one of the many stories of creation, of how we came to be. We learned the prophecies that come from these songs in the longhouse, the rules and history, and the philosophy and all the things that we learned. The Creator talks to us through [these]. Our people lived according to the laws—you can't go out and waste a resource. We brought these teachings forward down through [our] history. The Indians wanted to protect the lands—they fought to protect their lands, practice their sovereignty, and protect their game because it was important for them, realizing that someone was encroaching on their way of life. In our way, we wanted to live a good life and went through battles and finally made an agreement. The tribes had to give up title to the lands because of conquest or an agreement.

I went through a boarding school. I was punished for speaking my language. I was punished for wearing long hair. I was punished for singing my songs. And I learned that I was a heathen, a sinner, that I was no good. I could change my ways, but I could never change the color of my skin, the color of my blood. It made me wonder about the white man's ways. I listened to the preacher talk about Jesus, God, heaven, and hell. I went to the church, marched like a little kid. Because of that, I became influenced by the white man in the sense that maybe I was rebelling against the feds and against the white people for putting me through something like this.

The Corps of Engineers built a dam that flooded my people's fishing place. [They] built a series of dams on the Columbia—they were decimating our salmon populations, changing the ecosystem. In 1957, they flooded where I lived and where I fished, so I left. I can't fish no more. There are no more falls,

I won't hear the echoes off the walls no more. So I left my country for the big city. What a shock! White people coming up out of the ground like ants. Jeez! Walking down the street and pretty soon—*Shoo wawawa!* People coming out of the ground. What the hell? And they push you and they push you and walk all over you. I finally went down and explored and got caught up with people rushing into the ground, and I ended up on a subway. I didn't know where I was going. Finally, I had to ask a policeman where I was and where I could go. He gave me a map. I knew how to read English, [but] it took me a while to get back. The subway experience was something. I wrote home and told my grandmother, "White people, there are so many of them. They are like ants, they live underground. If you are in the way, they run over you."

I got to see the statues of those famous white people. And I said, "Huh, there are no famous Indians." Abe Lincoln, George Washington, Grant's Tomb. There is nothing there for the Indian. They never recognize the Indians and the history of the tribes. In the United States the Indian is overlooked. We are artifacts, we are curious people, we are like aliens. We are to be studied, and they find our remains in the ground. And they say, "Oh, look here, I found some bones. I found a basket, I found some shells, I found some fish bones. People lived here."

So our way of life was really affected. We had to try to get along according to the agreement. But we kept on talking: "Wouldn't it be nice if we could form our own Indian organization made of our conservation officers, our police officers, our technicians, our biologists, our hydrologists, the grassroots people, and our elders, our treaty people? They're the ones we need to get the idea to. Our elected people, they're just politicians. All they know is how to go to Washington to talk to Congress and not get anything done. Let's do it ourselves."

We sat down and listened to all the ideas, the visions, the strategies, management objectives, the CFRs [Code of Federal Regulations], everything. We looked at them [and said], "Let's get our own constitution and our own bylaws and our own people. We'll make it nonpolitical, a nontribal council, an organization where our people, as individuals, can speak up—not wait for the politicians. Let them become members, let them speak out. They are the ones who are working, they are the teachers."

We sat down and [thought] of the things that we could see as a need, never knowing that we would be sitting [here] in a big casino. We met in Canada, with another Indian organization, the Intertribal Timber Council. There was Buzz Cobell, John Smith, Herschel Mays, Levi George, Doug Dompier, and [me]. We sat down and dreamt everything we could. We knew it was just a dream list. "If we are going to dream, then let's dream and write it down." The dream list was this:

1. We're going to have a national organization, [and] there's going to be a headquarters.

2. We're going to have regions. We're going to have regional officers and regional office funding for education, [for] law enforcement. We would therefore become like a Bureau [of Indian Affairs], but Indian-run.

3. We would create our own training for enforcement officers, court systems, education, camps, and development of new programs for the cultural use of products like baskets [and] drums. Use of the materials from the animals. And even work it into the Indian colleges. Yeah, we were really dreaming that we could go anywhere in the United States and find a regional office.

4. And we would have a resource center with qualified elders, speakers, and technicians who had special skills and education and talents.

We made sure we had one token white man, Doug Dompier. [laughs] But he was very frank about it. He agreed: "Don't let no white man ever get into the circle. He will straighten your circle out and take over." And then we used the practice of prioritizing and got together again and again to really go over what we thought could actually happen. About nine months later, Buzz Cobell had a conference and after that we decided, let's do it. And we went through with it and set up as an organization—a nonprofit organization. About that time, Buzz took over the administrative duties.

Our people from all over the US and Canada can have one voice if we have a common problem, such as law enforcement or the cultural and social needs for feathers for our way of worship. We believe in the words of Chief Seattle, that we are only a strand in the web of life holding together. We are but a strand. We can hold things together and become a web of life that can maintain our way of life. Through voicing our concerns, we can protect our way of life, we can protect our natural resources. Wherever the salmon are, as long as the buffalo are still here, we'll still exist. Some of the treaty language says, "as long as the rivers flow, the sun shines, and the grass grows." They knew what they were talking about when they said those things. We are going to be here one hell of a long time.

JOHN ANTONIO

Officially, the society incorporated in 1983. Many people throughout Indian Country [who] are involved in fishery and wildlife resource management had similar thoughts about forming a national Indian fish and wildlife organization. We all had similar concerns regarding communication and sharing of ideas. When I started working with the Navajo Nation in 1976, we had gatherings to talk about natural resource issues that were specific to the Navajo Nation. Then we started talking with other tribes, because we felt it was important to share expertise. We knew there were other tribes, particularly Mescalero, White Mountain, and Zuni, that were involved in similar projects. It was a simple network we established—just by talking. We knew the people who worked with us from agencies, and so a lot of times we had cooperative projects.

So it was this simple concept of helping one another, communicating with one another, that led to building this coalition. It was the commonalities that tribes had in the management of fish and wildlife and other related natural resources that brought us together. We built a network of folks that you could depend on. We knew we were limited with staff, funding, and equipment, so it was only natural that we just started to borrow and share expertise, assistance, and equipment.

One of the big problems we saw was that there was a lot of inequity in funding for tribal fish and wildlife programs. You know, the BIA has a trust responsibility, but other agencies have trust responsibilities too. We've been trying for years to have parallel legislation for federal aid for fish and wildlife restoration. Federal aid funds are not available for tribes, even though tribal members pay the same excise tax on guns, ammunition, boats, fishing gear, things like that. That money goes into a Treasury account administered by the Fish and Wildlife Service. Because of the way the act is written, only states and territories can benefit. But as Indian people, we pay that same excise tax when we buy a box of shells, a rifle, or a fishing pole, or whatever and come back to our reservations. The problem is that no funds [make it to tribes]. We began addressing that issue, began to have a more unified approach, to write letters of support and testify before congressional subcommittees. So far, we haven't been able to change the laws, but it's something we're working on. The Indian Fish and Wildlife Resource Management Act of 1993 is a good example of how we were able to join together and successfully draft legislation with the help of Senator [Daniel] Inouye [of Hawai'i] and Congressman Bill Richardson from New Mexico.

In the beginning, some tribes were a little bit skeptical about the society. What was its function going to be? When you go to Congress, are you really representing all the tribes? We acknowledged that concern. Tribal councils have their own authority, their own powers, they can act on behalf of the Tribe. But if they agreed with us, with the concerns or proposals that we wanted to take up, then that helped us. So, for example, if the Hopi didn't agree with something, that was fine, that was respected, but if they gave support, like a tribal resolution, that would make our society's [message] more meaningful. Our purpose was to serve as a forum where the tribes could voice their views and develop some common strategies and take that forward.

Everybody was wondering, can we really make a difference? I guess the philosophy was power in unity. We were never concerned that people wouldn't take us seriously. It was more a matter of if we could establish our goals and objectives, and what our purpose was. The fact that we had guys like Senators Inouye, [Dennis] DeConcini, and [John] McCain, former Congressman Bill Richardson, and even Ben Campbell, who supported us more recently. Even agency heads, like Mollie Beattie, the former director of the Fish and Wildlife Service. She took the time to sit down and talk to us. To me, that showed them that we were serious about resolving issues, and they believed that we could make a difference. It proved that these guys wanted to make a difference. The society meant something. The neat thing about it is that when you get a core group [of] people who believe in the cause, that's something that can never die.

GERALD COBELL

I became involved with the society during its inception. I recall tribes were rallying to respond to the proposed National Fish Hatchery closures that would impact tribal fisheries nationwide, as well as several federal court cases relating to tribal jurisdiction over big game. I became acquainted with tribal rangers who were concerned about tribal fish and wildlife issues at different meetings in the Northwest when I attended a tribal funding and fishing rights workshop in Seattle taught by Alan Stay in about 1980 or 1981.

In July 1981, the Warm Springs Tribe of Oregon and the Bureau of Indian Affairs cohosted a tribal fish and wildlife workshop at the Kah-Nee-Ta Lodge facility on the Warm Springs Reservation. It was there that I became acquainted with Nathan "Eight Ball" Jim, who became a stalwart member of the society and one of my good friends. Eight Ball helped us a great deal over the years and is probably best remembered for his leadership and sense of humor, especially when he acted as our auctioneer during fundraising auctions following our annual banquets. Eight Ball represented the Warm Springs on the Columbia River Inter-Tribal Fish Commission, so he was able to grease the skids, so to speak, with the powerful Northwest fish and wildlife commissions and tribes, to convince them that the formation of the NAFWS was not competition for limited federal monies, but another ally in the fight to protect and preserve tribal fish and wildlife resources.

Nathan used his influence with the commission to obtain the help of their attorney, Rob Lathrop, to draft our charter and bylaws and eventually help us get incorporated. It's my hope that the society can create some kind of memorial so our good friend Eight Ball will not go unremembered. It was during these two meetings that discussions began regarding the development of a national Indian fish and wildlife organization, and I agreed to work with the Blackfeet Tribe to put on a national meeting in Montana. I was able to get the support of Earl Old Person, lifetime Chief and longtime Chairman of the Blackfeet Tribe.

The reason the NAFWS was formed was to facilitate a national awareness of issues and opportunities facing tribal fish and wildlife programs and to provide a focal point for dealing with federal agencies, conservation organizations, and other entities influencing tribal natural resource programs. Most of us felt that the National Congress of American Indians and other national Indian organizations were not responsive enough to meet the specialized needs of tribal fish and wildlife programs. We needed to develop our own organization. At this point, both the Mescalero Apache and White Mountain Apache Tribes were engaged in legal battles with their respective states for control of big game hunting within their reservations. State versus tribal jurisdiction was a huge issue.

A number of people played important roles in the formation of the NAFWS: John Smith and Steve Judd from Colville; Nathan Jim from the Warm Springs Tribe; Wes Martel from Wind River; John Antonio from the Pueblo of Laguna; Phil Stago from White Mountain Apache; Don Sandier from the Rural Alaska

Resources Association; Herschel Mays from the Confederated Salish and Kootenai Tribes; Ed Fairbanks from Leech Lake; Levi George, Wilfred Yallup, Joe Jay Pinkham, Bill Bradley, Vince Piel Jr., and the whole Yakama Nation contingent; Beverly Houten and others from Pyramid Lake Fisheries; Doug "The Old Dog" Dompier from the Columbia River Inter-Tribal Fish Commission; John Banks from Penobscot; and probably many others I've forgotten.

The society reached a major goal when we were able to get Mollie Beattie, director of the US Fish and Wildlife Service, to attend our annual conference. It provided proof that we had become a national Indian organization that commanded the respect of federal natural resource agencies.

One very special memento I have from my tenure as President is a beautiful bolo tie that belonged to Levi George, a good friend from the Yakama Nation and an avid supporter of the society, who is probably now welcoming Eight Ball on the other side. We were getting ready to hold our auction, and obviously many members had forgotten to bring something to donate. I was inspired to remove a ring or belt buckle or something and offer it up to be auctioned. Soon everyone at the head table took some prized possession off, and Eight Ball began auctioning it off. I can't remember what I paid for Levi's bolo tie, but it's priceless to me now.

DOUG DOMPIER

I became involved in the society as it was being formed. I was one of the original folks sitting in the lounge and afterwards in a suite for the rest of the evening discussing with other staff and some of our commissioners the need for such an organization. That was a time when we all smoked and didn't mind having a beer or two after the day was over. It would have been sometime in 1981.

Since I am not a tribal member, I have never held any formal office but have participated on numerous committees aimed at setting up both regional and national conferences. I have been a nonvoting member since we started. During the early years, [I wanted] to ensure that the bylaws were written so that only Native Americans could be elected to the board of directors. I was very concerned that if nontribal folks were elected, they would tend to dominate the society. That was not to be our role. I sometimes wish that the bylaws had been extended to exclude tribal members who worked for the federal government. There seems to be a conflict of interest in government workers holding such positions.

During the time the society was formed, we were in some terrible conflicts throughout the US. There were pockets of us in the Northwest and other parts of the country who were fighting for the tribes' treaty rights. Some of these battles escalated into national fights, such as the attempt by western states to try and make steelhead a national game fish and prevent the tribes from selling them. This [spurred the tribes] to meet and begin to know each other and [find out] what was happening in other parts of the country. We needed a forum in which tribal policy leaders could meet and talk and develop strategies. For the professional employees of the tribes, many who were being shunned by state and federal employees, it gave them the opportunity to meet with their counterparts. Many of those early visionaries who met all night in 1981 in Kah-Nee-Ta Lodge in [Warm Springs,] Oregon, are no longer with us. They were the tribal leaders who had the vision of what was needed.

One of the early challenges was how the tribes dealt with the different philosophies of the various regions. Our folks have been very strong on tribal rights, and because of our frequent conflicts with the state and federal agencies, we had a great distrust for them (and still do). This was not uniformly held in some of the other regions, so the tribes had to find ways to accommodate the differences.

RON SKATES

I am a member of the Cheyenne River Sioux Tribe of South Dakota. I started my career as a tribal game warden 27 [or] 28 years ago. Then I went to school and graduated and became a professional wildlife biologist and went to work for the US Fish and Wildlife Service. I've been with the service for 25 years.

In my career, I had a tremendous interest from a technician's and, later, a professional's standpoint in trying to help the tribes. In 1985 we had our first regional conference. I helped put that together from the Great Plains Region—each region had started having their own conferences. We had our first conference in Bairoil, Wyoming. One of the big things that's always been an issue for tribes in the Great Plains Region is the lack of dollars—and unstable governments. We felt we needed to come together and try to work as a group, try to assist each other, not only bringing the major issues to the forefront but trying to get funding through organizations like BIA or the Fish and Wildlife Service. Code development and things like that are a big issue because some of the codes and regulations the tribes have are basically unenforceable.

It was very interesting because our priorities as a society and as tribes are significantly different from those of the Fish and Wildlife Service. For example, the biggest focus in our region, the Plains, was law enforcement operation and training—help with regulations, game code development. [People said,] "Train us, certify us. Secondly, we need funding." Third was wildlife management—of big game particularly. Because we are originally inland nomadic tribes, we're more in sync with the big game than, for instance, the fisheries of the coastal and Great Lakes areas. The fourth priority was restoration and recovery of habitat. And fifth was [establishing] an equal partnership [with federal agencies].

Being a tribal member on a reservation, working for the fish and game department is extremely political. Your tenure and your career are probably not going to last too long unless you really play the political arena well. The politics change so often it's sometimes very difficult to keep any continuity going. I was very frustrated when I worked for my Tribe. But my heart was really into work[ing] with the tribes.

I've always felt that you can go out and work with other tribes much more effectively than you can work with your own Tribe. I've worked with my own Tribe through the service and been effective because I'm not in that day-to-day political arena. Because I knew who to contact, I could provide them things. I could get people there, help in getting some of their ponds stocked, and those kinds of things. I've been working with my Tribe my whole career. But I do it from my US Forest Service capacity, not in their day-to-day politics. I can do things to assist or find them sources where they can get funding or training or whatever it is [they need].

One dear friend of mine, whom I've known for a long time, has some problems with those of us in federal agencies. He's told me that right up front, and he's said that in front of the group. I respect that. I know where he's coming from. He's told me, "Don't take this too hard, that's just the way I believe." I said, "We all have our own thoughts, our own ideas and challenges. You've been effective in that arena. You've been able to wade through all the politics and stay alive, and I admire that. I think that's wonderful. But not all of us are able to do that. So how can we help?"

One time, I had [a request for a visit] from one of the tribes in Montana. They were having some problems developing a game code. A traditionalist segment of their membership said, "We don't want any management. We're doing just fine." I've been in some pretty heated situations, but before we could even show the video, one fellow stood up and said, "Who are you?" I introduced myself, and he said, "You're just one of those bureaucrats that's coming here to give us a bunch of grief, and we'd just as soon you leave the Reservation." In fact, he said, "You know, you come in here, and you tell us all the white man's policy and what we need to do. We've done just fine." This person was really on the attack. Finally, I just took my badge off and threw it on the table and said, "Okay, we're just Indian to Indian." I said, "You know I don't appreciate your comments here. I don't appreciate them because if you really cared about your resources, you'd have more left on this Reservation. You don't have much left because you don't care. That really bothers me. I grew up on the Reservation. If there was a bingo game or basketball in town, you'd all be there. Where are all our people when we're talking about fish and wildlife resources that are important to all of us? Where are they? Don't always throw the blame on us bureaucrats. As Indian people, we have a responsibility."

I thought I was going to get run off the Reservation. I used a little Lakota language to kind of offset that, [and] he got real quiet. Later, he came up and said, "You're right. How can I help?" We're still struggling with that Reservation today, but they've come a long way.

When you talk about traditional and contemporary, one of the things I've heard at many of these meetings is "It's our right to hunt and fish, we can take anything we want." The first thing I usually tell them is "No, it's not necessarily your right, it's a tribal right. If your tribal government wants to give you that authority, yeah, you can. But it's not an individual right. Let's make that clear." We already see things changing. We have the youth practicums, and we come together at some of the schools. Some of these young people, they don't want [the tribes] harvesting these animals. So I think the generations to come are going to move out of that old philosophy that this is an individual right. Those values have started to change.

ARTHUR "BUTCH" BLAZER

[The formation of the society] was a result of people getting together within their regions and making the decision, "Yeah, let's go for it." The biggest issue was the frustration among all of us of not having what we needed to properly manage our resources. Initially, the strong regions were the Southwest, the Pacific, the Plains, and the Great Lakes. Those four regions were the ones who really got the ball rolling. Eventually we got in the East, which later broke into the Northeast and the Southeast. Alaska was the hardest to get because of its isolation. None of us had money to travel up there, and they didn't have any money to travel down here.

When I first got out of school, I began working for my Tribe—this was back in 1975. One of the first things I walked into was the *New Mexico v. Mescalero Apache Tribe* case—a monumental hunting and fishing case which Mescalero won in the Supreme Court in 1983. It really made an impression on me. I had to testify in district court straight out of school. It was quite nerve-wracking, being out in front of a judge and a jury. We won that battle with the state of New Mexico when our attorneys got the state game commissioner to say on the stand that the Mescalero Apache Tribe managed their wildlife better than the state of New Mexico [did]. That's what won our case for us at the district level. And I am sure that is what carried the day all the way through the Supreme Court. It was a precedent-setting case and meant so much for Indians across the country. That got me thinking early on about the importance of tribes planning and putting a strategy together. That impressed on me that communication was important among those of us that do fish and wildlife management work. The same thoughts were occurring to other people in other regions, and it brought us together.

We struggled for quite a few years without any staff. But we eventually were able to get some money through the BIA to hire our first executive director, Dewey Schwalenberg. Later, some other very good people got elected to the board. We came together and agreed that we needed to develop a strategic plan and get it going.

We have had strong support from the Washington office [of the US Fish and Wildlife Service]. Gary Rankel has always been very supportive. He helped the society get the Intertribal Bison Cooperative [now the InterTribal Buffalo Council] going. The BIA has always been very helpful in running 638 contracts through them.

A good example took place several months ago when we had a meeting in Albuquerque, cosponsored by the society and the Fish and Wildlife Service, concerning the direction that the Indian Interim Fisheries Program is going to go. The service is looking to change direction from a sports fishing operation to a native fish operation, which is going to have a negative impact on recreational fishing on Indian lands. I really commend the service for this—they wanted to bring all the tribes together to hear how they felt about it early on, before any decisions were made. We helped them do that because we felt that was exactly what they should do. But when our bureau people in Washington, DC, found out what we were doing, the assistant secretary, Kevin Gover, called the director of the service and said, "What in the hell are you guys doing pulling all these tribes together? You're causing a lot of problems!" He didn't want it to happen, but it happened—[and] I'm glad that it did.

If you take a look at our Constitution, the mission statement is broken down into four or five components. One is education, with a strong emphasis on student education. Education has always stood out as our number one mission component because of that. So we got our first national practicum in 1991, and we started having regional practicums in the Southwest, which we've done on a consistent basis. When we started talking about a youth camp or a youth practicum, it was to bring professional Indian managers and tribal elders together with students. I think the traditional people see what we are trying to do, and you'd be surprised—they understand that it needs to happen. They were very supportive of the people who we brought in, and they learned a lot themselves. The kids weren't the only ones that were learning. That's why we have such a good time when we have those practicums: it's a learning experience for everyone.

There is also a real need out there in regard to law enforcement. This includes the tribal courts, tribal government, developing strong tribal codes, working with other federal agencies so they will support what is happening on our reservations in regards to poaching, stealing wood, those kinds of things. Probably the major impact that the society has had is in regard to providing additional conservation law enforcement specialized training, forensic training, and identification of endangered species. We haven't really become a force yet in trying to influence or help agencies change policy that will have a positive impact on Indian Country conservation law enforcement. But I can see it coming. We have to make sure that we pick our battles well because we are a small organization with a limited budget.

Early on, I was quite adamant about not becoming a political organization. I wanted it to stay a technical organization. This issue really came to mind when John Antonio was President of the society. He came to me after we had a discussion about the spotted owl issue in the Pacific Northwest. It was having an economic impact on certain tribes up there, and I think it was the Yakama who wanted John to write a letter on behalf of the society saying that what they were doing was wrong. I said, "No, we can't do that, because that's up to the Tribe." You go to battle for the owl because a Tribe wants the owl protected, and then down the road you have another Tribe that is heavy into timber. We are trying to protect the resources, but from a political and social aspect, we don't have the right as an organization to tell tribes how we think they should go. On my

[Mescalero Apache] Reservation, if our former Chairman, Wendell Chino, had to make a decision between the welfare of his people and the spotted owl, he would say, "Get rid of the spotted owl. I can't have my people starving because we can't harvest timber." I'm sure what he would like to do is have both. But if [he had] to choose between a resource and his people, his people would always come first—they always have. And that is a political decision that each Tribe has to make.

KEN POYNTER

I attended the first eastern regional society conference in the fall of 1987 at Cherokee, North Carolina. I had been working for the Passamaquoddy Tribe for a couple of years as a tribal biologist. I felt like I was working in a vacuum, trying to manage this land base. Lo and behold, I [went] to this society meeting, and I met other people who were doing the same thing and had encountered the same problems. They gave me suggestions on how to solve them myself. That was a tremendous help to me. By being involved with this organization and attending its meetings, I was able to develop a national network of my peers, who helped me do my job better. All of a sudden, there was a resource for me to tap into. I could get copies of documents that would assist me in the development of plans that I was doing for my Tribe. I was able to look at other tribal laws, issues, and management plans. I was able to find and talk with people about the always-present problem of inadequate funding. That was a major problem that my Tribe was having in regards to the dollars from the federal government and funding the program I was working under. It was that type of thing plus being actually able to pick up the phone and call somebody across the country who I had met at a society meeting and say, "Hey, this is what I'm doing, do you have any suggestions?" [We could] discuss projects, compare notes, etc. To me, that's what I was looking for.

The major thing that the tribes continue to address in tribal natural resource management, is inadequate funding. That's something that I was involved in, and then the society got involved. Right after starting as executive director, I wrote some testimony for an oversight committee on Capitol Hill. Prior to setting up the office in Colorado, we had looked at a Native American wildlife enhancement bill, which would have created a new and separate funding source for tribal natural resource management. We always met strong opposition in DC from a group called the International Association of Fish and Wildlife Agencies. They are extremely powerful. They receive funding through the federal aid program administered by the Fish and Wildlife Service, and they have been adamant about opposing tribal participation in that program. We tried to get a champion on Capitol Hill to help out and introduce the enhancement bill, but we were not successful. The bill would have created a specific additional funding source for tribal natural resource management for tribes. It's one thing to get a bill passed and turned into law, that's the easy part. The hard part is to get funding.

SALLY CARUFEL-WILLIAMS

My initial position with the society in 1991 was secretary-receptionist, and shortly after coming to work here, I took on the duties as membership coordinator. Patty Schwalenberg's mother and my mother had worked together for 15 years, and we developed a cooperative working relationship. Executive Director Dewey Schwalenberg left the organization in 1992. Ella Mumford took over as ED in April that year. One of Ella's first actions was to hire a bookkeeper, and David A. Lee was brought on staff. I took over duties as executive secretary and secretary to the board of directors in 1992.

At the board meeting in Tampa, Florida, in 1992, the board of directors decided to sponsor the second annual Native American Environmental Awareness Summer Youth Practicum. The first practicum had been held in Beulah, Wyoming, at the Ranch A Lodge owned by the Fish and Wildlife Service. It was coordinated by Clara Spotted Elk from Northern Cheyenne. Twenty-five Native American high school students from around the country attended, and their essays, written at the close of the practicum, were published in the society newsletter.

The second practicum started the practice of using society members, tribes, and contacts as instructors. We had instructors from the Great Lakes Indian Fish and Wildlife Commission, the Menominee Tribe of Wisconsin, the Bad River Band of Lake Superior Chippewas, the Oneida Tribe of Wisconsin, the Lac du Flambeau Band of Lake Superior Chippewas, Minnesota Chippewa Tribe Mille Lacs, and the Cheyenne River Sioux Tribe [CRST] environmental project. Jim Garrett from CRST drove 600 miles to teach and has served as an instructor since 1995. Most importantly, because the society values the traditional teachings of tribal elders, each host Tribe provided an opportunity for the students to interact with those elders. It was great! Each host Tribe also provided a feast.

[At the practicums], we learned some good traditional ways that we have incorporated into our program. Our national students really became aware of the connections tribal people have to homelands. However, travel costs were outrageous, and the program was moved back to Colorado. For a region and Tribe to host such an endeavor takes incredible commitment and manpower, so the board has had the practicum in Colorado ever since.

Because the practicum has evolved from a career conference type format to that of an academic program, it was important to have a certified teacher on staff. Great Lakes Region member Adrian "Dusty" Miller, who was currently teaching an environmental class at Native American Educational Services College in Kenosha, Wisconsin, took a strong interest in the practicum and came on staff in the summer of 1996 to serve as lead instructor and curriculum coordinator. Mr. Miller has been with the program ever since, and at the 1999 national conference [he] won the Chief Sealth Award.

During the practicum, students and staff spend time participating in classroom sessions, field education, recreational activities, [and] traditional methods and generally interacting with professional, cultural, and spiritual people.

Each student writes an essay to get into the program and writes an exit essay. The students also give an extensive presentation on their Reservation, Tribe, and community. In addition, students participate in a group project incorporating the components of the program in a mock scenario of a tribal council session. All in all, students participate in over 100 hours of on-task activities.

◆

The interviews with NAFWS founders in the previous section were conducted in 1999 and 2000. A lot has happened since then, including the arrival of a new cohort of sharp, young, talented tribal wildlife professionals and students. On 12 July 2021, via Zoom, editors Serra Hoagland and Steve Albert gathered some of the founders who were part of the original interviews (Butch Blazer, Buzz Cobell, and Sally Carufel-Williams), several students who were interning with the NAFWS that summer (Tamra Jones, Mathis Quintana, Samuel Chischilly, and Jovon Jojola), the NAFWS education coordinator (Ashley Carlisle), and the NAFWS executive director (Julie Thorstenson) to see where they thought the society and tribal wildlife management are headed. Some comments have been lightly edited for clarity and consistency.

Tamra: One thing that stands out to me is, what opportunities do I see in the future? [Being from Alaska], I need to become more educated about other tribes in the States. In Alaska there's so much opportunity that hasn't been available before for Native students, such as the number of scholarships. When you apply for a job, it stands out that you are Native. That in itself is amazing, and it's a very good thing. During the past four years of school, I felt very supported by my community as an Alaska Native. I'm going to cross my fingers that it continues to grow. I'm having a daughter here soon. I hope that she is able to have opportunities to be able to go to school and be a part of her culture. My language, Ahtna, has been dying out, and a lot of the elders that speak the language are passing away. It's really sad to see the language dying, but there are classes being offered to teach youth the Ahtna language. So that's what I think the future will hold. I'm very optimistic.

Mathis: I heard of the NAFWS through the American Indian Science and Engineering Society. I'm a member at Colorado State University. As for the future, one thing that really stood out to me was how much the NAFWS has grown. Just talking with Ashley and others, they told me about what it was like when they first started. I see NAFWS growing and getting stronger, with outreach to more tribes. We have been developing a contact list, and it was amazing to see how many tribes have natural resource programs. It's eye-opening to see how much impact we can have.

Samuel: The way I heard about the NAFWS was through my dad and my brother. The internship is giving me good experience with what is involved with a wildlife studies major, or the career I want to go into. I've been learning a lot through this initiative. I feel like my culture is declining in the number of people who practice it, the language, and some of our ceremonies. I feel like that's something we should work on: our cultures and our traditions.

Jovon: I first heard about the Native American Fish and Wildlife Society through the Southwestern Indian Polytechnic Institute. One of my advisors took the students in our program to a conference. Starting this internship, it's been eye-opening to see how many tribes there are within the US and how the NAFWS is trying to help them grow their wildlife and natural resources departments.

Ashley: Thinking back to my undergraduate years, I found out about NAFWS through a Google search when I was looking for scholarships. At first, I didn't know there was a national office. I was invited to the Southwest Regional Conference, and I saw that . . . there's some kind of national organization here. Later, I was so happy to see the posting for the job I have now, because this is exactly what I wanted to do. Coming into the office and talking with Julie [Thorstenson] and [NAFWS public information officer] Karen Lynch, and meeting the board of directors, I got to know how much history there is within NAFWS and see how within the past couple of years we [have been] starting to grow. Having this internship program is one of the things that is starting to really engage our youth, getting them to pursue opportunities. Finding the balance of that longevity, but also bringing in new students or emerging professionals, and being able to pass the torch, I guess you could say.

Tamra was talking about how language is dying. I think there's a gap between how there are a lot of Native people living more in the cities now, but they want to learn their language and be part of their culture. I think there hasn't been the opportunity to have access and the opportunity to do that. A lot of those gaps are being filled right now in terms of phone apps and online courses. I think that's really cool. Living here in Denver, it's hard to reach out to my roots in New Mexico, because I'm Navajo and, even though I grew up on the Navajo Reservation, I also grew up going to a private Christian school. There are a lot of students I have met here at CSU that grew up in Denver or Albuquerque, not on the Reservation, so there's that disconnect. I think an opportunity in the future could be apps or online services that [could] help bridge that gap. That can also happen at NAFWS with apps for learning about native plants and Traditional Ecological Knowledge. But there's also the issue of how some of these lessons and knowledge can't be shared like that.

Serra: One of the things I'm really interested in is workforce development. What do you all sense as far as your peers and your social group, the people who are of your same age and your cohort—do they generally have an interest in natural resources, or is natural resources kind of low on the interest list? Are most people interested in technology or business or law?

Tamra: As a college student, I'm surrounded by a lot of natural resource students, so that might skew my perspective a bit, but I think it is definitely something. Interest in natural resources and the environment is something that has been growing, especially in the last 20 years. But because of all the issues and all the challenges that have been introduced—which weren't seen as a problem in the past—now we can see that if we don't do anything to help manage them properly, then we're not going to have these [resources] available anymore. I think that realization has sparked an interest in natural resources, and how we can manage them. Obviously there's interest in business, economics, law, but these are interconnected with natural resources management. They're all so vital to being able to properly manage and work together with different people with different backgrounds.

Ashley: I'm surrounded with natural resource students in my social circle. Where I come from, there are a lot of high school students that are looking into college, and it seems like they're more steered towards agriculture—which is a part of natural resources—but a lot of that is really about 4-H . . . the main part of which is showing animals, raising livestock, baking, and cooking. I think having some kind of organization that is like 4-H but more natural resource–based [would be helpful]. We don't have Boy Scouts or Girl Scouts where I'm from.

Steve: You all have grown up with technology, you've never been without it. When the NAFWS was being formed, there was no such thing as the internet. How do you use technology? How do you see the interface between technology and environmental protection, and what do you think the potential is for your generation, who are so comfortable with it, using technology to help manage wildlife or other natural resources?

Tamra: I have wondered that myself because, thinking back to what you were just saying, you guys have grown up without it, and we have always been exposed to it. So [with] all the resources that I've had available if I need help with a problem or I want to understand something, it's so easy for me to just go to Google and be like, "What is the NAFWS?" and it pops up and I can read all about it. So I think the potential for technology to be used in natural resource management is very bright in the future, but I definitely hope that it doesn't get abused, because I know that kind of power can be abused. But I definitely think that technology can be a very good thing in natural resource management, and you can connect with people all over the world.

Mathis: I think there's a lot of potential with technology. I think it's just about getting tribes to know how to use it. I took a GIS [geographic information system] class, and I really liked how they did things, and I've seen how much potential that technology can have to describe things within natural resources.

Samuel: I believe the technology we have today is for the better. It allows us to do many more things than we could ever do, and faster. If we didn't have the technology we do now, knowing the landscape and where someone's land is would take weeks, if not months. The many scientific things that we take for granted wouldn't be accomplished. Even video calls and meetings online wouldn't be possible.

Julie: Tamra and I have a similar work history, which is kind of interesting. She's spending her summers working construction—I'm guessing to put herself through school. I told her that's how I put myself through school. I had opportunities to study abroad, but I had to pay tuition. I barely had enough tuition money to go to school, let alone go somewhere international. Now, NAFWS is able to offer opportunities to students from all over, and these guys are amazing. Just listening to them, I just get all jazzed up about the future.

We have four interns in four separate locations, and we're connecting. We can see each other and interact, and we can talk about things, and their exposure is definitely greater than [mine] when I was in college. We're able to expose them to more of Indian Country. I'm really happy that we have Ashley because she brings a lot of energy, and she also brings a lot of knowledge because she navigated through a lot of those things through her college career. For as young as she is, she took advantage of a lot of opportunities. She can help these students find scholarships and intern opportunities and connect with [each] other. A lot of times, [we] forget that some of our youth are really in a small world, and they don't know much outside of their reservation. I recently spoke to some junior high kids, and I asked them what state had the most tribes, and they said South Dakota, because that's what they know. So I think this is an opportunity, and hopefully they'll learn something about other tribes. And when they get ready to hit the career field, they won't be limited.

If we can somehow also reach out to high school students, that'd be good. That's where NAFWS helps fill that gap, by having our summer youth practicum, which is for incoming 10th- to 12th-graders, exposing them to natural resource careers and topics. We're also going to have profession-building workshops as part of the national summer youth practicum, exposing them to all the cool opportunities that are out there. We want to make college students and early professionals aware of opportunities. We're hopeful that once these guys all get together and are able to interact more, they'll continue to build that network and tell their friends about us, and we'll start seeing more people in the field. I get really excited when I listen to our interns talk and when I listen to Ashley talk. I'm excited to watch their careers develop and be able to say, "Hey, I knew them when."

Chapter 3

Connecting People, Science, and Culture

| *An Interview with* **SCOTT AIKIN**

The US Fish and Wildlife Service's Office of Native American Liaison includes a Native American programs coordinator and liaisons in each of the eight US Fish and Wildlife Service regions. Native American liaisons serve as points of contact within the service for tribal conservation issues, especially those related to fish and wildlife management. This interview with the current national Native American program coordinator was conducted by the editors of this volume, Serra Hoagland and Steve Albert, on 21 January 2021.

Steve: What do you think some of the big challenges are going to be in tribal wildlife management over the next 25 years?

Scott: Revisiting the topic of climate change will be a necessity. With regard to tribal communities, many of which are limited in terms of their footprint on the landscape, there's a dependency on maintaining cultural practices and connection to natural resources. Tribes have been in the forefront of bringing the understanding of traditional knowledge [in the climate change debate]. Senator [Lisa] Murkowski from Alaska was one of the first politicians to push for TEK [Traditional Ecological Knowledge] to be a part of our scientific approach.

As Native scientists, we can be more active in [facilitating] this. It's a very sensitive topic: [sacred] site locations and protection of habitat are a concern, as is giving knowledge out. Oftentimes, it's misused or taken advantage of. So how do we set up structures that allow for a free discussion about the history of landscape, the history of species movement and activity, but from a deep knowledge within a community? Mainstream scientists have yet to understand how to incorporate [these things]. It's incumbent on tribal communities to help define that.

Serra: How can the service or the Native American Fish and Wildlife Society help with that?

Scott: In society meetings, we've had this discussion through the years about incorporating TEK. When I was in Alaska about four years ago, there was a discussion by an elder describing a bird that was coming into downtown Fairbanks. The Gwich'in people didn't have a name for it, but by virtue of *not* having a name for it, [they knew] that migration patterns had changed. So trying to take scientific observation and reconnect it to human observation, cultural observation, and incorporating it in a sound and professional way, that's our task.

I also think a tribe can describe the impacts of climate change, [such as] the shifting of plant and animal communities. As we've discovered by pollen analysis over the last 20 years, there have been dramatic changes in plant communities. Animals are reacting to that. So how do tribes—with [the limitations of working within] a reservation boundary—meet their cultural needs? Some of those traditional lands or some of those [habitats] may have shifted.

Steve: You've talked before about the issue of cultural appropriation. How do you think tribes can engage with federal and state partners, yet still protect their privacy and [any] knowledge they may not want to share?

Scott: Tribes can represent the need to build empathy with the communities that don't have that cultural perspective. A federal agency is driven by a mission, [and] a state is driven by their [constituents] to do what's in their interests. Tribes generally say: "How are we living, protecting, and praying with the landscape?" Those are three different viewpoints. [With tribes,] there is a sense of fear associated with cultural appropriation. And yet we have to continue to step into that realm as Native people. In the next 25 years we have to figure out a way to meet as sovereigns in a process, a free flow of discussion, where [there is] respect and empathy.

Serra: I really liked what you said about building empathy. [Author] Robin Kimmerer brings up the similar point about allowing other people to become indigenous to place. When I read that at the time, I thought, "Yeah, that's great because then they'll build a better relationship to that place, have higher empathy for the environment." I couldn't quite pinpoint why I felt uncomfortable about it, but you're right that there's this looming sense of appropriation there.

Steve: What are your hopes for working with the new administration?

Scott: At the inauguration yesterday, President [Biden] said, "It's time to be bold," and I made note of it. I'm going to appropriate his comment, in a good way, and be bold for tribal communities as a voice with a willingness to mix in the national soup, if you will.

A question I've been asked is, "Don't all tribes have casinos? Aren't they doing well?" Less than 20% [of tribal casinos] are actually solvent, and of those, there are only a fraction that are making money. So, no, we're not breaking anybody's bank with casinos. But the discussion that comes up is, "Well, how come tribes don't help one another out more?" I do think that's a valuable question because, were tribes to help one another more, like the Native American Fish and Wildlife Society does, we [could] do a tremendous amount. For tribes, that really is in many ways a disservice to their tribal identity, their unique—my Nishnawbe—identity.

We have an understanding of our faith that fits into our daily life. We don't just manage natural resources. In fact, the term "natural resources" is on its surface an offensive term because it takes away that sacred reality of what it is. I'm helping the plant community, the animal community, the fish community. I can't look at them as a resource. Even though I have to use that term, I still have to look at them as a sacred reality that the Creator gives me, so that I can be the person I was created to be. And yet that's something that is attractive to the non-Indian community because it's a human experience. We don't have a [monopoly] of that human experience, even though we have to protect it.

Serra: I'm interested to hear your thoughts on tribes increasing their footprint: land buyback programs, etc. The Muckleshoot Tribe has purchased more forested lands. With the Cobell settlement, there's funding for that type of thing, [for example,] to help with fractionated allotments, etc. What do you think about that?

Scott: With Muckleshoot, their intent is to improve elk habitat so that they aren't beholden to the state requirements and limitations. Conversely though, for example with the Spokane Tribe, the service did a relocation from an area that had an excess amount of elk. But the Spokane Tribe radio-collared their elk [so we know that they] migrate off the reservation. So there's a benefit to the state. So where tribes are doing this sort of habitat management and improvement, and purchasing land through buyback, it's been helpful.

The initial idea of land buyback was for lands within exterior reservation boundaries. Muckleshoot was able to do that in a much broader sense. A lot of the acres that they bought back are not within the reservation boundary. They've actually bought private landholdings and are seeking to have those converted to trust lands. But the initial intent of land buyback was to purchase landholdings within [existing] reservation boundaries that had become taxable, taken out of trust status. At Potawatomi, they passed that threshold back in the early 2000s, where they actually had more than 50% of the reservation back into trust status. It was a huge milestone.

We'll still see echoes of that for the next maybe half a decade or so, maybe a decade. But when we buy back land, are we buying it back for a municipal purpose or for some broader purpose? It might be better to not limit ourselves to the reservation boundaries, even though that's where we have our biggest voice.

Steve: How do you protect Indigenous knowledge in an era of the internet where it just takes one person with a keyboard and an internet connection to really do some damage in terms of getting information out there that shouldn't be?

Scott: That's a concern. Tribes have some of the best access to GIS programming; ESRI did a tremendous job doing outreach to tribes and getting tribes GIS capabilities.[1] Our youth are so savvy with computers. Many tribes have an ability that exceeds state or federal GIS work. We have to come up with a mutually agreeable process of how to protect tribal data. If they want to protect it, we don't need to know *why* they want to protect it. We just need to talk about incorporating some of that data into the broader database.

Serra: You've touched on this a little bit, Scott, talking about youth and the next generation and their skills. Tell me a little more on what you think about the staffing and the general demographics of the workforce that will be working in tribal wildlife management.

Scott: We're at a challenging point in that we have a lot of folks that brought forward this effort from the '70s in natural resource [stewardship] within tribal communities. They're passing [away], and who's picking it up? What's the next level? Your viewpoint of 25 years is a good one. Not only is that the window [during which] we probably have to address climate impacts, it's also the window [in which] I think we can take on the capacity that we've been given. And I say that not only in the physical sense, but in the spiritual sense. We have a capacity within tribal communities that we didn't have 25 years ago. How do we teach our youth that there's a responsibility to carry that forward? I think youth are not given [an understanding of] the gravity of what it means to *not* go into these fields. It seems antithetical to how tribal communities view that in their ceremonies. But when they get to the office, they put on the natural resources hat. We can alter that internally within tribal communities to come up with a broader reality of what that means.

Many elders are not afraid to speak about a faith-based reality or a religious component, if you will. There *is* a spiritual connection to what we do, and I often start conversations and training within the Fish and Wildlife Service with a reminder that the reason we're so passionate about what we do is there's a spiritual connection. I would argue that passion comes from a gift, a gift given to you by the Creator. How do we bring our youth an awareness of that? If you have a gift and desire for this, what are the options for you? A lot of our youth are looking at business, at computers. I hope under this [Biden] administration, tribal lands can start to turn a corner, to be a place that isn't someplace to escape from, as many youths feel they need to do. Computers aren't the only reality. So, I'm [cognizant] of that when I'm encouraging youth to consider working for the federal government, because it's incumbent upon them to maintain their tribal identity.

We talk about standing on shoulders and being brought forward by our elders. For Native people, it's a common understanding that our elders have given us this opportunity. Well, what do we do with that opportunity? [We need] to really impress on our youth that this is an opportunity given, it's not an opportunity you are entitled to. There's an inherent responsibility to be willing to suffer with that, and you can go out and do something that takes you out of your comfort zone.

Working for the federal government is out of a comfort zone for a lot of people. I feel responsible for bringing a voice and a face into an agency that has an inherent value in terms of its mission to tribal communities. There's so much common ground. But big as that common ground is, it doesn't outweigh the fear and frustration [of tribes] and the historical abuses. When I did the invocation for the delisting of the bald eagle, my mom and dad were able to be there with me. My mom just couldn't believe that could be a possibility. We saw that yesterday [at President Biden's inauguration] with the National [Youth] Poet Laureate [Amanda Gorman].

Steve: Yeah, she was amazing.

Scott: That comes from a history of hope, from people who were provided that opportunity, and she took advantage of that opportunity [and was] able to give voice at a time that was so impactful. And she had the skills to do that. We've got youth on many levels that can do that.

Serra: One thing that concerns me a little is the trajectory of our overall population moving more into urban areas and losing that connection to the land. But even urban Native youth communities have a lot of those ties through their family and through their upbringing. So that gives me a little bit of reassurance. The other thing I wanted to ask you about, Scott, is you give so much, I don't know how you do it. I've never heard you say "no" to any of the times we ask[ed] for your help or when you talk to some of our Native students through The Wildlife Society or our professional development program. What advice do you give to future tribal wildlife professionals as to how to stay balanced?

Scott: I appreciate that. I very much respect and live a life that my grandparents and my mom gave me. I'm here not by chance, but by purpose—the purpose of my family and the Creator. Although [much of] my family has passed, I remember them, and I remember my spiritual connection [to them] through their teaching. So before I did this, I prayed, I asked for help. That's something we separate very much in society, but I think for Native people, we don't separate that. We have prayer. It's common to have prayer before a meeting. If there are tribal members, we have an invocation, we have a prayer, and the non-Indian community, though they don't understand the words, they're affected by it.

We might do it out of respect for our elders, but it's more than that. We are spiritual beings. Without that, then you're draining from your tank of energy and hope, and you can't fill your tank back up. The reason that we respect our elders is there's a wisdom they recognize. They have an understanding and a peace in the midst of chaos. Natural resource management is a nonspiritual act in many ways. But when we come as Native peoples to a gathering of natural resource managers, we bring the potential for a broader discussion of passion, of caring, of empathy, all of which are not just biological, but also spiritual.

We shouldn't act as though we have the answer, or as though we're different from others. I always tell my kids: we have the closest tie to a lived reality of all human beings. The lived reality is that we are part of a creation, we are part of a landscape, and we're integrally tied to that landscape. Just because we might have that echo and remembrance most closely as Native peoples, it doesn't mean that someone who's from another country [doesn't] have that too. Of course they do. It might have been forgotten, but if we have this remembrance intact, then we have a responsibility to help it be shared. So, [we should know] how to be bold, but also how also to be humble. [I] trust in the Creator that I've been placed here to do something today. Even though I might be tired, even though I might feel like I'm disconnected, I'll do my best—and I'll be asking for help. And something tremendous can come out of it. Oftentimes it does, and it fulfills me.

How do we reconnect with our spiritual being in the middle of that reality? Our innocence was taken, was abused in many ways, but it doesn't define us, and we can move beyond that with the teachings of our elders, who suffered in more ways than we have suffered. And yet they were able to overcome that from being committed in their spiritual understanding and being authentic. I'm always seeking help because it does take a lot of energy, but I never feel like I've expended too much. Not say[ing] that arrogantly, by any means. But every day I try to recognize I have a responsibility to be respectful of my family, who've given me this opportunity to be here, and respectful of the Creator for placing me in here at this time.

When I stood on the Jefferson Memorial and did that invocation [for the delisting of the bald eagle], it was the most surreal aspect of my life to be singing a prayer song that my grandfather had given me for the delisting. I'm a traditional dancer, and I have a bald eagle bustle as part of who I am. The Creator puts us there. So I made a strong statement with conviction that my grandfather assured me: you can stand outside and yell about issues, or you can step inside and talk with the people who are causing the issues. And the difference is you can see them and interact with them inside. They can only hear you if you're yelling outside. That's why I work for the federal government. In my almost 30 years, I've stepped down and come back in and stepped down and come back in, worked for my tribe. But I don't regret doing this because of [the] perseverance given to me by my family and my elders. That perseverance is being used in a good way to do the work that I do in the federal government.

NOTE

1. The Environmental Systems Research Institute is an international supplier of GIS (geographic information system) software, web GIS, and geodatabase management applications.

Chapter 4

Who Stands for the River?

| WINONA LADUKE

I stood by the Shell River. Looked at her good. She's a beauty. Deep and wide here, right before she joins with the Crow Wing. For generations, she was a path of Anishinaabe from Lake Superior to Hudson Bay. The old people used to travel by river, using the power of the water to move their *jiimaanan*, canoes. Where I live by Shell Lake, the river is small, but she's mighty. She comes out with a rush from the lake, brilliant with the shine of the sun on the water. *Waasabiik*.

The boss of the Shell is the heron, the great blue heron, *zhashagi*. The Shell River and the lake remind us of the story of long ago—the story of the people who followed a shell in the sky.

The river, she meanders south from the lake, crosses near the first set of agricultural fields, and then runs back into the woods. Then comes more of the irrigators, center pivot irrigators, and fields of sprayed potatoes for RD Offutt, for the big McDonald's. *The river continues.*

She meanders southeast toward towns like Menagha, or Minikaag ("the place where the blueberries grow"), Minnesota, and then joins up with the Crow Wing River. *Nishtuau*. Big rivers, and rivers getting bigger. But before that, if you go down by Duck Lake, you can see the river crossing. That's where Enbridge, the Canadian oil pipeline company, shoved a pipeline under the river. That pipe was installed in July of 2021 and now can push up to 915,000 barrels of oil a day underneath the Shell. The Shell River gets crossed 4 times by Enbridge's Line 3; 4 out of the 65 river crossings. We will cross our fingers for her.

After all, she's not as remote a river crossing as the Willow or Moose Rivers. All of those rivers are put at risk by Enbridge Line 3. The Willow already had a frac-out—saw it with my own eyes, and saw that if I walked to shore, the Minnesota Department of Natural Resources officer would arrest me. That's how it works when the oil pipeline company pays for the law enforcement. Enbridge paid over $8 million to 96 law enforcement agencies in Minnesota.[1] The Minnesota DNR received the most ($2.1 million) to protect Enbridge's right to move oil through Minnesota. That's the same agency that is in charge of monitoring the corporation. It's a total conflict of interest. And thus far, Enbridge has brought

Minnesota 28 frac-outs and 3 aquifer piercings. The company secured 5 billion gallons of water from the state in the summer of 2021, the single largest allocation of water in the history of Minnesota. No hearings, no environmental impact statement.

Enbridge's Kalamazoo spill lasted 17 hours before Enbridge closed the valve. This pipeline gushes about 20,000 gallons a minute in a leak. New pipes are not better than no pipes. Ugh.

It's like that Joni Mitchell song, "You don't know what you got 'til it's gone / They paved paradise / put up a parking lot." We don't get out much these days. It's a beautiful world out there, and she's worth protecting.

This is how it works: First they cut the forests, then they dig the trenches, blow through a few burial mounds, and some bears' dens. And then the pipe comes, and the drill. It's brutal. It's a rape, *and we are witnesses*.

Oil is not good for water. And it's not good for mussels. Not good for the mussels of the Shell River. The Shell has the largest inland mussel population on the upper Mississippi. Bitumen, the tar sands stuff, goes to the bottom.

The north, Giiwedinong, is a beautiful land full of life. Some of those relatives have four legs, some have wings, and some are in the waters. That's to say that this is where the wild things are. There are bears, wolves, otters, beavers, butterflies, and frogs—all things which belong here. We are their voice. This is where we live, and it's worth something to us.

Indigenous peoples represent 4% of the world's population, but live with 75% of the world's biodiversity. Indakiingimin, the land to which we belong. The skies are dark, and the stars are brilliant, even the northern lights, *waawaate*, where the spirits are dancing. That's something you can't see everywhere.

How about, we keep things out of where the wild things live? Things like pipelines and mines, even 5G. The whole world doesn't need 5G. Sometimes, just listen to the ice crack on the lake, or maybe hear a wolf howl, that's good enough.

I'd like to think that the Minnesota Department of Natural Resources was looking out for the wild things, and the rivers maybe, but no, sadly, that's not the case. The Minnesota DNR first allocated permits to Enbridge for 636 million gallons of water discharged, sucked from lakes, and more, and approved the destruction of endangered species. Then, in June of 2021, the DNR gave Enbridge an even better deal—a $150 permit amendment fee and 5 billion gallons of water allocated to the corporation. And that's during the worst drought in known history.

That's crazy stuff. And for all of it, the DNR gets to keep more than $2 million. That's for the taking of endangered species—paid to the state which issued the permits. That's compensation. In 2021, the Minnesota DNR looked a lot more like mercenaries than water and wildlife protectors. It's clear that the 1855 treaty territory needs to be protected. Tribal people need to protect the water and the rice. Our people need to stand up for those who cannot speak.

Who stands up for the wild things? Who stands up for the Shell River, the Mississippi, the Willow River, and the rice? Someone needs to.

A global movement called the Rights of Nature has taken hold in constitutions, reaffirming the rights of rivers and mountains. In July 2018, Bangladesh became the first country to grant all of its rivers the same legal status as humans.[2] From now on, its rivers will be treated as living entities in a court of law. In Minnesota in 2019, the White Earth Ojibwe recognized the right of wild rice to continue in perpetuity. New Zealand has recognized both a river and a mountain, and the Bolivian and Ecuadoran governments have recognized Mother Earth as an entity. Ecuador's Constitution recognizes that "Nature or Pachamama, where life is reproduced and exists, has the right to exist, persist, maintain and regenerate its vital cycles, structure, functions and its processes in evolution." In December of 2021, the Ecuadorian Constitutional Court overturned mining permits in a protected area. The seven justices of the court found that mining activities pursued by a state mining company and its Canadian partner threatened the Los Cedros protected area's right to exist and flourish.[3]

The Rights of Nature movement challenges the legal framework which characterizes nature as a passive resource to be owned and used and replaces it with a means for nature to protect itself.

Nature has rights, and that includes the right to exist uninterrupted by 5G and pipelines, the right to be free of oil and garbage. We stay in our place and they, the wild things, stay in their world. That way, we don't get in trouble, like viruses crossing and such. It turns out that if you mess with the wild things, you can get a pandemic—this coronavirus came from bats in China. Leave the wild things alone. The United Nations reaffirms that the encroachment on biodiversity is a core cause of health, ecological, and economic disaster. Next time, leave them alone.

In the meantime, Enbridge drilled the Shell River. The tribes and citizens tried to stop the oil in every regulatory hearing, in the courts, and with our bodies. In July of 2021, I was arrested with six other women on the Shell River, all charged with trespassing and obstruction. I spent three nights in jail because of Enbridge, while the corporation, after 28 frac-outs and 3 aquifer piercings, faces no charges. The company's destruction continues, now pushing through Wisconsin and Michigan for its Line 5, the pipe which will ultimately bring the oil back to Canada for refining.

But this is the end of the fossil fuel era, and I am going to bet on the river, not the pipe, and stand my ground. The courts continue to shut down pipelines; the Keystone XL pipeline is the latest, and the Trans Mountain in Canada is also imperiled. Line 3, however, never got to federal court. There was no federal environmental impact statement, only a good deal of money paid by a Canadian corporation to capture the regulatory systems of the state. It's time for Indigenous legal systems and regulatory systems to protect the land from which we come, the land and waters to which we belong. When we transform and rematriate our legal and governance systems, we will find that the rights of the river are greater than the rights of the oil company. She tells us about life.

Let's remember the Shell. She's a beauty, I wish someone would protect her.

NOTES

1. See https://twitter.com/MNSnarkDept/status/1523046832020021251/i. Accessed 15 July 2022.
2. R. Chandran, Fears of evictions as Bangladesh gives rivers legal rights. Reuters, 4 July 2019. https://www.reuters.com/article/us-bangladesh-landrights-rivers/fears-of-evictions-as-bangladesh-gives-rivers-legal-rights-idUSKCN1TZ1ZR. Accessed 28 July 2021.
3. See https://insideclimatenews.org/news/03122021/ecuador-rights-of-nature/. Accessed 15 July 2022.

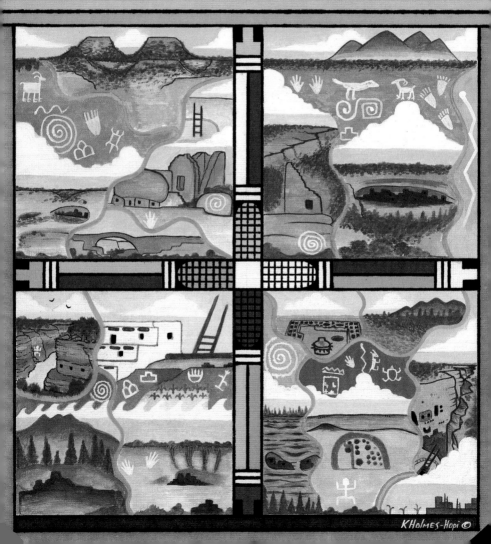

Part II

Legal Issues

Despite the enshrinement of tribal sovereignty in the US Constitution, tribes have had to constantly fight to maintain this right of self-determination and the ability to legally manage their own lands and waters. In the chapters that follow, several prominent legal scholars and wildlife professionals describe some battles; the intersection of tribal, federal, and state law; and the implications for co-management in the coming decades.

Chapter 5

The Importance of Meaningful Federal-Tribal Consultation in Land and Natural Resource Management

| MICHAEL C. BLUMM *and* LIZZY PENNOCK

INTRODUCTION

Tribal knowledge of the environment is vast but often untapped or ignored by federal officials when making decisions that affect tribal land and natural resources of cultural significance. Although the United States has long had a government-to-government relationship with federally recognized tribes, it has often failed to live up to its end of the bargain. Tribal calls for meaningful consultation, or any consultation at all, have often gone unheeded.

Consultation is significant to tribes for several reasons. Tribes frequently have a distinct land management approach, which nontribal officials do not adequately understand. Policies and regulations formed without tribal consultation or the consideration of tribal values or rights can force a management scheme on tribes inconsistent with their needs, historical resource management programs, and legal rights. Tribes place considerable cultural, religious, and historical significance on places and resources that other land managers often do not recognize or protect. Moreover, when tribes must defend their rights and resources after being left out of federal decision-making, the result is often significant expenditures of funding, time, and legal resources that otherwise would be unnecessary.[1]

The government-to-government relationship between tribes and the US government arises out of the trust doctrine, and government-to-government consultation is a substantial aspect of that relationship. The federal government's trust responsibility to Indian tribes emerged from early federal-tribal treaties, executive orders, statutes, the US Constitution, and various Supreme Court opinions.[2] In his 1831 *Cherokee Nation v. Georgia* decision, Chief Justice John Marshall declared that Indian tribes were "domestic dependent nations" that "look[ed] to [the federal] government for [their] protection."[3] *Cohen's Handbook of Federal Indian Law* considers *Cherokee Nation* to have laid the foundation for analogizing the government-to-government relationship as a trust relationship with an accompanying "federal duty to protect tribal rights to exist as self-governing entities."[4] The federal government's interpretation of its trust responsibility skewed toward "a nearly absolute and unreviewable congressional plenary power" in the 19th and 20th centuries, but the modern trust doctrine purports to recognize tribal self-determination and sovereignty.[5]

Consultation arises out of the trust duty and other contexts, like those prescribed by the National Historic Preservation Act (NHPA) of 1966.[6] If consultation in any context is to be meaningful, however, federal agencies must treat tribes as more than the public or a stakeholder to consult before an activity can occur. To adequately address tribal concerns and perspectives, federal officials must understand tribal cultures, histories, and legal rights. In order to treat tribes as sovereigns, federal officials must understand Indian law and the unique status of tribal governments in US law, including the government-to-government relationship under the federal trust obligation.

Meaningful tribal consultation can prevent federal agencies from making uninformed decisions affecting culturally significant tribal lands and resources, and those consultations may come in various packages.[7] For example, the intertribal coalition that successfully petitioned President Barack Obama to proclaim Bears Ears National Monument described the desired relationship as one of ongoing "collaborative management."[8] According to the coalition, collaborative management harmonizes Western science with traditional knowledge founded on Native cultural values and should engage tribes from beginning to end.[9] Another successful consultation resulted in the 1997 Joint Secretarial Order on American Indian Tribal Rights, Federal-Tribal Trust Responsibilities, and the Endangered Species Act.[10] The order's consultation called for high-level government officials and tribal representatives in highly structured negotiations that made time for presentations on and understanding of the relevant cultural, historical, and legal issues, enabling federal negotiators to understand tribal experiences and backgrounds.[11]

In this chapter we contend that the current practice of tribal consultation in land and resource management for culturally significant tribal lands is often "too little, too late."[12] But federal agencies can remedy this injustice by incorporating the three essential elements of meaningful consultation that we suggest: (1) early and consistent tribal engagement, (2) face-to-face interactions, and (3) a deep understanding by federal officials of tribal cultures and land management practices. These elements of meaningful consultation can assure robust tribal participation in land and resource management and make it more likely that tribes will substantively influence management decisions. We also analyze various consultation arrangements—some that have achieved successful consultation and others that have failed—both of which provide lessons for the future.

We conclude that meaningful consultation requires face-to-face negotiations from the beginning to the end of decision-making processes, and incorporating tribal perspectives, knowledge, and rights. Through meaningful consultation, the federal government can begin to fulfill its trust obligation to honor the government-to-government relations with tribes. If decision-making incorporates tribes' perspectives and their knowledge of land and resource management for culturally significant tribal lands, the result will be better federal land management.

THE NIXON ANNOUNCEMENT OF A GOVERNMENT-TO-GOVERNMENT RELATIONSHIP

After two centuries of turbulent tribal policy, in 1970 President Richard Nixon announced the federal government's commitment to encouraging tribal self-determination and fostering a government-to-government relationship with tribes.[13] The announcement was a landmark, officially ending the termination era and declaring a "new direction of Indian policy aimed at Indian self-determination."[14] The announcement moved the federal government away from both termination and paternalism, creating a policy in which it put Indian people at the helm of decision-making related to them. Nixon suggested that achieving the "new and balanced" federal-tribal government-to-government relationship meant that both governments must "play complementary roles" when it came to "Indian problems."[15]

Nixon's announcement seemingly quashed the notion that the US government might not have a responsibility to consult with tribes as sovereigns. He made it clear that the only question for the federal government to consider was *how* to carry out that responsibility and how to make Indian self-determination an enduring national policy.[16] This balanced relationship in which the federal government and tribal governments have complementary roles cannot exist if the federal government does not consult meaningfully with tribes in environmental decision-making affecting culturally significant tribal lands and resources.

CONSULTATION UNDER THE EXECUTIVE ORDERS AND THE NATIONAL HISTORIC PRESERVATION ACT

Several presidents after Nixon reaffirmed the federal policy of fostering a government-to-government relationship with tribes.[17] Together, Congress and the executive branch developed the federal policy of Indian self-determination through several executive orders and section 106 of the NHPA. Section 106 and its implementing regulations allow tribes to sue federal agencies for failure to adequately consult with them as the regulations require.[18] Executive orders require consultation in a broader context, applying to all "regulatory policies that have tribal implications."[19] Executive order consultation, however, is unenforceable in court because its scope is limited to "internal management of the executive branch."[20]

Section 106 of the NHPA

Section 106 and its implementing regulations form the principal statutory requirement for tribal consultation and outline mechanisms for federal-tribal consultation. This section requires US officials with jurisdiction over a federal "undertaking" to account for the effects on "any district, site, building, structure or object" listed in the National Register of Historic Places before spending federal money or granting a license.[21]

Section 106 regulations direct federal officials to begin consultation "at the early stages of project planning" to identify historic properties that the undertaking may affect and to assess alternatives to "avoid, minimize or mitigate any adverse effects" of the undertaking on historic properties.[22] Federal agencies must give the Advisory Council on Historic Preservation a reasonable opportunity to comment on the undertaking and "involve the consulting parties . . . in findings and determinations made during the section 106 process."[23] If the undertaking has the potential to affect a historic property that is religiously or culturally significant to any federally recognized tribe, that tribe is a "consulting party." However, several procedural and administrative elements have hampered implementation of section 106. The law's scope is limited, since consultation is based on the "scale of the undertaking" and the federal government's involvement—which are relative terms. Further, section 106 applies only to properties and sites listed on or eligible for the National Register of Historic Places.

Additionally, the federal government's efforts often fall short of ensuring meaningful tribal engagement under section 106, regularly resulting in rote and less than rigorous consultation—that is, box-checking. For example, archaeological surveys conducted by private entities sometimes fail to recognize tribally significant resources. Complicating this matter is the fact that tribes sometimes intentionally withhold information from the federal government for National Register listing because they do not want to make cultural resources publicly accessible. If a tribe chooses to withhold information detailing the locations or the attributes of culturally or religiously significant resources, it runs the risk of their destruction through project development. In some cases, even when tribes have willingly disclosed information, the federal government has failed to consider tribal concerns before granting project approval. Instead of engaging in meaningful back-and-forth consultation, federal agencies sometimes document "every contact or communication with a tribe, no matter how inconsequential" as proof that consultation took place.[24] Those contacts can include one-way communications like mailing to a tribe a notice of an agency's intent to prepare an environmental impact statement, a tribal member speaking at a public meeting, and federal agencies treating tribes interchangeably, counting communications with one as communications with all.

Although these actions might be considered to be consultation, they are not *meaningful*. Federal agents are unlikely to grasp tribal knowledge and perspectives and apply them in decision-making by merely mailing notices and attending public meetings. Because each tribe is unique, no individual tribe can be overlooked, and the appropriate authorities within each tribe must be consulted. Even when tribal consultation complies with section 106, it does not automatically satisfy the government-to-government obligation. Whether compliance with section 106 fully satisfies the government-to-government consultation obligation depends on the level of tribal participation in the resolution of adverse effects.

The Clinton Executive Orders

Throughout his tenure, President Bill Clinton used executive orders (EOs) to strengthen the US government's commitment to a government-to-government relationship with tribes. Several Clinton administration executive orders purported to "establish regular and meaningful consultation and collaboration" with tribal governments. EO 12875 in 1993 declared a policy to protect the "American people" from the consequences of "unfunded federal mandates" on tribal governments.[25] But the "regular and meaningful consultation" directed by EO 12875 placed consultation with state, local, and tribal governments on the same level, even though the federal government's responsibility toward tribes is distinct.

EO 13175 in 2000 recognized as "fundamental" the "unique legal relationship" between the federal government and tribal governments: the federal government has, over time, "establish[ed] and define[d] a trust relationship with Indian tribes."[26] EO 13175 promoted tribal "self-government," "sovereignty[,] and self-determination" by requiring federal agencies through "regular and meaningful consultation" in a government-to-government framework to carry out the "complementary roles" that Nixon's announcement had envisioned 30 years before. EO 13175 directed federal agencies to establish an "accountable process" to ensure that tribal officials have an opportunity to contribute "meaningful and timely" input when agencies develop regulatory policies that have tribal implications, and to consult tribal officials early in the development process.

These EOs both incorporate elements essential to meaningful consultation but leave much to be desired. EO 13175 recognized the value of "early" consultation and back-and-forth communication in which tribal officials provide concerns that the agencies must consider before issuing a policy or regulation. The EOs, however, do not define "meaningful consultation." EO 13175 left the details of how to engage in meaningful and timely consultation up to the federal agency. Moreover, that executive order does not require an agency to act on tribal concerns, but merely to summarize them and the agency's response. The agency discretion granted by the EOs for fashioning consultation does not "protect tribal rights to exist as self-governing entities" because it relegates tribes to participating in whatever process the agency decides.

TRIBAL CONSULTATION CASE STUDIES

We analyze the consultation processes via three case studies: the secretarial order on the Endangered Species Act; the proclamation (and diminishment) of Bears Ears National Monument in Utah; and oil and gas lease sales in Chaco Canyon, New Mexico. The case studies are not exhaustive but represent several aspects of the tribal consultation process, each discussing whether federal agencies engaged in consultation, and, if so, whether it was meaningful or amounted to box-checking.[27] Procedural aspects of consultation include the rank of the federal official that engages in consultation and whether an agency consults tribes directly or groups them in with the public or nontribal stakeholders. Substantive aspects include tribes' identification of places and resources of cultural significance and co-management frameworks between tribal governments and state or federal governments

An early example of meaningful consultation that preceded Clinton's EOs grew out of Judge Robert Belloni's historic decision in *Sohappy v. Smith*, later consolidated into *US v. Oregon*.[28] After ruling that Columbia River treaty tribes were entitled to a "fair share" of the salmon harvest allocation because treaty language expressly assured them "a right of taking fish in common with" white settlers, Judge Belloni called for meaningful tribal participation in fishery management.[29] Despite the *Sohappy* decision, Oregon was still discriminating against tribal fishers as late as 1975, so Belloni ordered the tribes and states to cooperate on developing a comprehensive fish management plan.[30] Belloni's order thus laid the groundwork for decades of meaningful negotiations, which have resulted in a series of management plans governing salmon harvest under a co-management framework. The management plans, requiring the concurrence of both the states and the tribes, are a significant and substantive result of the negotiations and may serve as a general model for co-management in other contexts.[31]

Both the example set by Judge Belloni and the case studies below reveal that there is no one definition of "meaningful consultation." Our analysis of several agency consultation efforts shows that meaningful consultation must include, at a minimum, face-to-face discussions and early and consistent tribal engagement. The case studies also show that meaningful consultation arises when federal agencies—or the federal judiciary—adequately understand tribes' cultures and their land management practices.

The Secretarial Order on Tribal Rights and the Endangered Species Act

The process that led to the Joint Secretarial Order on Tribal Rights and the Endangered Species Act (ESA) in 1997 is a prominent example of meaningful consultation. Like the collaboration Judge Belloni ordered in *Sohappy*, this consultation happened before Clinton's EOs requiring "regular and meaningful consultation" in 1998 and 2000, and it demonstrates several essential elements of meaningful consultation. The order attempted to harmonize federal law with "tribal rights to manage their resources in accordance with their own beliefs and values."

Tribes came together in the mid-1990s to discuss how to protect tribal interests in light of the ESA because its enforcement often disregarded tribal sovereignty and resource management practices. A group of tribal resource managers and lawyers organized tribal efforts on a national scale to develop a tribal consensus on ESA implementation in Indian Country, beginning at a workshop in February 1996.

The tribal consensus reflected a desire "to avoid ESA conflicts through good, cooperative tribal land management."[32] Once participants settled on this central message, they began drafting a proposal calling for a nationwide joint secretarial order that would establish working relationships between tribal governments and federal agencies for ESA implementation. The tribes soon presented their proposal to the Department of the Interior, triggering the consultation process.

Consultation on the secretarial order demonstrated how to effectively implement several essential elements of meaningful consultation, especially highlighting several crucial procedural aspects. First, Secretary of the Interior Bruce Babbitt had a special counsel who briefed him on the issues and the nature of the tribal position. Second, Babbitt met face-to-face with tribal leaders who presented the national tribal consensus, instituting a year and a half of negotiations. Babbitt's efforts to understand the tribal position allowed him to begin the process with an understanding of the issue's importance to tribes. As a result, he appointed an appropriate negotiating team that included high-level federal representatives.

Unlike the type of consultation that would emerge later under the executive orders, which gave federal agencies enormous discretion in creating the consultation process, the "structure and protocols" of the several two-day negotiating sessions for the secretarial order were "carefully negotiated" between tribal and federal representatives.[33] The negotiating sessions devoted substantial time for those involved to develop a deep understanding of the "cultural, historical, and legal background" and the implications of the intersection of the ESA with Indian wildlife management.[34] These two elements are essential because tribal members are the appropriate source for instructing agencies as to how best to interact with them, and because agency officials may not effectively apply that information without understanding the context from which it arises.

The final secretarial order that emerged from these consultations was, according to Professor Charles Wilkinson, "a sensible harmonizing of Indian law and the ESA" and can serve as a positive example of a government-to-government relationship in which both sides are respected as sovereigns.[35] The secretarial order called for "extensive cooperation between tribes and federal administrators,"[36] requiring federal agencies to provide scientific, technical, and informative assistance for tribal development of conservation and management plans for ecosystems on which ESA-listed or listing-eligible species depend.

Bears Ears National Monument

Indigenous people have called the Bears Ears region in southeastern Utah home for many thousands of years. The area is dominated by a pair of culturally significant buttes (resembling the ears of a bear), which are surrounded by largely undeveloped US public lands.[37] Bears Ears is, according to the Native American Rights Fund, "one of the densest and most significant cultural landscapes in the United States."[38] Looting, vandalism, and development for resource extraction have long threatened the integrity of Bears Ears.[39]

In 2010, several tribes with deep ties to the area embarked on legal efforts to protect the land, and they soon formed the Bears Ears Inter-Tribal Coalition. Despite repeated requests, tribes were excluded from land management planning by members of Utah's congressional delegation and local governments.[40] In response, the coalition sent a proposal to President Obama in 2015, requesting that he proclaim the Bears Ears National Monument and establish a framework for collaborative federal-tribal management of the monument. Obama responded by establishing the monument and a tribal advisory commission in December 2016. Efforts to lobby the Biden administration to reinstate the monument boundaries that President Donald Trump severely diminished were ultimately successful.[41] President Biden signed a proclamation on 8 October 2021 to restore the boundaries established under Obama, and it retained an additional 11,200 acres added by Trump.[42] Significantly, the US government and the five tribes of the Bears Ears Inter-Tribal Coalition signed a "one-of-a-kind agreement" to share management responsibilities for the monument.[43]

Background

In 2011, Navajo Nation President Ben Shelly met with Secretary of the Interior Ken Salazar to request national monument protection for Bears Ears. In 2013, the Navajo Nation began to work with the newly formed nonprofit organization Utah Diné Bikéyah (UDB) to represent tribal interests in the debate over management of Utah public lands, including Bears Ears. UDB engaged the local community through town hall meetings, hosted numerous tribal gatherings focused on land protection strategies, and developed a sophisticated map of Bears Ears, highlighting the cultural significance of specific lands.[44]

The San Juan County commissioners and members of Utah's congressional delegation were simultaneously working on land management plans, although they failed to meaningfully engage the tribes despite tribal efforts to participate.[45] The Navajo Nation and UDB submitted their proposal to the county. The commissioners engaged in a series of meetings with UDB in early 2015, achieving little. By August 2015, the county was urging the state legislature to pass a bill that would open culturally significant areas for resource extraction.[46]

At the same time, Utah Congressmen Rob Bishop and Jason Chaffetz were pushing the Public Land Initiative (PLI) through Congress. UDB and the Navajo Nation shared an early version of their proposal with federal officials involved in the PLI process and visited Washington, DC, to meet with the congressmen supporting the PLI, but the tribes never received a single substantive response. In 2016, Bishop released the PLI, which would protect 1.39 million acres of Bears Ears without any tribal management.[47]

With little hope for a version of the PLI that would protect tribal interests, the tribes began working on a separate proposal to protect Bears Ears by using a presidential proclamation under the Antiquities Act, maintaining the campaign that the Navajo Nation had begun in 2011.[48] The Hopi, Navajo, Uintah and Ouray Ute, Ute Mountain Ute, and Zuni tribal governments united to form the Bears Ears Inter-Tribal Coalition in July 2015, which drafted the Bears Ears proclamation proposal.

The Bears Ears Inter-Tribal Coalition Proposed Collaborative Management
In October 2015, the coalition submitted a comprehensive land management proposal to President Obama, requesting that he proclaim 1.9 million acres of land surrounding Bears Ears as a national monument under the Antiquities Act. The proposal called for collaborative management of the lands within the proclamation boundaries. The coalition's proposed version of collaborative management would combine Native traditional knowledge and cultures with existing federal public land practices and be more than mere consultation with federal agencies. It would require long-term, active engagement by the tribes in managing the conservation of Bears Ears.

Under the collaborative management proposal, an administrative commission with eight members would oversee management of the monument. The commission would have one person from each of the five tribes in the intertribal coalition and one person from each of the three federal agencies with land management authority over public lands within the proposed monument boundaries.[49] Through "joint decision-making," the commission would oversee the development of the governing management plan and would form policy.

The proposal outlined a collaborative management framework that would fuse Western land management and tribal knowledge.[50] Because integrating traditional knowledge with existing federal land management practices is a centerpiece of collaborative management, the coalition proposed an institute for combining traditional knowledge with Western science. A monument manager would report to the commission and oversee operational staff experienced in both traditional Native American values and knowledge and Western science and public land management.

Collaborative management, as proposed, would replace the consultation required of federal agencies in similar contexts, like under Clinton's EOs. While EO-based consultation often "becomes merely a box to be checked that allow[s] federal agencies to proceed on the projects which they prefer,"[51] the coalition's proposed framework would ensure long-term joint decision-making through the commission and the monument manager. In all respects, the intertribal coalition sought a deep fusion of Western and tribal land management practices.

Obama's Bears Ears Proclamation

In December 2016, during the last weeks of his second term, President Obama proclaimed the Bears Ears National Monument, although the size fell short of what the tribes had proposed, preserving only 1.35 million acres. However, Obama did create the Bears Ears Commission "to ensure that management decisions affecting the monument reflect tribal expertise and traditional and historic knowledge."[52] The commission would have an elected representative from each of the five tribes but no federal officials. The commission would partner with the federal agencies for general decision-making, and federal agencies would "carefully and fully consider integrating" the commission members' knowledge and expertise.[53] The management plan developed under the proclamation would also create a framework for ongoing meaningful engagement between the commission and the federal agencies.

The Obama administration's proclamation set a high bar for what was possible for a government-to-government relationship, reflecting a respect for tribal cultures and Bears Ears' importance to them. Like the secretarial order, the proclamation demonstrated that substantive results could ensue when the federal government develops a deep understanding of tribal culture and values, because it can create a meaningful role for tribal governments to contribute to public land management, especially of resources that are culturally significant to tribes.

Although the proclamation did not adopt every aspect of the coalition's proposal, the collaboration it outlined went well beyond what President Clinton's executive orders required. EO 13175 called for "regular and meaningful consultation" but gave federal agencies considerable discretion in fashioning that process. The Bears Ears management proposal would enable tribes to instruct federal agencies in how to engage them meaningfully. EO 13175 required only summary impact statements detailing tribal concerns and describing the extent to which an agency addressed them. The Obama proclamation, on the other hand, required agencies to provide a "written explanation of their reasoning" if they "decide[d] not to incorporate specific recommendations" submitted by the commission.[54] Thus, agencies could no longer merely list tribal concerns before moving forward with a project.

Using tribal involvement to design a consultation framework is redolent of what the Nixon announcement called for in 1970, when it recognized that federal programs and funding would be more effective "if the people who are most affected by these programs are responsible for operating them."[55] By applying Indigenous knowledge of the culturally significant tribal lands and resources within Bears Ears, the government could further its responsibility to tribes and achieve a balanced relationship between governments.[56] The Bears Ears management scheme envisioned a genuine government-to-government relationship between sovereigns.

Trump, Biden, and the Future of Bears Ears
Within a year of Obama's proclamation, the Trump administration used the Antiquities Act to reduce the Bears Ears boundaries by more than 85%, splitting the remaining 15% into two segments.[57] In early 2020, Trump's Interior Department promulgated a management plan that would allow drilling, mining, and grazing on lands that the administration had removed from protection.[58] Several groups representing the interests of the five tribes in the coalition filed lawsuits challenging this action.[59] Ongoing litigation was stayed on 21 January 2021, however, per President Biden's EO 13990, which directed the secretary of the Interior to conduct a 60-day review of the Trump administration's proclamation—in consultation with the attorney general, several other agency secretaries, and tribal governments—to determine whether the Biden administration could restore the boundaries established by the Obama administration.[60]

By March 2021, the coalition reported that it had had consultations with the Departments of the Interior and Agriculture.[61] Its members had a face-to-face meeting with Secretary of the Interior Deb Haaland, the first Native American cabinet secretary, in April 2021.[62] Substantive results of the ongoing consultation flowed from the reestablishment of the Bears Ears Commission and co-management framework, either as proposed by the coalition or as proclaimed by Obama.

The coalition urged Secretary Haaland to recommend that Biden reestablish the monument at the originally proposed 1.9 million acres and, in the interests of expediency, advocated for executive branch action (rather than legislative).[63] Executive action restoring or enlarging the monument as proclaimed by Obama, however, invited potential litigation from the monuments' opponents, who argued that the Obama-era monument was too large and who were encouraged by a statement by Chief Justice John Roberts in which he questioned the scope of presidential authority under the Antiquities Act.[64] In December 2021, the Utah attorney general's office hired a law firm to research potential litigation strategies to challenge the monument's restoration, meaning that there may be litigation.[65]

The Chaco Canyon Oil and Gas Leases

Chaco Canyon, much of which is part of Chaco Culture National Historical Park, and the surrounding land in the San Juan basin in northwestern New Mexico supported a sprawling mecca of Indigenous life for hundreds of years and remains important to the Navajo Nation and more than 20 Pueblo tribes.[66] Culturally significant tribal sites in the basin are at risk from private companies seeking to drill for oil and gas. Tribes in the area allege that federal land managers with authority over oil and gas drilling have consistently failed to adequately consult them concerning the management of Chaco Culture National Historical Park and the surrounding area.[67]

Consultation on Chaco Canyon's Resource Management Plan

A resource management plan (RMP) published by the Bureau of Land Management (BLM) in 2003 authorized nearly 10,000 oil and gas wells in the San Juan basin, of which about 4,000 have already been drilled. The RMP, encompassing 4.2 million acres of land, including over 675,000 acres of Navajo Nation trust surface land and 210,000 acres of allotments held by individuals of the Navajo Nation, governs land and resource management in the basin, including decision-making processes for oil and gas development. BLM began the process of amending the 2003 RMP in 2014, and in 2016, BLM and the Bureau of Indian Affairs (BIA) announced a joint effort to analyze resource and land management in the area for both public and tribal lands adjacent to Chaco Canyon.[68]

BLM began the NHPA section 106 consultation process because Chaco Canyon is a qualifying property.[69] When BLM began the process to amend the RMP in 2014, however, it did not consider tribal lands, even though the RMP governs almost 1 million acres of trust lands and tribal member–owned allotments. BLM began the 2016 RMP amendment consultation process by seeking public comments. Between October 2016 and February 2017, the consultation consisted of meetings with interested stakeholders and the public.

Meanwhile, BLM was still auctioning lease sales under the 2003 RMP, which was developed without adequate tribal consultation. Specifically, BLM planned to auction 4,500 acres of land for oil and gas development in March 2018. The Greater Chaco Coalition, an ad hoc group formed by tribal, environmental, and local community groups, protested lease auctions, claiming that tribal consultation was inadequate. These protests caused BLM to cancel the sale, and the bureau acknowledged its failure to adequately survey the area for cultural resources.

Even after announcing an expanded analysis in 2016 that would include tribal lands, in which the Interior Department touted its "commitment to ensuring that the region's rich cultural and archaeological resources are protected," BLM consistently failed to directly engage tribes.[70] After committing in 2016 to working with Native American leaders and integrating traditional knowledge in the management of culturally significant tribal lands, and after admitting its failure in 2018 to consult with tribes when canceling the lease sale, BLM released the draft RMP amendment in February 2019. The amendment's "preferred alternative" approved over 3,050 new wells in the planning area, just 33 wells short of the plan's maximum development alternative.

Tribes alleged that BLM's consultation for the RMP amendment draft again failed to directly engage them. Public review began in February 2020, just days before New Mexico's COVID-19 stay-at-home orders went into effect.[71] Those orders meant that the public meetings in May 2020 would be virtual.[72] The Navajo Nation and the Pueblos repeatedly requested that BLM prolong the public process for the RMP amendment until there could be in-person, face-to-face meetings instead of online.[73] In response, BLM added "four additional 'virtual' open houses" in August 2020, during which no public comments became part of the official record.[74]

The Navajo Nation had filed a lawsuit in 2017 alleging, in part, that BLM had failed to consult tribes about the effects of issuing oil and gas leases near Chaco Culture National Historical Park and had failed to analyze the indirect effects the wells would have on the park.[75] However, the courts ruled that BLM did not violate the NHPA, finding its analysis adequate for historic sites potentially affected by oil and gas drilling, since the park itself was not slated for leasing. Employing what might be classified as a "soft glance" review, the district court explained that it was "not tasked with determining if [BLM] correctly decided whether an oil well . . . altered a historic site" under the NHPA, but merely whether the bureau had "followed the proper procedures."[76] Documentation supporting the agency's findings "need not be a topic treatise or even an essay," the court reasoned, but must provide only "some explanation."[77] Thus, the court held that the bureau did not violate the NHPA, a determination the 10th Circuit upheld in 2019.[78]

Most tribes viewed the district court's deference to the bureau's consultation as judicial box-checking, illustrating the court's willingness to rubber-stamp BLM's treatment of the section 106 process as purely procedural. Further, tribes argued in both Chaco consultation processes that BLM failed to engage in meaningful consultation with the Navajo Nation and the Pueblos. Failing to provide even cursory consultation, BLM did not engage the tribes early in the decision-making process, as directed by the Clinton administration's EO 13175. When BLM canceled the lease sale in 2018—perhaps the only substantive result from this process—it conceded that it had originally approved the sale even though "concern" remained "about the proximity to Chaco [Canyon] of some of the leases and uncertainty about cultural impacts."[79]

By 2020, the Bureau of Land Management had conducted no new cultural resource studies, but gave $434,000 to the Navajo Nation to conduct an ethnographic study to identify such resources in this region. The study remained incomplete as of April 2022. However, on 15 November 2021, the Biden administration proposed to "ban oil and gas drilling within a 10-mile radius of the Chaco Culture Heritage Withdrawal Area" for 20 years, withdrawing close to 336,000 acres of federal lands from leasing under mineral leasing laws.[80] According to the Navajo Nation, which advocated for a 5-mile buffer, this proposal went forward "without proper tribal consultation," while the All Pueblo Council of Governors supported the proposal.[81] BLM accepted comments on the proposal until May 2022, and its analysis was in progress as of the writing of this chapter.

The Bureau of Land Management also failed to provide meaningful consultation by refusing face-to-face interactions before the Biden administration's proposal regarding lease sales in the Chaco Canyon region. Tribes lacked the funding and human resources to adequately participate in the RMP amendment because they were fighting the disproportionate effects of COVID-19 in their communities.[82] An agency cannot "ensure meaningful and timely input by tribal officials" if the tribes lack the capacity to review the documents.[83] While tribes responded to a health emergency, BLM moved to quickly approve the amendment, authorizing nearly 3,000 new gas and oil wells.[84]

The federal government's process concerning the oil and gas leases in Chaco Canyon was not only a failure to consult, but a display of disrespect, prioritizing the approval of oil wells over the health and interests of tribal members who were at disproportionate risk during a global pandemic.[85] In March 2021, the Biden administration placed an indefinite moratorium on new oil and gas lease auctions, meaning that the 3,000 oil well leases proposed under the latest RMP alternative could not be sold, but on 29 June 2022 the Biden administration resumed oil and gas leasing on public lands.[86] Should the Bureau of Land Management's proposal to withdraw close to 340,000 acres of lands around the Chaco Cultural Heritage Area from oil and gas leasing be approved, however, leasing will be banned on those lands regardless of the leasing resumption on other public lands.

CONCLUSION

Federal agencies can technically meet the consultation requirements under the NHPA and the government-to-government consultation prescribed in executive orders without consulting meaningfully with tribes. This does not mean that the US government lacks an obligation to go above these bare minimum legal requirements. To fulfill its trust obligation to engage in government-to-government relationships with tribes and to "protect tribal rights to exist as self-governing entities," the US government must engage in meaningful consultation. Incorporating the essential elements of meaningful consultation is necessary for a government-to-government relationship between sovereigns. A partnership in which the federal government treats tribes as respected sovereigns cannot exist if federal agencies leave tribes out of the decision-making processes that affect their culturally significant lands and natural resources. When the US government merely engages in box-checking consultation and damages culturally significant tribal lands and resources, it is acting inconsistently with its trust obligation.

The case studies discussed in this chapter expose the lengths to which tribes must go to ensure that the federal government adequately considers their interests. Even when agencies are willing to hear their voices, the legal, administrative, financial, and personnel resources required are often beyond the means of many tribes. In the two case studies in which meaningful consultation was eventually achieved—Bears Ears and the secretarial order—the tribes shouldered the burden of making the federal agencies, which have a trust responsibility to protect tribal interests, understand the value and importance of their lands, cultures, and traditions. To require tribes to regularly perform this labor for every consultation process just to get a seat at the table is not sustainable.

Even if the United States did not have consultation obligations, the government's best interests are served by meaningfully consulting with tribes in land and natural resource management decision-making that affects properties with cultural importance to the tribes. At minimum, federal agencies can avoid litigation and project delays that occur when tribes assert that their rights are being ignored, as in the Chaco Canyon case. Federal agencies have much to gain from understanding and incorporating unique tribal knowledge and expertise in land management—as does the public. One way to help federal land managers gain this understanding would be to establish institutes that promote the use of Western scientific knowledge, Indigenous knowledge, and local knowledge in management decision-making, as suggested in the Bears Ears proposal. These institutes could provide access to the tribal knowledge needed to ensure meaningful tribal participation in federal decision-making. In the words of Russell Attebery, Chairman of the Karuk Tribe, "[N]obody knows Indian [C]ountry like the people who live there."[87]

NOTES

1. Letter from Mark Ingersoll, Chairman of the Confederated Tribes of Coos, Lower Umpqua, and Siuslaw Indians, to Larry Roberts, Acting Assistant Secretary of Indian Affairs, 30 November 2016, https://www.bia.gov/sites/default/files/dup/assets/as-ia/raca/pdf/idc2-055648.pdf, at 3 ("[W]e have been forced to devote extraordinary amounts of staff time, legal resources, and scarce funding over the past decade in an effort to compel the Federal Energy Regulatory Commission [FERC] to simply do what is required of them by the National History Preservation Act . . . , and by FERC's federal trust responsibility.").

2. See COHEN'S HANDBOOK OF FEDERAL INDIAN LAW, s. 5.04[3](a) (Nell Jessup Newton ed., 2012) [hereinafter COHEN TREATISE] ("Today, the trust doctrine is one of the cornerstones of Indian law."). The Supreme Court cases that had a considerable role in defining the trust relationship are Johnson v. M'Intosh, 21 US 543 (1823), Cherokee Nation v. Georgia, 30 US 1 (1831), and Worcester v. Georgia, 31 US 515 (1832).

3. Cherokee Nation v. Georgia, 30 US 1, 17 (1831).

4. COHEN TREATISE, *supra* note 2, at s. 5.04[3](a).

5. *Id.* ("In the late 19th and early 20th centuries, courts relied on [early Supreme Court decisions] to justify broad exercise of power to dispose of tribal property and alter the relationships of tribes to the federal government, even without tribal consent") (discussing an order by the secretary of the Interior "reaffirming the federal trust responsibility's application to all Interior agencies and bureaus"). See also US Dep't of the Interior, Secretarial Order No. 3335, Reaffirmation of the Federal Trust Responsibility to Federally Recognized Tribes and Individual Indian Beneficiaries (20 August 2014), https://www.doi.gov/sites/doi.gov/files/migrated/news/pressreleases/upload/Signed-SO-3335.pdf, 4 ("During the last few decades, the trust relationship has evolved" into today's "Era of Tribal Self-Determination.").

6. National Historic Preservation Act, 16 USC §§ 470a *et seq.*

7. See Presidential Memorandum on Tribal Consultation, 2009 DAILY COMP. PRES. DOC. (5 November), https://obamawhitehouse.archives.gov/the-press-office/memorandum-tribal-consultation-signed-president [hereinafter Obama's Consultation Memo] ("Consultation is a critical ingredient of a sound and productive Federal-tribal relationship.").

8. Charles Wilkinson, *At Bears Ears We Can Hear the Voices of Our Ancestors in Every Canyon and on Every Mesa Top: The Creation of the First Native National Monument*, 50 ARIZ. ST. L. J. 317, 326 (2018) (describing "collaborative management" as a "deeper tribal-federal relationship" than merely "co-management," where tribes are involved in the entire land management decision-making process).

9. Proposal from the Bears Ears Inter-Tribal Coalition to President Barack Obama, at 28 (15 October 2015), http://www.bearsearscoalition.org/wp-content/uploads/2015/10/Bears-Ears-Inter-Tribal-Coalition-Proposal-10-15-15.pdf [hereinafter Bears Ears Proposal].

10. US Departments of Commerce and Interior, Joint Secretarial Order on American Indian Tribal Rights, Federal-Tribal Trust Responsibilities, and the Endangered Species Act, No. 3206 (5 June 1997).

11. Charles Wilkinson, *The Role of Bilateralism in Fulfilling the Federal-Tribal Relationship: The Tribal Rights–Endangered Species Secretarial Order*, 72 WASH. L. REV. 1063, 1077–79 (1997) ("[Meetings] were designed to allow the tribal side to explain some of the many unique and varied circumstances that apply when federal laws are sought to be extended into Indian country.").

12. Mark Ingersoll Letter, *supra* note 1, at 7.

13. Special Message to the Congress on Indian Affairs, 1 PUB. PAPERS 564 (8 July 1970) [hereinafter Nixon Announcement].

14. *Id.* at 1. The termination era saw a "harsh attack on tribal sovereignty and cultures," where the federal government's goal was "pro-assimilation and anti–special rights for Indians." The government "abrogated express treaty rights" and "unilaterally ended the government-to-government relationships" that the United States had with over 100 tribes. Carole Goldberg, *President Nixon's Indian Law Legacy: A Counterstory*, 63 UCLA L. REV. 1506, 1510 (2016).

15. Nixon Announcement, *supra* note 13, at 2.

16. *Id.*

17. Presidents Clinton, George W. Bush, and Obama reaffirmed the federal commitment to the government-to-government relationship. Memorandum on Government-to-Government Relationships with Tribal Governments (23 September 2004); and Presidential Memorandum on Tribal Consultation, 2009 DAILY COMP. PRES. DOC. (5 November 2009), https://obamawhitehouse.archives.gov/the-press-office/memorandum-tribal-consultation-signed-president

18. See, e.g., Quechan Tribe of the Fort Yuma Indian Rsvr. v. US Dep't of Interior, 755 F. Supp. 2d 1104, 1108 (S.D. Cal. 2010) (explaining that "[t]he Court's review of agency action under [the NHPA] is governed by the Administrative Procedures Act," and a failure to engage in adequate consultation violates § 706(2)(D) of the APA because it is an "agency action" "without observance of procedure required by law").

19. See, e.g., Exec. Order No. 13175, 65 Fed. Reg. 67,249 (6 November 2000) [hereinafter EO 13175], at § 5(a).

20. *Id.* at § 10.

21. 16 USC § 470f. See also *What Is the National Register of Historic Places?* NAT'L PARK SERV., https://www.nps.gov/subjects/nationalregister/what-is-the-national-register.htm#:~:text=The%20National%20Register%20of%20Historic%20Places%20is%20the,of%20the%20Nation%27s%20historic%20places%20worthy%20of%20preservation (last visited 18 April 2021).

22. 36 CFR § 800.1(a).

23. *Id.* See also *id.* at § 800.2(a)(4) ("The Council issues regulations to implement section 106, provides guidance and advice on the application of the procedures in this part, and generally oversees the operation of the section 106 process.").

24. *Id.* at 7.

25. Exec. Order No. 12875, 58 Fed. Reg. 58,093 (26 October 1993).

26. EO 13175, *supra* note 19, at § 2(a).

27. Other consultation case studies we considered included the Jordan Cove liquefied natural gas project and the memorandum of agreement to remove four dams on the Klamath River. For more information on the Jordan Cove project, see Mark Ingersoll Letter, *supra* note 1, at 3–6; and Ted Sickinger, *Feds Uphold State Denial on Jordan Cove LNG's Coastal Zone Permit, Another Roadblock for the Controversial Project*, OREGON

Live (9 February 2021), https://www.oregonlive.com/environment/2021/02/feds-uphold-state-denial-on-jordan-cove-lngs-coastal-zone-permit-another-roadblock-for-the-controversial-project.html. For more information on the Klamath dam removal process, see Memorandum of Agreement Implementing the Klamath Hydroelectric Settlement Agreement for Dam Removal (November 2020), at 1, http://www.klamathrenewal.org/wp-content/uploads/2020/11/Klamath-MOA.pdf; Jamie Parfitt, *Fight over Klamath River Dam Removal Project Goes to Federal Regulators*, KDRV News (16 February 2021), https://www.kdrv.com/content/news/Fight-over-Klamath-River-dam-removal-project-goes-to-federal-regulators-573806011.html; and Konrad Fisher, *The Klamath River's Advocates Succeed on Their Second Try with New Agreement for Largest-Ever Dam Removal*, Waterkeeper All., https://waterkeeper.org/magazines/volume-13-issue-1/klamath-river-dam-removal/ (last visited 27 April 2021).

28. The federal government began taking action on behalf of the tribes to protect treaty fishing rights in the late 1960s, including by representing individual treaty fishermen in state criminal prosecutions. See Michael C. Blumm & Cari Baermann, *The Belloni Decision and Its Legacy*: United States v. Oregon *and Its Far-Reaching Effects after a Half-Century*, 50 Envtl. L. 347, 366 (2020). Tribal activists, including Sohappy, sued Oregon state officials in 1968, "challenging the state's restrictions on treaty fishing and seeking to stop the state's arrests of treaty fishermen" (*id.* at 364). In the same year, the United States initiated the US v. Oregon suit to similarly protect tribal treaty rights for salmon harvest, and due to "the overlap of treaty rights issues, Judge Belloni consolidated the two cases" in 1969 (*id.* at 366–67).

29. Sohappy v. Smith, 302 F. Supp. 899, 911–12 (D. Or. 1969). See also Blumm & Baermann, *supra* note 28, at 352, and *id.* at 366 (describing the Columbia River treaty tribes as including the tribes of the Yakama, Umatilla, and Warm Springs Reservations and the Nez Perce Tribe).

30. Blumm & Baermann, *supra* note 28, at 374.

31. See Blumm & Baermann, *supra* note 28, at 385. Note, however, that this apparent tribal veto in the co-management framework came as a result of a federal court's interpretation of management as necessary to satisfy express treaty rights.

32. Wilkinson, *supra* note 11, 1074.

33. *Id.* at 1077.

34. *Id.* at 1078.

35. *Id.* at 1081.

36. *Id.* at 1082.

37. Elouise Wilson, Mary R. Benally, Ahjani Yepa, & Cynthia Wilson, *Women of Bears Ears Are Asking You to Help Save It*, New York Times (25 April 2021) ("We are among the Women of Bears Ears—Indigenous women who support our families and communities in the protections of ancestral lands. . . . From these Southwestern lands, twin buttes arise; they are known as Bears Ears."). See also *Native American Connections*, Bears Ears Inter-Tribal Coalition, https://bearsearscoalition.org/ancestral-and-modern-day-land-users/ (last visited 26 April 2021) ("Several southwestern tribes trace their ancestry to the ancient peoples who populated the [Bears Ears] region since time immemorial.").

38. Native American Rights Fund (NARF), *Protecting Bears Ears National Monument*, https://www.narf.org/cases/bears-ears/ (last visited 15 April 2021).

39. Dean B. Suagee, *Tribes Call for Collaborative Management of Bears Ears National Monument*, The Hill (10 June 2016), https://thehill.com/blogs/congress-blog/judicial/283078-tribes-call-for-collaborative-management-of-bears-ears-national.

40. Wilkinson, *supra* note 8, 317, 327. See also NARF, *Protecting Bears Ears*, *supra* note 38.

41. https://www.bearsearscoalition.org/bears-ears-national-monument-restored/ (last visited 15 July 2022).

42. https://www.whitehouse.gov/briefing-room/statements-releases/2021/10/07

/fact-sheet-president-biden-restores-protections-for-three-national-monuments-and-renews-american-leadership-to-steward-lands-waters-and-cultural-resources/ (last visited 15 July 2022).

43. https://www.whitehouse.gov/briefing-room/statements-releases/2021/10/07/fact-sheet-president-biden-restores-protections-for-three-national-monuments-and-renews-american-leadership-to-steward-lands-waters-and-cultural-resources/ (last visited 15 July 2022).

44. Bears Ears Inter-Tribal Coalition, *Proposal from the Bears Ears Inter-Tribal Coalition to President Barack Obama* (15 October 2015), http://www.bearsearscoalition.org/wp-content/uploads/2015/10/Bears-Ears-Inter-Tribal-Coalition-Proposal-10-15-15.pdf, 15 ("[UDB] has interviewed and surveyed thousands of people; held eight Town Hall meetings; obtained over 15,000 statements of support; held five annual gatherings of Tribes at Bears Ears to discuss land protection strategies; interviewed dozens of elders and medicine men; developed sophisticated GIS [geographic information system] data and many maps displaying that data; and obtained 24 resolutions of support from many Navajo chapter houses and Tribes."). See also NARF, *Protecting Bears Ears, supra* note 38; and *Interactive Map*, BEARS EARS INTER-TRIBAL COALITION, https://bearsearscoalition.org/interactive-map/ (last visited 18 April 2021).

45. Bears Ears Proposal, *supra* note 9, at 15.

46. NARF, *Protecting Bears Ears, supra* note 38 ("One month" after a series of meeting with UDB, Navajo Nation, and Ute Mountain Ute Tribe, the San Juan County commissioners "urged the Utah State Legislature to pass HB 3931, which undermined the Bears Ears proposal by designating large areas of the region as 'Energy Zones' that would be fast-tracked for grazing, energy, and mineral development.").

47. NARF, *Protecting Bears Ears, supra* note 38.

48. Antiquities Act, 54 USC § 320301 (1906).

49. Anna Brady, *Through Bears Ears, Tribes Lead the Way for True Collaboration over Utah's Public Lands*, UNIV. OF UTAH S. J. QUINNEY COLL. OF L. (9 November 2015), https://law.utah.edu/through-bears-ears-tribes-lead-the-way-for-true-collaboration-over-utahs-public-lands/.

50. *Id.* at 31.

51. Wilkinson, *supra* note 8, 326.

52. Proclamation No. 9558, 82 Fed. Reg. 1139, 1144 (28 December 2016) (establishing Bears Ears National Monument).

53. *Id.*

54. *Id.*

55. Nixon Announcement, *supra* note 13.

56. *Id.* at 2.

57. Proclamation No. 9681, 82 Fed. Reg. 58,081 (4 December 2017) (modifying the Bears Ears National Monument). See Hopi Tribe v. Trump, Case 1:17-cv-02590-TSC, at *14 (D.D.C., 20 March 2019) ("Shortly [after Trump decreased the Bears Ears National Monument] . . . , Plaintiffs sued, alleging that President Trump's Proclamation was not authorized by the [Antiquities] Act, and violates the United States Constitution.").

58. See Michael Doyle & Jennifer Yachnin, *Biden's Legal Team Has Done Its Bears Ears Homework*, E&E NEWS (5 February 2021), https://www.eenews.net/stories/1063724513.

59. Three cases were filed against the Trump diminishment: Hopi Tribe v. Trump, Case No. 17-cv-2590 (TSC), Utah Diné Bikéyah v. Trump, Case No. 17-cv-2605 (TSC), and Nat. Res. Def. Council v. Trump, Case No. 17-cv-2606 (TSC), which were consolidated before the federal district court in DC.

60. Exec. Order No. 13990, 86 Fed. Reg. 7037 (20 January 2021) (Protecting Public Health and Environment and Restoring Science to Tackle the Climate Crisis), at § 3(a) [hereinafter EO 13990].

61. *Bears Ears Inter-Tribal Coalition Supports Review of Monument Boundaries*, NATIVE AM. RTS. FUND (17 March 2021), https://www.narf.org/bears-ears-update/.

62. Jennifer Yachnin, *Haaland's Utah Trip Beset by Bears Ears Lobbying*, E&E News (8 April 2021). See also *Secretary Deb Haaland*, US Dep't. of the Interior, https://www.doi.gov/secretary-deb-haaland (last visited 13 April 2021); and Michael Doyle, *Problems, Opportunities Aplenty Await Haaland at Interior*, E&E News (22 February 2021) ("A voice like mine has never been a Cabinet secretary or at the head of . . . [I]nterior. . . . [I]t's profound to think about the history of this country's policies to exterminate Native Americans and the resilience of our ancestors that gave me a place here today") (quoting Haaland).

63. Yachnin, *Haaland's Utah Trip*, supra note 62. See also Jennifer Yachnin, *Tribal Leaders: Bears Ears Can't Wait for Legislative Fix*, E&E News (21 April 2021).

64. Utah's governor "warned . . . that his state could file its own legal challenge if Biden opts to restore the monument ahead of congressional action," arguing that "the Antiquities Act provides a limit on the size" of protected sites, and Chief Justice Roberts "recently appeared to invite new challenges to the law" (Yachnin, *Tribal Leaders*, supra note 63). For the Roberts statement, see Jennifer Yachnin, *Chief Justice Roberts Invites Antiquities Act Challenges*, E&E News (24 March 2021).

65. See EO 13990, *supra* note 60.

66. Jonathon Thompson, *Drilling Chaco: What's Actually at Stake*, High Cnty. News (13 April 2015), https://www.hcn.org/articles/drilling-chaco-whats-really-at-stake (describing Chaco Canyon as "the center of a large society that extended hundreds of miles beyond the canyon[']s walls" to many historical sites "concentrated in the central San Juan Basin"). See also Arlyssa Becenti, *Feds Proceed with Chaco Drilling Plan while Tribes Distracted by Pandemic*, Navajo Times (4 June 2020), https://navajotimes.com/coronavirus-updates/feds-proceed-with-chaco-drilling-plan-while-tribes-distracted-by-pandemic/ (explaining that both the Navajo Nation and Pueblo tribes consider Chaco Canyon their ancestral home).

67. Thompson, *supra* note 66 (describing unprotected sites as a "prime target for oil and gas drillers"). Several structures, including Pueblo Bonito, are protected from oil and gas drilling as part of the Chaco Culture National Historical Park, but the surrounding areas are not. Oil and gas drilling adjacent to protected areas has negative effects since drilling creates light and noise pollution. See *id*. See also Joey Keefe, *Groups Blast Trump Administration Plans for More Drilling at Chaco Canyon*, New Mexico Wild (26 September 2020), https://www.nmwild.org/2020/09/26/groups-blast-trump-administration-plans-for-more-drilling-at-chaco-canyon/ (quoting the executive director of the New Mexico Wilderness Alliance, who described the Bureau of Land Management's "consultation" process during the COVID pandemic as "shameful" and "compounding a tragic history of disrespect and broken trust").

68. Press Release, US Dept. of Interior, Interior Department Announces Broader Plan to Review Management of Lands in Northwestern New Mexico (20 October 2016), https://www.doi.gov/pressreleases/interior-department-announces-broader-plan-review-management-lands-northwestern-new.

69. Diné Citizens Against Ruining Our Env't v. Jewell, 312 F. Supp. 3d 1031, 1051 (D.N.M. 2018).

70. See Rebecca Sobel, *Greater Chaco Coalition Responds to BLM's Broken Promises*, WildEarth Guardians (2 March 2020), https://wildearthguardians.org/pressreleases/greater-chaco-coalition-responds-to-blms-broken-promises/.

71. *Greater Chaco Coalition Demands BLM Respect Tribes and Communities, Echoes Request to Postpone Drilling Plan*, Frack Off Chaco (27 September 2020), https://www.frackoffchaco.org/blog/press-release-9-17-2020 [hereinafter Coalition Demands BLM Respect].

72. Becenti, *supra* note 66 (noting that the five virtual public meetings held by the regional BLM office "weren't ideal for affected tribal members, either because many [were] without internet/broadband connection" or "were busy with community obligations" regarding COVID-19).

73. Letter from Daniel Tso, Chairman of the Navajo Nation Health, Education and Human Services Committee, to Tim Spisak, Director of Bureau of Land Management, New Mexico State Office (13 August 2020), https://drive.google.com/file/d/1u_Zdp7ssx aDbfF0fcS6TmBw-AQoTIIkx/view.

74. Tso questioned the "immediately, and indefinitely, suspect" RMPA process. He explained that "the Navajo Nation was still in the midst of an extreme human health emergency" and that the Navajo Nation could not be expected to engage in "meaningful consultation" because it could not be in person, tribal members lacked internet access, and they needed translation into the Navajo language (*id.*). See also Coalition Demands BLM Respect, *supra* note 71.

75. Diné Citizens Against Ruining Our Env't, 312 F. Supp. at 1031, 1051, 1081.

76. *Id.* at 1100.

77. *Id.* at 1101.

78. *Id.* at 1109, *aff'd*, Diné Citizens Against Ruining Our Env't v. Bernhardt, 923 F.3d 831, 850 (10th Cir. 2019).

79. John R. Moses, *Zinke Places Chaco Canyon Drilling Leases on Hold, Pending Cultural Review*, FARMINGTON DAILY (2 March 2018), https://www.daily-times.com/story/news /local/navajo-nation/2018/03/02/chaco-drilling-leases-hold-pending-zinke-cultural -review/389984002/ (quoting Secretary of the Interior Ryan Zinke).

80. https://www.navajonationcouncil.org/wp-content/uploads/2022/04/Chaco _Consultation_2022.04.29.pdf (last visited 15 July 2022); and https://eplanning.blm.gov /eplanning-ui/project/2016892/510 (last visited 15 July 2022).

81. https://www.navajonationcouncil.org/wp-content/uploads/2022/04/Chaco _Consultation_2022.04.29.pdf (last visited 15 July 2022); and https://www.durango herald.com/articles/proposed-drilling-ban-around-chaco-canyon-gets-support-from -all-pueblo-council-of-governors/ (last visited 15 July 2022).

82. Becenti, *supra* note 66.

83. EO 13175, *supra* note 19, at § 5(a).

84. Becenti, *supra* note 66.

85. See, e.g., Liz Mineo, *For Native Americans COVID-19 Is the Worst of Both Worlds at the Same Time*, HARVARD GAZETTE (8 May 2020), https://news.harvard.edu/gazette /story/2020/05/the-impact-of-covid-19-on-native-american-communities/ ("As of April 30[, 2020], the Navajo [N]ation had the third-highest per capita rate of COVID-19 in the country, after New Jersey and New York. Worsening the situation, Native Americans appear to have a higher risk of serious complications.").

86. Donald McGahn II, Jeffery Schlegel, David Stringer, & Charles Wehland, *Biden Administration Announces Moratorium on New Federal Oil and Gas Leases*, JONES DAY (24 March 2021), https://www.mondaq.com/unitedstates/oil-gas-electricity/1050810 /biden-administration-announces-moratorium-on-new-federal-oil-and-gas-leases; https://biologicaldiversity.org/w/news/press-releases/lawsuit-challenges-bidens -resumption-of-oil-gas-leasing-on-public-lands-2022-06-29/ (last visited 15 July 2022).

87. Oregon Humanities, *A Conversation on the History and Future of Settlement and Water Use in the Klamath Basin*, YOUTUBE (15 March 2021), https://www.youtube.com /watch?v=rzmo2qYSgG0 (Russell Attebery at 15:54, discussing meaningful consultation).

Chapter 6

An Introduction to Indian Reserved Water Rights

VANESSA L. RAY-HODGE and DYLAN R. HEDDEN-NICELY

There has been a lot said about the sacredness of our land which is our body, and the values of our culture which is our soul. But water is the blood of our tribes, and if its life-giving flow is stopped, or it is polluted, all else will die and the many thousands of years of our communal existence will come to an end.
—Governor Frank Tenorio, San Felipe Pueblo, 1976

Mní Wičhóni (Lakota)—Water Is Life—reverberated across the United States in 2016 as the Standing Rock Sioux Tribe fought to protect its waters, land, and other cultural and natural resources from the existential threats posed by the construction of the Dakota Access Pipeline (DAPL), a 1,168-mile-long crude oil pipeline running from North Dakota to Illinois. Tribes across the United States traveled to support the No DAPL movement. It had struck a chord in Indian Country and highlighted the struggle that many Indian tribes face in protecting tribal waters and other natural resources. The Standing Rock Sioux Tribe continues to pursue avenues to protect their water and other resources from DAPL. And they are not alone, as the fight by Indian tribes to secure and protect their water and other resources is not new.

The federal government has long encouraged non-Indian development of land and particularly water resources across the United States. Many early federal water-related projects authorized the taking of Indian lands to build dams and flood control projects that primarily benefited non-Indian communities and devastated Indian communities.[1] The US Supreme Court had recognized the existence of Indian reserved rights in 1908.[2] Despite that, a 1973 report by the national water commissioner found that

> more than 50 years elapsed before the Supreme Court again discussed significant aspects of Indian water rights. During most of this 50-year period, the United States was pursuing a policy of encouraging the settlement of the West. . . . With the encouragement, or at least the cooperation, of the Secretary of the Interior—the very office entrusted

with protection of all Indian rights—many large irrigation projects were constructed on streams that flowed through or bordered Indian Reservations.... With few exceptions the projects were planned and built by the Federal Government without any attempt to define, let alone protect, prior rights that Indians might have had in the water used for the projects.... In the history of the United States Government's treatment of Indian tribes, its failure to protect Indian water rights for use on the Reservations it set aside for them is one of the sorrier chapters.[3]

The failure of the United States in advancing and protecting Indian water rights has left tribes far behind their non-Indian neighbors on many socioeconomic levels. As water development continued in the second half of the 20th century and into the 21st, water-related litigation increased significantly in scope and complexity. At the same time, non-Indian and tribal uses of water became increasingly intertwined and competitive as tribes began to obtain the resources necessary to develop some of their own water rights. The United States, as trustee for Indian tribes, holds Indian reserved water rights in trust for the benefit of tribes and is generally required to be involved when those rights are being litigated or otherwise determined. As a result, litigants often seek to require the participation of the United States in order to quantify federal and Indian water rights, but many have found that they cannot require federal participation because the United States has immunity from lawsuits unless that immunity has been expressly waived by Congress.

In 1952, Congress passed the McCarran Amendment, which has been interpreted by the Supreme Court as providing a waiver of the United States' immunity from suit in actions seeking to join the United States as a defendant in general stream adjudications.[4] In interpreting the scope of the waiver, however, the Court held that the waiver is limited and that any general stream adjudication seeking to come under the McCarran Amendment must include certain procedural requirements to effectuate the waiver. For example, the Court ruled that the amendment only provides a limited waiver of sovereign immunity for purposes of joinder to (1) comprehensive general stream adjudications in which (2) the rights of all competing claimants are adjudicated.[5] The waiver does not subject the United States to private suits to decide priorities between the United States and a particular claimant. Thus, all water users in a river system or basin must be joined in a state proceeding for the McCarran Amendment to allow the involuntary joinder of the United States.

The McCarran Amendment does not, however, give state courts exclusive jurisdiction over federal reserved water rights.[6] Federal courts can still adjudicate these water rights claims, but they may abstain from taking cases where there is an ongoing state general stream adjudication. Since the mid-20th century, due in part to the McCarran Amendment, more tribes have been forced into, or actively sought, litigation to define, quantify, and protect their water rights. In virtually all of the litigation involving Indian water rights, tribes have faced staunch opposition from other water users, who actively seek to limit the amount and uses of water to which tribes are entitled to sustain their homelands. This has created a tremendous amount of hostility among tribes, their non-Indian neighbors, and states when it comes to water rights, especially in areas of the West where there are increasing demands on limited water supplies.

GENERAL OVERVIEW OF WATER RIGHTS

The establishment and protection of water rights continue to be critically important issues to tribes from a cultural and spiritual standpoint, as well as to economically sustain reservation communities and the natural resources on which tribes and their members depend.

In eastern states, water rights are often based on a riparian water law system under which the owner of property adjacent to a stream or lake is entitled to a reasonable use of that water for beneficial purposes.[7] Riparian users generally have a duty to share water.[8] In contrast, across the West, most state water rights are based on what is known as the prior appropriation doctrine under which water can be diverted for a beneficial use or purpose.[9] A "beneficial purpose" is defined under state law, but usually includes irrigation, livestock watering, and various domestic, municipal, commercial, and industrial uses.

In prior appropriation (western) states, a water right has two main attributes. The first is a priority date based on the date water was first used or diverted for a beneficial purpose. This establishes the water user's priority vis-à-vis other users in a basin. The second attribute is the quantity to which a user is entitled for a particular purpose. In these states, individuals or entities with an early priority date have senior rights over users with later (junior) rights. What this means practically is that in times of water shortages, an individual or entity with an early (senior) priority date is entitled to receive their full quantity of water for beneficial purposes before users with junior rights. This has been summarized by the saying "first in time, first in line" (or "first in right").

Water rights in prior appropriation states can be lost by nonuse, which can come from forfeiture (failure to comply with statutory conditions for use of water) or abandonment (relinquishment of the right), which are both usually governed by state statutory law. Indian tribes, however, are not generally subject to state laws or regulations in this regard, and as discussed below, Indian reserved water rights exist outside state law.[10]

ESTABLISHMENT OF THE RESERVED WATER RIGHTS DOCTRINE

For more than a century, the Supreme Court has held that when the United States withdraws land from the public domain and reserves that land for a federal purpose, the United States by implication reserves unappropriated water necessary to accomplish the purpose of the reservation. This is known as the reserved rights doctrine, also referred to sometimes as the Winters doctrine or Winters rights, as it was first outlined in the landmark Indian case *Winters v. US*.

The *Winters* case involved a water rights dispute between the United States on behalf of the Gros Ventre and Assiniboine Tribes of the Fort Belknap Reservation in Montana and non-Indians outside the Reservation. The Fort Belknap Reservation was created based on an agreement between the tribes and the United States on 1 May 1888, which reserved a tract of land "as an Indian reservation as and for a permanent home and abiding place" for the tribes.[11] In the court case, the United States sought an injunction prohibiting Henry Winters and other non-Indians from diverting water from the Milk River and its tributaries for irrigation upstream from the Reservation, because insufficient water was reaching Reservation lands that the Bureau of Indian Affairs (BIA) and tribes wanted to develop for the pursuit of agriculture by irrigation.

The United States argued that "all of the waters of the river are necessary for [current uses] and the purposes for which the [R]eservation was created, and that in furthering and advancing the civilization and improvement of the Indians, and to encourage habits of industry and thrift among them, it is essential and necessary that all of the waters of the river flow down the channel uninterruptedly and undiminished in quantity and undeteriorated quality."[12] However, the 1888 agreement establishing the Fort Belknap Reservation did not specifically mention water rights. Non-Indians, like Winters, argued that they lawfully had begun appropriating water under state law by 1900, and thus their rights were senior to the claims of the United States and the tribes.

Despite the fact that the non-Indian irrigators had put the water to use first and under state law would have had senior rights based on beneficial use, the US Supreme Court in *Winters* held that the tribes had superior rights to the water under federal law. In reaching this finding the Court found that the case turned on the meaning of the 1888 agreement establishing the Reservation. The Court noted that in establishing the Reservation for the tribes, "[t]he lands were arid, and, without irrigation, were practically valueless . . . and 'civilized communities could not be established thereon.'"[13] Thus because a purpose of the Reservation when established was to promote agriculture and because water was needed in order for the tribes to sustain irrigated agriculture within the Reservation, the Court found that the tribes retained federally protected water rights, explaining that the "power of the government to reserve the waters and exempt them from appropriation under state law is not denied, and could not be."[14] Even though non-Indians upstream from the Reservation had been irrigating before the tribes did, the tribes were entitled to water for irrigation as a matter of federal law with at least an 1888 priority date even though the water had not previously been put to use. This kind of federal water right has since become known as a reserved water right.

In *Winters*, the Supreme Court did not place a limit on the amount of reserved water to which the tribes were entitled. Rather, the Court enjoined the interference by non-Indians with the water that the BIA and tribes needed to develop agriculture. In that respect, the decree in *Winters* was open-ended because, while tribes were found to have senior reserved water rights, those rights were not fully determined.

Arizona v. California, decided in 1963, is the only major US Supreme Court decision other than *Winters* explaining the nature and extent of Indian reserved water rights.[15] The case began in 1952 when the state of Arizona filed a suit in the original jurisdiction of the US Supreme Court against California and Nevada to determine its rights to water from the Colorado River. The United States intervened, asserting, among other things, reserved water rights to the mainstem of the Colorado River and its tributaries in the lower Colorado River basin for several tribes. Ultimately, the Supreme Court addressed only the reserved water rights for five Indian reservations along the mainstem.[16]

The Supreme Court referred the case to a special master (a subordinate official appointed to gather evidence, find facts, state conclusions of law, and recommend a decree), who held lengthy hearings on the issues presented.[17] The special master concluded, without much discussion, that the five reservations were established so that the "Indians would settle on the Reservation land and develop an agricultural economy."[18] The special master also "found both as a matter of fact and law that when the United States created these reservations or added to them, it reserved not only land but also the use of enough water from the Colorado [River] to irrigate the irrigable portion of the reserved lands."[19] Overall, "the aggregate quantity of water which the Master held was reserved for all the reservations is about 1,000,000 acre-feet," to be used on approximately 135,000 irrigable acres of land.[20]

In confirming the special master's conclusion, the Supreme Court rejected arguments that because most of the five Indian reservations at issue were established by executive order, they did not have federal reserved water rights. In doing so, the Supreme Court made clear that Indian reservations established by executive order must be treated the same as reservations created by treaty or congressionally ratified agreements.[21] Relying on *Winters*, the Court found that "water from the river would be essential to the life of the Indian people," and therefore the United States intended to "reserve waters necessary to make the reservation livable."[22] These rights, the Court found, are "present perfected rights" that vested as of the date that the reservation was established.[23] Thus, the Supreme Court found that the five tribes had federally protected water rights to the Colorado River to make their reservations livable and suitable for agriculture.

In deciding the quantity of reserved water to which the five tribes were entitled, the Court agreed with the special master, who found that the United States reserved "the use of enough water from the Colorado River to irrigate the irrigable portion of the reserved lands."[24] The Supreme Court found the special master's calculation of 1 million acre-feet for all five reservations to be reasonable because it was "intended to satisfy the future as well as the present needs of the Indian Reservations."[25]

The Court rejected Arizona's arguments that "the Master ha[d] awarded too much water" and that the quantity of water for each reservation should be based on its reasonably foreseeable needs, which the Court construed to mean the number of Indians.[26] Instead, where Indian reservations were set aside with an agricultural purpose, the Court concluded, "the only feasible and fair way by which reserved water for the reservations can be measured is irrigable acreage."[27] Irrigable acreage, the Court decided, yielded an equitable result because it provided certainty to non-Indians by capping the quantity to which the tribes would be entitled while simultaneously providing certainty to the tribes by ensuring they would always have the water necessary to irrigate their lands.[28] Notably, the reserved water rights for the five tribes included water for domestic and stock-related uses.[29]

Both *Winters* and *Arizona v. California* establish that preemptive federal law embodied in the treaties, executive orders, and agreements establishing Indian reservations exempt tribes from principles of state water law—including the principle of prior appropriation—that would otherwise limit tribal water use and thereby thwart the future use of water by tribes necessary to fulfill the purposes for which their reservations were created.[30] Accordingly, federal law supplants state law principles of prior appropriation by recognizing federally protected water rights for Indian reservations that are not based on or measured by actual beneficial use. Instead, federal law secures for tribes reserved water rights to fulfill the purposes for which the Indian reservation was created. The amount of water reserved must also be sufficient to ensure that tribes have enough water to meet their present and future needs.

THE RESERVED RIGHTS DOCTRINE EXTENDS TO NON-INDIAN FEDERAL RESERVATIONS AND HAS IMPLICATIONS FOR INDIAN RESERVED WATER RIGHTS

In addition to addressing Indian water rights, the Supreme Court extended the reserved rights doctrine to include non-Indian reservations in *Arizona v. California*, and the scope of Indian reserved water rights has been colored by non-Indian reserved water rights cases ever since. Of particular importance are two US Supreme Court cases: *Cappaert v. US* (hereinafter *Cappaert*) and *US v. New Mexico* (hereinafter *New Mexico*).[31]

In *Cappaert*, the United States sought to secure reserved water rights to ensure that Devils Hole in the Death Valley National Monument in Nevada had enough underground water to maintain a pool at levels that would preserve the federally endangered Devils Hole pupfish (*Cyprinodon diabolis*). The Cappaert family owned a multimillion-dollar commercial ranch near the monument and had begun pumping groundwater in 1968 from a basin or aquifer that was also the monument's source of water.[32] Water measurements taken by the United States since 1962 showed that since the Cappaerts had begun pumping, the water in Devils Hole had dropped to levels that negatively impacted the ability of the pupfish to spawn in quantities that would prevent extinction.

The 1952 presidential proclamation establishing Devils Hole specifically mentioned the "underground pool": "said pool is a unique subsurface remnant of the prehistoric chain of lakes which in Pleistocene times formed the Death Valley Lake System," and "that this subterranean pool is an integral part of the hydrographic history of the Death Valley region is further confirmed by the presence in this pool of a particular race of desert fish . . . which is found nowhere else in the world."[33] The proclamation set aside a 40-acre tract as part of the Death Valley National Monument to protect the pool. The Court held that, given the purposes for which Devils Hole was established, when President Harry Truman reserved the monument he impliedly reserved sufficient water to permit "a peculiar race of desert fish . . . , which is found nowhere else in the world . . . to spawn."[34]

In finding a reserved water right for Devils Hole, the Supreme Court explained that the reserved rights doctrine "is based on the necessity of water for the purpose of the federal reservation" and that "the United States can protect its water from subsequent diversion, whether the water diversion is of surface or groundwater."[35] The Court held that the proclamation establishing Devils Hole reserved "only that amount of water necessary to fulfill the purpose of the reservation, no more."[36] Thus the injunction must be tailored "to minimal need, curtailing pumping only to the extent necessary to preserve an adequate water level at Devil's [sic] Hole" sufficient to allow the fish to spawn and survive.[37] *Cappaert* thus introduced the concept that reserved water is implied where necessary to fulfill a particular purpose of a federal reservation, at least in a non-Indian context.

In the *New Mexico* case, the US Supreme Court developed what is known as the primary/secondary purposes test for determining reserved water rights on non-Indian federal reservations. Here, the US claimed reserved rights for the Gila National Forest in a general stream adjudication of the Rio Mimbres, and the Court had to determine "what quantity of water, if any, the United States reserved . . . when it set aside the Gila National Forest in 1899."[38] The Court noted that "each time" it had applied the reserved rights doctrine, the Court had "carefully examined both the asserted water right and the specific purposes for which the land was reserved, and concluded that without the water the purposes of the reservation would be entirely defeated."[39] This "careful examination" is required because reserved rights are implied, and because "with respect to the allocation of water . . . [Congress] has almost invariably deferred to the state law."[40] The Supreme Court then held that in reserving lands for a national forest, the United States reserved only the amount of water "necessary to fulfill the very purposes [or primary purposes] for which" the forest was established, and where water is "only valuable for a secondary use [or purpose] of the reservation" the United States must acquire that water under state law "in the same manner as any other public or private appropriator."[41]

In *New Mexico*, the Supreme Court was presented with two statutes involving national forests and had to determine the extent to which national forests have reserved water rights. The first statute, the Organic Administration Act of 1897 (hereinafter Organic Act), initially defined the purposes of national forests as the conservation of water flow and the furnishing of a continuous supply of timber. Later, Congress enacted the Multiple-Use Sustained-Yield Act of 1960 (hereinafter MUSYA), which broadened the purposes of national forests to include outdoor recreation, range, timber, watershed, and wildlife and fish purposes. In *New Mexico*, the United States claimed water rights to fulfill the purposes of both the Organic Act and the MUSYA.

However, the Supreme Court did not agree that the United States could claim reserved water rights under both statutes, finding that the United States was entitled to water rights based solely on the narrow purposes set out in the Organic Act. The Court found that the legislative history of the Organic Act confirmed that the purpose of national forests was to protect forest growth and favorable forest conditions for economic purposes. In contrast, the Court found that the purposes of the MUSYA would require "significant amounts of water for purposes quite inconsistent with" the goals of the Organic Act.[42] The legislative history of the MUSYA further indicated that it was intended to be "supplemental to, but not in derogation of, the purposes for which the national forests were established" in the Organic Act.[43]

Based on the two acts, the Court held that Congress only reserved as a matter of federal law the rights to water for national forests based on the primary purposes set forth in the Organic Act. Any water rights to fulfill the additional supplemental purposes of national forests established in the much broader MUSYA must be fulfilled by acquiring state-based water rights. As a result, the United States did not have a reserved right to water for instream flows under the MUSYA and would have to establish any such instream flow rights based on state law. This primary/secondary purposes test can thus be a limitation on reserved water rights—at least with respect to non-Indian federal reservations.

Winters and *Arizona v. California* establish a solid foundation for Indian tribes to seek reserved water rights for present and future purposes, whereas *Cappaert* and *New Mexico* appear to create some limiting principles to the scope of reserved water rights, albeit in non-Indian contexts. However, unlike Devils Hole in the Death Valley National Monument and the Gila National Forest, Indian reservations were established to provide permanent homes and to sustain human beings. As a result, the applicability of *Cappaert* and *New Mexico* to tribal reserved water rights is heavily disputed.

COURTS HAVE DEVELOPED VARIED APPROACHES TO DETERMINING THE SCOPE OF INDIAN RESERVED WATER RIGHTS

Winters, *Arizona v. California*, and other early reserved water rights cases were initiated and prosecuted solely by the United States with minimal participation or input by the interested tribes. Consequently, it was the United States—along with its values and prerogatives—that defined the purposes for which these reservations were set aside. For the United States, the primary objective in setting aside Indian reservations was often to encourage agriculture-based economies wherein tribal members would leave their traditional ways and adopt Euro-American practices and values. As a result, much of the early reserved water rights litigation centered on the quantification of water rights necessary to support irrigated agriculture. But of course, the purposes of a reservation cannot be defined by the United States alone. This requires analysis of the mutual intent of both the United States and the tribes whose homelands are reserved. And since *Winters* and *Arizona v. California*, tribes have become increasingly more involved in litigation involving their reserved water rights, often arguing for a broader and more flexible approach to quantifying their reserved water rights. Courts grappling with questions regarding the scope of reserved water rights for Indian tribes have taken varying approaches over the years, often ending in conflicting results for Indian tribes from state to state.

Wyoming was the first state to conclude a McCarran Amendment determination that fully adjudicated the rights of Indian tribes after *Winters* and *Arizona v. California*. In the *Big Horn I* decision,[44] the Wyoming Supreme Court had to determine the reserved water rights of the Eastern Shoshone and Northern Arapaho Tribes on the Wind River Reservation. The court first considered whether Congress intended to reserve water when the Wind River Reservation was created. The court found that the special master and the district court did not "blindly presume a reserved water right existed."[45] Rather, they both engaged in a "careful examination" of the "asserted water right and the specific purposes for which the land was reserved, and concluded that without the water the purposes of the [R]eservation would be entirely defeated."[46]

Next, the court sought to determine the purposes of the Reservation because "the amount of water impliedly reserved is determined by the purposes for which the [R]eservation was created."[47] Both the special master and the trial court found the tribes' treaty unambiguous and thus determined the Reservation's purpose "from the four corners of the treaty."[48] The special master found that the purpose of the Reservation was to provide "a homeland" and included a broad reserved water right for irrigation, livestock, fisheries, wildlife, aesthetics, and mineral, industrial, domestic, commercial, and municipal uses. The trial court disagreed and concluded that the purpose of the Reservation was solely agricultural.

On appeal, the Wyoming Supreme Court had "no difficulty" in "finding that it was the intent at the time to create a reservation with a sole agricultural purpose."[49] The court noted that the Ninth Circuit had "applied the specific purpose test" from *New Mexico* in at least one other Indian water rights case, although it "did not mandate that a single purpose for the reservation be found."[50] The court also recognized that the primary/secondary distinction "has been drawn into question because the standards governing non-Indian federal reserved water rights differ from those governing Indian reserved water rights."[51] The court did not decide whether the primary/secondary distinction applied, but proceeded to consider the treaty creating the Wind River Reservation.

In determining the agricultural purpose for the Wind River Reservation, the court placed significant weight on certain portions of the tribes' treaty while ignoring other important parts that contemplated broader activities, such as education, hunting, and timber harvest. In particular, although one article of the treaty referenced the Reservation as a "permanent homeland," the court discounted the significance of this and focused on numerous other provisions that encouraged agriculture. Those other provisions included a reference to "agricultural reservations," authorization for allotments for farming purposes, the provision of seeds and implements for farming, higher annual stipends for Indians engaged in farming, and a $50 prize for the "ten best Indian farmers."[52] The court found the import of all these provisions was that the "treaty does not encourage any other occupation or pursuit."[53]

In addition, the court upheld reserved rights for municipal, domestic, commercial, and livestock uses on the basis that they were traditionally subsumed "within the agricultural reserved water award."[54] The court also cited the Special Master's Report from *Arizona v. California*, which "indicated that PIA [practicably irrigable acreage] was the measure of water necessary for agriculture and related purposes."[55]

In stark contrast to the *Big Horn I* decision, the state of Arizona has taken a vastly different approach to Indian reserved water rights. In *In re Gen. Adjudication of All Rights to Use Water in the Gila River Sys. & Source*,[56] known as the *Gila V* decision, the Arizona Supreme Court took a broad approach in determining the purposes of an Indian reservation, setting forth the "permanent homeland" standard of reserved water rights for Indian reservations. *Gila V* concerned a state general stream adjudication under the McCarran Amendment involving multiple tribes with reservations in the Gila River system. The Arizona Supreme Court initially looked to both *Winters* and *Arizona v. California* "to determine the manner in which water rights on Indian lands are to be quantified."[57]

The court noted that in *Winters*, the "purpose for creating the Fort Belknap [R]eservation was to establish a permanent homeland for the Gros Ventre and Assiniboine Indians."[58] The Arizona Supreme Court found that since *Winters*, the US Supreme Court "has strengthened the reserved rights doctrine" beginning with *Arizona v. California*.[59] Then came *Cappaert*, which rejected an equitable balancing of interests but also found that the allocation for a federal reservation must be tailored to the "minimal need" of the reservation. The court noted that *Cappaert*'s "minimal need" limitation "makes good sense because federally reserved water rights are implied" and "uncircumscribed by the beneficial use doctrine."[60] Finally, the Arizona Supreme Court discussed *New Mexico* and its primary/secondary distinction. The court stated that the distinction's "application to federal Indian reservations [wa]s one of the issues before us today."[61]

In examining the applicability of *New Mexico*, the Arizona Supreme Court found that "it is necessary to distinguish between Indian and non-Indian reservations" when applying the reserved water rights doctrine.[62] The court found that reserved water rights for non-Indian federal reservations "are narrowly quantified to meet the original, primary purpose of the reservation."⁶³ The primary purpose of a non-Indian federal reservation "is strictly construed after careful examination."⁶⁴ Indian reservations, however, are different because Indians "have reserved water rights for 'future needs and changes in use.'"⁶⁵ Moreover, unlike other federal reservations, the US government has a trust responsibility to Indians. Given those significant differences, the court held that the primary/secondary distinction did not apply to Indian reservations.

As to the purposes of the various Indian reservations at issue, the tribes and the United States argued that federal case law had "preemptively determined" that every Indian reservation was established as a permanent homeland.⁶⁶ The non-Indians argued that each case must analyze the specific tribe's treaty or relevant documents "to determine that reservation's individual purpose."⁶⁷ Rather than decide whether federal case law "preemptively determined" the issue, the Arizona Supreme Court "agree[d] with the [US] Supreme Court that the essential purpose of Indian reservations is to provide Native American people with a 'permanent home and abiding place' that is a 'livable' environment."⁶⁸

The Arizona Supreme Court found that the case-by-case consideration of historical documents to determine each reservation's purpose "is highly questionable" for multiple reasons.⁶⁹ First, the court noted that many reservations were pieced together over time, which could lead to arguments that each alteration had a different purpose, which is "inconsistent with the concept of a permanent, unified homeland."⁷⁰ Second, the court emphasized the fact that the intent to reserve water rights is implied, and the "historical search for a reservation's purpose tends to focus only on the motives of Congress," which leaves tribal intent "out of the equation."⁷¹ Finally, the court recognized that other water rights-holders are not constrained to "use water in the same manner as their ancestors in the 1800s."⁷² The permanent homeland concept allowed this "flexibility and practicality."⁷³

As for quantification, the court found that *Cappaert's* "minimal need" limitation still applied to Indian reservations, as did the "need for individualized, fact-based quantifications."⁷⁴ However, the court rejected the PIA standard "as the exclusive quantification measure for determining water rights on Indian lands" because it forces some tribes to "pretend to be farmers" in a modern era when most agricultural projects are marginal and risky enterprises.⁷⁵ Instead, the court found that a "balancing of a myriad of factors" was the proper approach to quantifying Indian water rights.⁷⁶ The court offered a list of nonexclusive factors that a court should consider for quantification, such as (1) the tribe's history and traditions relating to water use; (2) preservation of the tribe's culture; (3) the tribe's geography, topography, and natural resources (including groundwater); (4) the tribe's economic base; (5) the past water use on the reservation; and (6) the present and projected population of the reservation.

No other court to date has directly applied *Gila* V. At least one federal district court has found that the "homeland purpose theory adopted in Gila River V is contrary to the primary purpose doctrine under federal law."⁷⁷ In 2017, the Ninth Circuit "considered" *New Mexico's* application to Indian reserved water rights in *Agua Caliente Band of Cahuilla Indians v. Coachella Valley Water District*.[78] That case was a federal action brought by the Agua Caliente Band of Cahuilla Indians to establish and quantify its reserved water rights to groundwater underlying its Reservation in Palm Springs, California. The United States intervened in the case as trustee on behalf of the Band. The case was separated into phases, and the first phase was to determine whether the Band has reserved rights to groundwater. The district court determined that the Band did have reserved rights to groundwater, after which several water districts appealed the ruling. Before reaching the question of whether the Band's reserved water rights extended to groundwater, the Ninth Circuit first considered whether the United States reserved water for the Band when establishing its Reservation, which turned on whether the water was "necessary to accomplish the purpose of the [R]eservation."[79] The Ninth Circuit noted that *New Mexico* "established a 'primary-secondary use' distinction," and although it previously had found that *New Mexico* is "not directly applicable to Winters doctrine rights on Indian reservations," it "clearly establishes several useful guidelines."[80]

The water agencies argued that *New Mexico* stands for the proposition that water is reserved "only if other sources of water then available cannot meet the [R]eservation's water demands."[81] The water agencies argued that the Band did not need a federal reserved water right to prevent the purposes of the Reservation from being defeated because the Band could obtain adequate water for those purposes under state law. In rejecting those arguments, the Ninth Circuit found that *New Mexico* is "not so narrow."[82] Instead, the Ninth Circuit noted that the relevant question is whether "the purpose underlying the [R]eservation envisions water use."[83] The Ninth Circuit also found that *New Mexico* "remains faithful" to and "did not alter the test envisioned by *Winters*."[84] The Ninth Circuit explained that *New Mexico* simply "added an important inquiry related to the question of *how much* water is reserved," but "[i]t did not, however, eliminate the threshold issue—that a reserved right exists if the purposes underlying a reservation envision access to water."[85]

Then, in determining "the primary purpose of the Tribe's [R]eservation and whether that purpose contemplates water use," the Ninth Circuit considered "the document and circumstances surrounding [the Reservation's] creation, and the history of the Indians for whom it was created."[86] The executive orders establishing the Band's Reservation expressly set the land aside for "the permanent use and occupancy of the Mission Indians" and for "Indian purposes."[87] These executive orders were prompted by government reports that recognized a need to secure the Band "permanent homes, with land and water enough."[88] Although the Ninth Circuit found that these circumstances were "imprecise," circuit precedent supported the idea that the "general purpose, to provide a home for the Indians, is a broad one and must be liberally construed."[89] And the creation of the Agua Caliente Reservation was no different than the reservations considered in other cases: "Water is inherently tied to the Tribe's ability to live permanently on the [R]eservation. Without water, the underlying purpose—to establish a home and support an agrarian society—would be entirely defeated.... Thus, we hold that the United States implicitly reserved a right to water when it created the Agua Caliente Reservation."[90]

Although the Ninth Circuit found a homeland purpose for the Agua Caliente Band, it is not yet clear how that homeland purpose will be applied in determining the quantity of water to which the Band is entitled or how *New Mexico* will factor into that determination.

In 2019, the Idaho Supreme Court waded into the scope of the reserved rights doctrine when it was called upon to determine the Coeur d'Alene Tribe's entitlement to reserved water rights.[91] The state of Idaho initiated that case as a general stream adjudication and joined the United States as a party pursuant to the McCarran Amendment. The Tribe later intervened in the case, and together the Tribe and the United States argued that the purpose of the creation of the Coeur d'Alene Reservation was to set aside a permanent homeland for the Coeur d'Alene people. That homeland contemplated water for a series of consumptive uses, including agricultural, domestic, commercial, municipal, and industrial, as well as a series of nonconsumptive uses to support the Tribe's traditional subsistence lifeway, including water rights to maintain Coeur d'Alene Lake, and instream flows to support the tribal fishery and seeps, springs, and wetlands within the reservation necessary for tribal hunting, gathering, and cultural practices.

The state of Idaho and several other objectors contended that (1) the *New Mexico* primary/secondary purposes test should apply to the Coeur d'Alene Tribe's water rights; and (2) the primary purpose for the Coeur d'Alene Reservation was solely to support the Tribe's agricultural and domestic needs. The district court concluded that *New Mexico* did apply but that the primary purpose of the Reservation was to support not only the Tribe's domestic, municipal, and agricultural needs, but also its traditional hunting and fishing practices. The parties appealed, setting the applicability of *New Mexico* as a primary issue for the Idaho Supreme Court.

The Idaho Supreme Court found that *New Mexico*'s primary/secondary purposes test does not apply to the water rights appurtenant to Indian reservations. In reaching that conclusion, the court highlighted several important distinctions between non-Indian and Indian reserved water rights. First, the court pointed out that "the two rights have different origins."[92] While non-Indian reserved water rights are "created by the document that reserves the land from the public domain . . . [A]boriginal-Indian reserved water rights exist from time immemorial and are merely recognized by the document that reserves the Indian land."[93] Likewise, the court found the federal government's role as trustee over reserved water rights important, underscoring that "in the context of Indian reservations, the government, as trustee of such lands, must act for the Indians' benefit."[94] Similarly, the court emphasized the importance of the Indian canons of construction, citing its own precedent recognizing that treaties between the United States and tribes are a bargained-for exchange and have special rules regarding their interpretation. Based on these factors, the court concluded that the "purposes behind the creation of an Indian reservation should be more broadly construed and not limited solely to what may be considered a 'primary' purpose."[95]

The court turned next to the applicability of the homeland purpose theory. In addition to those already mentioned, the court's main criticism of the primary/secondary purposes distinction was that it "runs counter to the concept that the purpose of many Indian reservations was to establish a 'home and abiding place' for the tribes."[96] Indeed, after analyzing much of the same case law that it used to dismiss the applicability of *New Mexico*, the court found that "[t]he homeland purpose theory gains support from precedent of the U.S. Supreme Court, other jurisdictions, and this Court."[97] Consistent with its overall reliance on the canons and federal policy supporting tribal self-determination, the Idaho Supreme Court then agreed with its sister courts that "the essential purpose of Indian reservations is to provide Native American people with a 'permanent home and abiding place,' that is, a 'livable environment.' "[98]

Nonetheless, the court was careful to contrast its holding with the conclusions of the Arizona Supreme Court in *Gila V*, which found a homeland purpose as a matter of law for all reservations and eschewed reliance on historical documents and circumstances. For its part, the Idaho Supreme Court found that although the "[l]anguage in *Winters* suggests a homeland purpose theory may arise in certain reservations," the Arizona court's approach "deviates from case law and the established interpretation framework regarding examination of formative documents to determine the purposes for an Indian reservation."[99] Accordingly, the Idaho Supreme Court directed that courts "are constrained to use the formative documents and circumstances to determine the reservation's purposes."[100] Notwithstanding this subtle but important admonition, the Idaho Supreme Court expressly found that a homeland purpose "should be recognized when established by the formative documents."[101] Examination of the documents and history relevant to the creation of the Coeur d'Alene Reservation ultimately led the court to find that "the Reservation's creation demonstrate[s] a homeland purpose consisting of the following uses: domestic, agriculture, hunting and fishing, plant gathering, and cultural."[102]

Taken together, the myriad approaches taken by courts throughout the West demonstrate no clear convergence on a singular method to determining the scope of the reserved rights doctrine. This underscores another problem with the McCarran Amendment's deference to state court general stream adjudications: the differing approaches taken by the state courts have led to confusion about the doctrine, which has caused disparities between tribes. Those disparities are based not on the text and circumstances of an individual tribe's treaty, but rather on policy choices made by state court judges about the relative value of reserved rights versus the water rights of often politically powerful non-Indians.

Nonetheless, the Idaho Supreme Court's decision will hopefully guide the way for future courts as they continue to refine the scope of the reserved rights doctrine. The Idaho court's analysis cut through much of the confusion caused by other decisions and instead came to an analytically sound conclusion rooted deeply in the foundations of federal Indian law and treaty interpretation as laid down by the US Supreme Court. More important, the Idaho Supreme Court's approach (for the most part) does honor to the agreements the United States forged with the tribes, all of which sacrificed so much in exchange for the singular goal of preserving some small fragment of their Aboriginal domain as their permanent homeland.

INDIAN RESERVED WATER RIGHTS HAVE BEEN FOUND TO EXTEND TO WATER FOR HUNTING, FISHING, AND GATHERING

Winters, *Arizona v. California*, and the other early reserved water rights cases were brought by the United States on behalf of the tribes. As a result, the federal perspective that agriculture was the main purpose for Indian reservations took center stage during these cases. Not surprisingly, tribes have a markedly broader view of the purposes underlying their homelands. For them, their reservation was set aside as a permanent homeland where tribal people can continue to practice their traditional ways of life while also adopting new ways to develop the tribal economy. That necessarily requires water rights to consume water for irrigation, domestic, municipal, commercial, and industrial uses, but it can also require water to remain in situ to ensure the continued vitality of the plants and animals with which tribes have coexisted since time immemorial.

Reserved water rights to maintain traditional subsistence activities have "no corollary in the common law of prior appropriation," which almost always requires that "some diversion of [the] natural flow of a stream [take place] to effect a valid appropriation."[103] Instead, these water rights are considered to be nonconsumptive: "the holder of the right is not entitled to withdraw water from the stream. . . . Rather, the entitlement consists of the right to prevent other appropriators from depleting the streams below a protected level in any area where the non-consumptive right applies."[104] Despite the importance of these other needs, for a host of socioeconomic and political reasons it was not until the 1970s that tribes were finally positioned to broadly assert their rights to water for uses beyond irrigation and domestic use. That effort was initiated by the Confederated Tribes of the Colville Reservation in *Colville Confederated Tribes v. Walton*.[105] Although they relied on all fish native to their region, the people of the Colville Reservation were traditionally salmon fishers.[106] Indeed, salmon are central to the way of life for the Confederated Tribes and have long been of "economic and religious importance to them."[107] Not surprisingly, when the time came to set aside a reservation for the Confederated Tribes, they demanded a homeland that bordered the Columbia River, which at the time was home to the largest salmon fishery in the West. That way of life, however, was devastated in the middle of the 20th century when the United States erected Grand Coulee and other dams within the Columbia River basin, which extirpated the salmon fishery throughout the Colville Reservation.

Although the Confederated Tribes have never given up their traditional salmon fishing, they resolved to supplement their dwindling fishery through the development of a replacement fishery composed of Lahontan cutthroat trout (*Oncorhynchus clarkii henshawi*). Those fish were brought in to live in Omak Lake, but in order to spawn they would need to migrate up No Name Creek, which feeds into Omak Lake.[108] Although No Name Creek provided ideal spawning habitat under natural conditions, it became apparent that non-Indian land and water use was decreasing water flows and quality (raising turbidity and temperature) in No Name Creek below the threshold necessary for the trout to successfully spawn and return to Omak Lake. The tribes initiated a stream adjudication to determine, among other legal rights, their right to maintain the flows in No Name Creek in order to protect their trout fishery.

Defendant William Boyd Walton argued that the sole purpose for the creation of the Colville Reservation was to develop and maintain an agricultural economy for the tribes and their members. Although the Ninth Circuit agreed that "one purpose for creating this [R]eservation was to provide a homeland for the Indians to maintain their agrarian society . . . it was not the only purpose for creating the [R]eservation."[109] The court's conclusion was based on its analysis of the tribes' intent in setting aside the Colville Reservation. It found that "[t]he Colvilles traditionally fished for both salmon and trout. Like other Pacific Northwest Indians, fishing was of economic and religious importance to them."[110] Accordingly, the court concluded that "preservation of the tribe's access to fishing grounds was one purpose for the creation of the Colville Reservation."[111]

Matters were complicated, however, by the fact that the Colvilles were claiming a water right for a fishery that did not exist at the time the Reservation was set aside. Instead, the argument went, if the Tribes had intended to reserve instream water rights it would have been to support their historic fishery on the Columbia River that existed at the time the Reservation was created. But because that fishery had been destroyed, the Tribes' water right had likewise been extinguished. The court rejected this argument, concluding that since reserved water rights may be used in "any lawful manner . . . subsequent acts making the historically intended use of the water unnecessary do not divest the Tribe[s] of the right to the water."[112] The court concluded that the Tribes' water rights that were set aside to maintain its historical fishery continued to exist and were available to be repurposed to support the Tribes' new replacement fishery.

For such a groundbreaking decision, the Ninth Circuit spent surprisingly little time in *Walton* analyzing the Winters doctrine to determine whether its scope could encompass nonconsumptive reserved instream flow water rights. That analysis came just a few years later in the Ninth Circuit's decision in *US v. Adair*.[113] That case involved the water rights of the Klamath Tribe, which set aside a Reservation in 1864 in what is today south-central Oregon.[114] In exchange for the cession of its territory outside that Reservation, the Tribe reserved, among other things, the "exclusive right of taking fish in the streams and lakes included in said [R]eservation, and of gathering edible roots, seeds, and berries within its limits."[115] The Klamaths argued that they necessarily were entitled to a right to water in sufficient quantities to maintain these subsistence rights. Like the non-Indian parties in *Walton*, however, the state of Oregon and related objectors in *Adair* argued that a reservation can only have one "primary purpose" and that "the intent of the 1864 Treaty was to convert the Indians to an agricultural way of life."[116] Therefore, these parties reasoned, the only water rights the Tribe was entitled to were those necessary to support the Reservation's agriculture. The court disagreed, concluding that US Supreme Court precedent did not "require . . . us to choose between . . . activities or to identify a single essential purpose which the parties to the 1864 Treaty intended the Klamath Reservation to serve."[117]

In overruling the state and non-Indian objections, the Ninth Circuit reaffirmed that the Winters doctrine is not limited to agriculture but instead extends to whatever water rights are necessary to fulfill the purposes for creating the reservation in question. The important question became then "whether securing to the Indians the right to hunt, fish, and gather was a . . . purpose of the Klamath Reservation."[118] That, according to the court, turned on "an analysis of the intent of the parties to the 1864 Klamath Treaty."[119]

The 1864 treaty expressly mentioned only the right to fish and gather; it failed to make mention of the Tribe's rights to hunt or trap, which gave rise to the argument that hunting and trapping were not proper purposes in the creation of the Klamath Reservation. Importantly, however, the court did not limit itself to the text of the 1864 treaty to discern the intent of the parties. Instead, it concluded that its analysis of the intent of the parties must be driven by both the text of the 1864 treaty and the circumstances surrounding its creation. Luckily, those circumstances had already been thoroughly examined by the federal courts in two cases analyzing the scope of the Klamaths' hunting rights.[120] Those courts had received voluminous evidence leading to the inescapable conclusion that the Klamaths' subsistence way of life depended not only on fishing and gathering, but also on hunting and trapping.[121] The court likewise noted that the United States "was probably aware that [these subsistence rights] held the greatest promise for sustaining the Klamath on their [R]eservation."[122] Collectively, the historical importance to the Tribe of all of these activities, as well as the United States' awareness of that importance, led the court to conclude that "one of the 'very purposes' of establishing the Klamath Reservation was to secure to the Tribe a continuation of its traditional [subsistence] lifestyle."[123]

Adair also clarified the priority date for water rights appurtenant to traditional subsistence rights. The state and non-Indian objectors argued that these rights should be treated like consumptive reserved water rights to support irrigated agriculture and awarded a priority date of no earlier than the date of the establishment of the Reservation. The Ninth Circuit rejected this argument, siding instead with the argument of the Tribes and the United States that "a pre-[R]eservation priority date is appropriate for tribal water uses that predate establishment of the [R]eservation."[124] In so doing the court married reserved subsistence water rights to the "fundamental principles of prior appropriations law—that priority for a particular water right dates from the time of first use."[125] In this case, the "Tribe had lived in Central Oregon and Northern California for more than a thousand years. . . . Within its domain, the Tribe used the waters that flowed over its land for domestic purposes and to support its hunting, fishing, and gathering lifestyle."[126]

The court recognized that the Tribes' eons-long "use and occupation of land and water" had long ripened into an Aboriginal right "to all of its vast holdings."[127] And after a thorough rendition of the Indian canons of construction as well as the reserved rights doctrine, the court concluded that it was not possible that either the Tribes or the United States would have understood the 1864 treaty or the creation of the Klamath Reservation to include a relinquishment of the Tribes' right to the use of the water they had always used on the land they reserved as their permanent home. As a result, the court concluded, the water rights appurtenant to the Tribes' traditional subsistence rights "necessarily carry a priority date of time immemorial. The rights were not created by the 1864 Treaty, rather, the treaty confirmed the continued existence of these rights."[128]

INDIAN RESERVED WATER RIGHTS MAY EXTEND OFF-RESERVATION TO PROTECT TRADITIONAL WAYS OF LIFE

Both *Walton* and *Adair* addressed the question of water rights appurtenant to on-reservation subsistence rights and established a strong foundation for Indian tribes to assert reserved water rights within their reservations to support their traditional ways of life (hunting, fishing, gathering, and cultural and spiritual purposes).[129] However, many tribes, particularly in the northwestern and upper midwestern portions of the United States, hold rights to hunt, fish, and gather off their reservation as well. This gives rise to an important but largely unsettled question of whether tribes may hold reserved water rights to protect these off-reservation rights (for a fuller discussion of off-reservation hunting and fishing rights, see chapter 8 of this volume).

The federal judiciary has only peripherally addressed this question in a case called *Kittitas Reclamation District v. Sunnyside Valley Irrigation District*.[130] That suit involved an effort by the Yakama Nation to compel the release of sufficient water from Cle Elum Dam to preserve salmon redds (egg-laying sites) that had been deposited in the Cle Elum River downstream from the dam and were being threatened by low flows caused by irrigation diversions. Although the sites were located off the Yakama Reservation, the Nation's fishing rights persist in the Cle Elum River pursuant to its 1855 treaty, which reserved "the right of taking fish at all usual and accustomed places, in common with the citizens of the Territory."[131]

Nonetheless, the water of the Cle Elum River had previously been adjudicated in a case that did not involve the water rights of the Yakama Nation.[132] Ever since, the adjudicated water rights of the Cle Elum River had been administered according to a consent decree (an agreement or settlement reached without admission of guilt by any party) by a water master. Because time was short, the Yakama Nation did not move to adjudicate its water rights in the Cle Elum River, which would have caused the federal court to address directly the question of off-reservation reserved water rights. Instead, the Nation requested that the water master (responsible for measuring, monitoring, reporting, and enforcing surface water diversions) maintain the flow in the river, which would have required the master to curtail a number of non-Indian adjudicated water rights. The master, in turn, requested instructions from the district court responsible for administering the consent decree, which ordered the release necessary to protect the redds. Without adjudicating the rights of the Yakama Nation, the Ninth Circuit affirmed. In so doing, the court at least implicitly recognized that tribes have the right to sufficient water, even off-reservation, necessary to maintain their subsistence rights.

The federal court's implicit holding was made explicit by the Washington Supreme Court when it found that the Yakama Nation held off-reservation water rights sufficient to maintain the fishery in its "usual and accustomed" fishing locations.[133] In response to *Kittitas* and the Yakama attempt to commence a federal court adjudication, the Washington State Department of Ecology commenced a general stream adjudication in Washington state for the Yakima River basin in a case that has since become known as *Acquavella*. The United States was joined to that adjudication pursuant to the McCarran Amendment and brought claims on behalf of the Nation for, among other things, on- and off-reservation instream flow water rights to protect aquatic habitat.[134]

The Washington State Court, affirmed by the Washington Supreme Court, expressly found that Congress "recognized . . . there was . . . a specifically definitive fishing right expressly reserved by the Indians. Therefore, the fishery rights in the 1855 Treaty are a . . . purpose of the Treaty."[135] However, the court went on to determine that those water rights had been "diminished" through the settlement of a series of claims the Yakama Nation had made to the Indian Claims Commission for the destruction of its fishing rights throughout its Aboriginal territory. The court did not conclude that the Nation's fishing rights had been extinguished because "[n]o factual evidence on the petition was ever presented to the [Indian Claims Commission]. From the facts as we know them, from the documented history noted herein, the Tribe would not and could not have been able to prove the complete destruction of the fishery prior to 1946. Thus, it appears that the real gravamen of the cause of action was for the diminution of the treaty fishing rights."[136]

Although the court's decision that the Nation's water rights had been diminished was based on dubious and problematic reasoning, it is noteworthy for its innovative approach to the quantification of the Nation's instream flow water rights. The court began by acknowledging that the Nation continued to be entitled to sufficient water to maintain fish life in the river.[137] However, the court found "it would be inappropriate for the Court to set specific, discrete quantifications . . . for all times and conditions" because of the "ever changing circumstances" that present themselves within the fishery each year.[138] Indeed, the court found that the success of the fishery depends on several "variables that may enter into the determination, on an annual basis."[139] Accordingly, the court tasked the Yakima River System Operations Advisory Committee with determining the flows necessary to "maintain fish life in the river" each year.[140] The committee also was charged with balancing the other water use needs in the basin and considering factors such as "water quality, climatic and temperature changes, changes in substrate locations within the stream, etc."[141]

In the 2019 Idaho Supreme Court case discussed above, the court was asked to determine the context of the Coeur d'Alene Tribe's entitlement to reserved rights.[142] The court determined that reservation of the Tribe's continued right to hunt, fish, and gather was a component part of the overall homeland purpose for the creation of the Coeur d'Alene Reservation. Accordingly, the Idaho Supreme Court affirmed the finding of the district court that the Tribe was entitled to instream flow water rights for fish-bearing streams within the Reservation.

However, the Tribe's instream flow claims to support their fishery transcended the Reservation's boundaries. The primary native fish in this region are bull trout (*Salvelinus confluentus*) and west-slope cutthroat trout (*Oncorhynchus clarkii*). These fish are adfluvial, meaning they live in Coeur d'Alene Lake but must return to the lake's tributary streams to spawn. Most of these spawning grounds are located outside the Reservation, so the Tribe and the United States argued that without sufficient water in these off-reservation streams, the Tribe's on-reservation fishery would be destroyed. The parties further reasoned that the Tribe was entitled to a reserved right in the off-reservation streams to support its on-reservation fishing rights.

The Idaho Supreme Court began its analysis of off-reservation instream flows by expressly recognizing their existence.[143] The court then recognized that "appurtenance is not dependent upon a 'physical relationship' with the land."[144] Instead, it adopted the Ninth Circuit's definition of "appurtenant," which is not driven by "the location of the water" but instead "has to do with the relationship between reserved federal land and the use of the water."[145] The court also recognized that "there has been some recognition by the United States Supreme Court in granting reserved water rights in water that were not directly located on tribal-owned land."[146] Taken together, these factors led the Idaho Supreme Court to conclude that reserved water rights may be sourced from both on- and off-reservation land, so long as the water is necessary to fulfill the purposes of the reservation.

Nonetheless, a bare majority of the justices found that although the Tribe may have been entitled to reserved water rights off-reservation, those water rights had been ceded by the Tribe in an agreement in 1889, which took place subsequent to the creation of the Coeur d'Alene Reservation in 1873.[147] The court had begun its analysis with a thorough explanation of the Indian canons of construction. The court had remained faithful to those canons until its analysis of the off-reservation instream flow claims, where it concluded, "This is not a situation where there is ambiguity."[148] The Tribe having explicitly relinquished its "right, title, and claim" to lands outside the Reservation constituted a voluntary relinquishment of any claim to off-reservation water rights, even those that would now arguably benefit an on-reservation purpose.[149]

The court backfilled its textual conclusion with a truncated version of the history and circumstances surrounding the 1889 cession agreement. Despite reaffirming throughout its decision that treaties and agreements between the Tribe and the United States should be interpreted "'in the sense the Indians themselves would have interpreted' or understood them," the Idaho Supreme Court cited just three statements in the historical record—all of which were made by federal officials—to support its conclusion that the Tribe unambiguously ceded its off-reservation water rights.[150]

Chief Justice Roger Burdick and Justice Joel Horton dissented from the court's decision regarding off-reservation instream flows, taking the majority to task for "inexplicably abandon[ing] the canons of construction" it had relied on throughout the case.[151] To these justices, neither the text nor the circumstances surrounding the 1889 agreement demonstrated that the Tribe "clearly, voluntarily, and unambiguously ceded their right to instreams flows that support the fish in waters outside of the Reservation."[152]

On the textual side, the dissent criticized the majority for focusing on the "right, title, and claim" language in the 1889 agreement while glossing over "the language that qualifies it—'to all lands.' The Tribe ceded 'right, title, and claim . . . to all lands outside the reservation.'"[153] Unlike the majority, which interpreted "land" to also mean "water," to Justice Burdick it "seem[ed] plain that when the parties said 'land,' they meant land, and when the parties said 'waters,' they meant water."[154]

The dissent's textual analysis was buttressed by circumstances surrounding the 1889 agreement, stressing that "it's worth remembering what exactly the Majority claims that the Tribe gave up. To the Tribe, water was of paramount importance . . . [and t]he fishery was the lifeblood of the Tribe."[155] Reservation of the waterways was of the essence to the Tribe as it negotiated the original agreement setting aside the Coeur d'Alene Reservation. As a result, these justices concluded, the Tribe's instream flow water rights should "protect the fish habitat extended as far as the fish do."[156]

Taken together, *Kittitas*, *Acquavella*, and *In re* CSRBA demonstrate that tribes are entitled to instream flow water rights, regardless of location, if they are necessary to fulfill the purposes for which a reservation was created. Unfortunately, they likewise demonstrate the many obstacles—whether they be the Indian Claims Commission, subsequent cession agreements, or other factual or legal circumstances—that tribes may confront when making these claims. As a result, any tribe looking to claim reserved water rights off-reservation must make a thorough examination of their history, from before the creation of the reservation through the present, to hedge against a negative outcome in their case. However, with caution, care, and significant effort, Indian tribes can use reserved water rights to protect the plants, animals, and environment that make up their homelands.

CONCLUSION

As water demands in the West increase, Indian tribes face increasing pressures to resolve their water rights claims in order to protect their water resources and ensure sustainable homelands for their members. This includes the ability of tribes to provide clean drinking water and engage in economic development, but also to use their water to preserve their traditional ways of life. With the exception of the *Arizona v. California* and *Big Horn I* cases, none of the Indian water rights cases we have discussed were complete adjudications, meaning that the tribes involved did not have finality with respect to their water rights. Unfortunately, litigating tribal water rights can take decades and strain tribal financial resources. And despite the long-standing existence of the Winters reserved rights doctrine and early recognition by the US Supreme Court that tribes have command and use of all of their waters, there is always uncertainty with respect to whether a tribe will be successful in obtaining all of their due water rights. In lieu of litigation, many tribes have explored settlements, where they can negotiate for uses and quantities of water that align with their cultural values. Settlement is not always faster than litigation, but it allows an opportunity for tribes, states, and non-Indian water users to come together and develop creative solutions.

NOTES

Quoted in US Geological Survey, National Water Summary on Wetland Resources (1996), https://www.govinfo.gov/content/pkg/CZIC-gb624-n37-1996/html/CZIC-gb624-n37-1996.htm.

1. See, e.g., Pub. L. 102-575, § 3501-11, 106 Stat. 4600 (30 October 1992) (providing additional equitable compensation to the Three Affiliated Tribes and the Standing Rock Sioux Tribe for the taking of their lands for the Garrison Dam and the Oahe Dam and Reservoir).

2. Winters v. US, 207 US 564 (1908).

3. NATIONAL WATER COMMISSION, WATER POLICIES FOR THE FUTURE: FINAL REPORT TO THE PRESIDENT AND TO THE CONGRESS OF THE UNITED STATES BY THE NATIONAL WATER COMMISSION 474–75 (1973).

4. 43 USC § 666 (1952); Colorado River Water Conservation Dist. v. US, 424 US 800 (1976). But see Dylan R. Hedden-Nicely, *The Legislative History of the McCarran Amendment: An Effort to Determine Whether Congress Intended for State Court Jurisdiction to Extend to Indian Reserved Water Rights*, 46 ENVTL. L. REV. 845 (2017).

5. See, e.g., Colorado River Water Conservation Dist., 424 US 800.

6. See *id.* at 809; Arizona v. San Carlos Apache Tribe of Arizona, 463 US 545 n. 10 (1983).

7. See, e.g., 78 Am. Jur. 2d Waters § 55 (February 2021).

8. See, e.g., ANTHONY DAN TARLOCK, LAW OF WATER RIGHTS AND RESOURCES § 3:10 (July 2020).

9. There are some states that rely on both the prior appropriation and riparian rights doctrines. California is one example.

10. "With the adoption of the Constitution, Indian relations became the exclusive province of federal law." County of Oneida v. Oneida Indian Nation, 470 US 226, 234 (1985).

11. 207 US at 565.

12. *Id.* at 567.

13. *Id.* at 576.

14. *Id.* at 577.

15. 373 US 546 (1963).

16. Those five were the Chemehuevi Indian Tribe, Cocopah Indian Tribe, Fort Yuma Quechan Indian Tribe, Colorado River Indian Tribes, and Fort Mojave Indian Tribe.

17. Arizona v. California, 373 US at 551.

18. Simon H. Rifkind, Special Master, Report at 260, Arizona v. California, 373 US 546 (1963).

19. 373 US at 596.

20. *Id.*

21. *Id.* at 598.

22. *Id.* at 599.

23. *Id.* at 600.

24. *Id.* at 596.

25. *Id.* at 600. One acre-foot of water—the amount of water needed to cover one acre of land with one foot of water—is approximately 325,900 gallons.

26. *Id.* at 598.

27. *Id.* at 601.

28. 373 US at 601. This type of quantification for irrigation has become known as the practicably irrigable acreage (PIA) standard.

29. Special Master's Report at 262 (the quantity of water reserved is the amount sufficient "to irrigate all of the practicably irrigable lands in a Reservation and to supply stock and related domestic uses").

30. Winters, 207 US at 576; Arizona v. California, 373 US at 597–98.

31. Cappaert v. US, 426 US 128 (1976); US v. New Mexico, 438 US 696 (1978).

32. Cappaert, 426 US at 133. Although the decision mentions underground water, the Supreme Court declined to decide in *Cappaert* whether the reserved rights doctrine extends to groundwater because it concluded that "water in the pool is surface water" (*id.* at 142).

33. *Id.* at 132.

34. *Id.* at 132–33.

35. *Id.* at 143.

36. *Id.* at 141.

37. *Id.*

38. 438 US at 698.

39. *Id.* at 700.

40. *Id.* at 701–2.

41. *Id.* at 702.

42. *Id.* at 713.

43. *Id.*

44. *In re* Gen. Adjudication of All Rights to Use Water in the Big Horn River Sys., 753 P.2d 76 (Wyo. 988) (Big Horn I).

45. *Id.* at 91.

46. *Id.* at 90.

47. *Id.* at 94.

48. *Id.*

49. *Id.* at 96.
50. *Id.*
51. *Id.*
52. *Id.* at 97.
53. *Id.*
54. *Id.* at 99.
55. *Id.*
56. 35 P.3d 68 (Ariz. 2001).
57. *Id.* at 72.
58. *Id.*
59. *Id.*
60. *Id.* at 73 n. 1.
61. *Id.*
62. *Id.*
63. *Id.*
64. *Id.* at 74.
65. *Id.* at 73 (citation omitted).
66. *Id.* at 74.
67. *Id.*
68. *Id.* (citations omitted).
69. *Id.*
70. *Id.*
71. *Id.* at 75.
72. *Id.* at 76.
73. *Id.*
74. *Id.* at 76 n. 5.
75. *Id.* at 78–79.
76. *Id.* at 79.
77. US v. Washington, 375 F. Supp. 2d 1050, 1063 (W.D. Wash. 2005), *vacated pursuant to settlement sub nom.* US *ex rel.* Lummi Indian Nation v. Washington, No. C01-0047Z, 2007 WL 4190400 (W.D. Wash., 20 November 2007).
78. 849 F.3d 1262 (9th Cir. 2017).
79. *Id.* at 1268.
80. *Id.* at 1269 and n. 6 (cleaned up).
81. *Id.* at 1269.
82. *Id.*
83. *Id.*
84. *Id.* at 1269–70.
85. *Id.* at 1270.
86. *Id.* (quoting Colville Confederated Tribes v. Walton, 647 F.2d 42, 47 [9th Cir. 1981]).
87. *Id.*
88. *Id.* at 1270 n. 7.
89. *Id.* at 1270.
90. *Id.*
91. *In re* CSRBA Case No. 49576, Subcase No. 91-7755, 448 P.3d 322 (Idaho 2019).
92. *Id.* at 343.
93. *Id.*
94. *Id.* at 344.
95. *Id.* at 346.
96. *Id.* at 344 (quoting 207 US at 565).
97. *Id.*
98. *Id.* at 345 (quoting 35 P.3d at 74).
99. *Id.* at 344, 347.

100. *Id.* at 348.
101. *Id.*
102. *Id.* at 348.
103. US v Adair, 723 F.2d 1394, 1410–11 (9th Cir. 1983).
104. *Id.* at 1411.
105. 460 F. Supp. 1320 (E.D. Wash. 1978). See also Dylan R. Hedden-Nicely, *The Historical Evolution of the Methodology for Quantifying Federal Reserved Instream Water Rights for Indian Tribes*, 50 ENVTL L. REV. 207 (2020).
106. Colville Confederated Tribes v. Walton, 647 F.2d 42, 45 (9th Cir. 1981).
107. *Id.* at 48.
108. Hedden-Nicely, *supra* note 105, at 215.
109. Walton, 647 F.2d at 47–48.
110. *Id.* at 48.
111. *Id.*
112. *Id.*
113. 723 F.2d 1394 (9th Cir. 1984).
114. *Id.* at 1397–98.
115. Treaty with the Klamaths, 16 Stat. 707 (1864).
116. Adair, 723 F.2d at 1409–10.
117. *Id.* at 1410.
118. *Id.* at 1409.
119. *Id.*
120. Klamath and Modoc Tribes v. Maison, 139 F. Supp. 634 (D. Oregon 1956); Kimball v. Callahan, 493 F.2d 564 (9th Cir. 1974).
121. Adair, 723 F.2d at 1409. See also Maison, 139 F. Supp. at 636–37.
122. Adair, 723 F.2d at 1409 n. 15.
123. *Id.* at 1409.
124. *Id.* at 1412.
125. *Id.* at 1414.
126. *Id.* at 1413.
127. *Id.*
128. *Id.* at 1414.
129. The Idaho Supreme Court in *In re* CSRBA, similarly and more recently determined that the Coeur d'Alene Tribe had reserved rights for traditional purposes. See, generally, 448 P.3d 322.
130. 763 F.2d 1032 (9th Cir. 1980).
131. Treaty with the Yakama, 12 Stat. 951 (1855).
132. See Kittitas, 763 F.2d at 1034.
133. In the Matter of the Determination of the Rights to the Use of the Surface Waters of the Yakima River Drainage Basin, 850 P.2d 1306 (Wash. 1993).
134. Washington Dept. of Ecology v. Acquavella, Memorandum Opinion Re: Motions for Partial Summary Judgment, Case No. 77-2-01484-5 (Wash. Super. Ct., 29 May 1990).
135. *Id.* at 45.
136. *Id.* at 51–52.
137. Washington Dept. of Ecology v. Acquavella, Amendment to Memorandum Opinion Re: Motions for Partial Summary Judgment, Case No. 77-2-01484-5, 60–61 (Wash. Super. Ct., 22 October 1990) (hereinafter Acquavella, Amended Memo Op.).
138. *Id.* at 59.
139. *Id.* at 58.
140. *Id.* at 59. See US Bureau of Reclamation, Interim Comprehensive Basin Operating Plan for the Yakima Project, Washington, § 2002 (2002).
141. Acquavella, Amended Memo Op. at 58.
142. 448 P.3d 322 (Idaho 2019).

143. *Id.* at 354.
144. *Id.* (quoting Joyce Livestock Co. v. US, 156 P.3d 502, 514 [Idaho 2007]).
145. *Id.* at 354.
146. *Id.* Specifically, the court pointed to the US Supreme Court's recognition in Arizona v. California of reserved water rights to the Cocopah Reservation, which at the time of the Court's decision was understood to be near but not physically touching the Colorado River. See 373 US at 344–45.
147. 448 P.3d at 355.
148. *Id.* at 356.
149. *Id.*
150. *Id* at 355.
151. *Id.* at 364–65.
152. *Id.* at 363.
153. *Id.* at 364.
154. *Id.*
155. *Id.* at 363.
156. *Id.* at 367.

Chapter 7

The Promise of Intertribal Wildlife Management

BETHANY R. BERGER

As the contributions to this book show, Indigenous governments play an ever-larger role in managing wildlife in ecologically and culturally sustainable ways. In this chapter I focus on an understudied aspect of this management: much of it is not by individual tribes, but by intertribal entities. Far more than ad hoc collaborations, these organizations are often decades old, with recognized legal attributes similar to those of the tribes that create them.

WHY INTERTRIBAL MANAGEMENT?

Intertribal management responds to a core challenge facing wildlife conservation and Indigenous peoples: going it alone does not work. Ecosystems ignore reservation and other legal boundaries. Effective management of the habitat in one's control has little impact if species depend on travel outside the habitat or are dependent on species or conditions that exist outside the habitat. In addition, federal law often recognizes rights to the same wildlife resources in multiple tribes, and traditional use patterns may create shared use of resources as well. The small size of many tribes also limits the political, administrative, and scientific resources of individual tribes working in isolation, and the often fierce opposition to tribal management by private, state, and federal parties creates particular dangers from conflicts and inconsistency between tribes.

Ecosystems Ignore Reservation Boundaries

Crossing geopolitical boundaries may be common or necessary to a species' life cycle, and activity outside tribal lands can have profound impacts on habitat within it. The United States' history of dispossession of Native peoples of their lands means that some key species exist only outside areas protected by tribal legal rights. For the Indigenous peoples of the Pacific Northwest, for example, salmon were "not much less necessary . . . than the atmosphere they breathed"

(*Washington v. Wash. State Commercial Passenger Fishing Vessel Assoc.*, 443 US 658, 680 [1979], quoting *United States v. Winans*, 198 US 371, 380–81 [1905]). Salmon are anadromous, migrating from the Pacific Ocean through freshwater rivers to reach their spawning grounds. While treaties guarantee these tribes rights to fish for salmon both on their reservations and at all "usual and accustomed" places outside them, actions anywhere along this route may deprive all tribes of the resource. Tribes must work together to protect the entire habitat and ensure that no one entity or person, tribal or otherwise, monopolizes it.

Similarly, Lakes Huron, Michigan, and Superior are spread over three states and are divided into three separate treaty areas involving separate tribes. Fish populations cross those treaty boundaries within the lakes and sometimes between them. Threats to those populations likewise cross treaty and lake boundaries. The Atlantic lamprey (*Petromyzon marinus*) crossed into the Great Lakes through the Welland Canal in Ontario; quagga mussels (*Dreissena bugensis*) and zebra mussels (*Dreissena polymorpha*) arrive on transatlantic ship ballasts; and invasive water plants do the same (Matheny 2021). Addressing the threat in just one place won't work.

A 2019 US Supreme Court case illustrates the challenges posed by migration of land animals across legal boundaries. In 2014, Clayvin Herrera and other Crow Tribe members shot several bull elk (*Cervus elaphus*) they had pursued from the Crow Reservation in Montana onto the Big Horn Reservation in Wyoming (*Herrera v. Wyoming*, 139 S. Ct. 1686, 1693 [2019]). Wyoming charged Herrera with hunting out of season and without a state hunting license. The Supreme Court ultimately held that the Crows retained their treaty rights to hunt in the area, but treaties with the neighboring Northern Cheyenne Tribe also preserve their rights to hunt in the same area (Treaty with the Northern Cheyenne and Northern Arapaho, 15 Stat. 655, Art. 2 [1868]). An intertribal approach to cross-border hunting and management can prevent conflicts between these two treaty-protected rights.

Rights Are Often Shared between Tribes

Rights are often shared between tribes as a matter either of law or of culture and tradition. Although the United States today recognizes 574 Indian tribes and Alaska Native villages, modern tribal status often conceals traditional interrelationships and divisions. The seven modern Sioux Tribes of South Dakota, for example, were created by the abrogation of treaties that preserved far more land for united Lakota, Nakota, and Dakota peoples. Elsewhere, as with the Confederated Tribes of the Colville Reservation and the Eastern Shoshone and Northern Arapaho Tribes of the Wind River Reservation, federal officials located unrelated tribes on a single reservation. Intertribal management may help alleviate some of the artificiality of contemporary tribal definitions.

Federal treaties and statutes protecting off-reservation hunting and fishing rights sometimes reserve those rights to multiple tribes without differentiating among them. This is true, for example, of the rights administered by the Northwest Indian Fisheries Commission, the Columbia River Inter-Tribal Fish Commission, the Great Lakes Indian Fish and Wildlife Commission, and the Chippewa Ottawa Resource Authority. The Marine Mammal Protection Act (1972) exempts "any Indian, Aleut, or Eskimo who resides in Alaska and who dwells on the coast of the North Pacific Ocean or the Arctic Ocean" from prohibitions on taking whales for subsistence purposes (16 USC § 1371[b]), but multiple Alaska Native villages must share a strict quota on subsistence whaling set by the International Whaling Commission (Ikuta 2021).

Even where tribes do not share resources as a matter of law, they may do so as a matter of culture and tradition. For example, the San Francisco Peaks near Flagstaff, Arizona, are sacred for the Navajo Nation and the Hopi, Havasupai, Hualapai, and White Mountain Apache Tribes, a place where Indigenous people gather plants for religious and medicinal purposes. By joining together to protest the use of recycled sewage effluent for snowmaking on the mountains, the tribes were able to rally public outcry and gain an initial injunction from the courts, although it was later reversed (*Navajo Nation v. US Forest Service*, 479 F.3d 1024 [9th Cir. 2007], *reversed* 535 F.3d 1058 [2008]). In a story with a hopefully happier ending, the Navajo Nation and the Hopi, Ute, and Zuni Tribes formed a coalition that led to the creation of Bears Ears National Monument and secured a significant voice in its management (Wilkinson 2018:325, 331). Although President Donald Trump slashed the monument's size to allow resource exploitation, President Joe Biden has significantly restored it. The unity of the intertribal coalition for the creation and Indigenous management of Bears Ears was a powerful factor in its relative success.

Pooling Resources Is Necessary

Because many tribes are relatively small, few have the sole ability to fund the expert and administrative work that effective management requires. Of the 574 federally recognized tribes, only 14 have a resident population over 10,000 (Office of the Assistant Secretary—Indian Affairs 2014:table 4). By pooling resources, tribes can muster the resources to play a significant role in the multifaceted demands of wildlife management.

This chapter alone cannot provide a comprehensive list of all intertribal wildlife management organizations; one publication says there are "scores" (Wilkinson 2020). Here, I discuss three of the most prominent.

The Northwest Indian Fisheries Commission

The Northwest Indian Fisheries Commission (NWIFC) is the oldest contemporary intertribal wildlife management organization and one of the largest. The commission supports natural resources management for the 20 federally recognized tribes in what is now western Washington.[1] Founded to support salmon co-management, its mission has expanded to support shellfish, non-anadromous groundfish, and land-based wildlife management, and more broadly to protect the habitat of the entire Puget Sound basin and coastal waters.

7.1. Member tribes of the Northwest Indian Fisheries Commission.
COURTESY OF THE NORTHWEST INDIAN FISHERIES COMMISSION

NWIFC formed in 1974 in the wake of US District Court judge George H. Boldt's landmark decision in *United States v. Washington* (384 F. Supp. 312 [W.D. Wash. 1974]; Boldt Decision) in which the court affirmed that the tribes retained their treaty rights to fish at "usual and accustomed places" outside their reservations, that they had rights to up to half of the harvestable catch, and that these rights were not subject to general state laws (*id.* at 333–34, 343), but there were conditions. These conditions included regulations that would protect the resource, personnel to enforce the regulations, and regular reporting to the state on the number of fish caught by tribal treaty fishers (*id.* at 341). Judge Boldt held that only two of Washington's treaty tribes, the Yakama and the Quinault, met those conditions at that time (*id.* at 333). Many of the remaining tribes were quite small. For each tribe to meet Boldt's conditions would have taken a substantial portion of their financial and personnel resources.

On 1 May 1974, three months after Boldt's decision, leaders from the Lummi, Quinault, Makah, Squaxin Island, Muckleshoot, and Skokomish Tribes met to discuss forming a commission (Ott 2011). Such an intertribal organization had been discussed since the 1960s, but the need to implement the self-governance requirements of the Boldt Decision added urgency (*id.*). The federal government also pressured the tribes to consolidate decision-making and management in a single intertribal entity (Hollowed 2021). Tribes successfully resisted this pressure, but the commission helped placate federal demands by acting as a coordinating body for the tribes.

Rather than unified intertribal management, the model adopted by the commission might be described as sharing resources combined with almost complete tribal independence. This is clearly reflected in the NWIFC Constitution, which in enumerating the "powers" of the commission first lists a limitation: "No Commission vote or action may be taken that will supersede the fishery rights and responsibilities or legal rights and legal interests of any member Tribe or Treaty Drainage/Regional Area that has objected" (NWIFC Const., Art. IV, § 1). The statement of purposes similarly emphasizes that its services are "supplemental and supportive" to tribes that "express a need" for them and that while the commission is a "coordinating body . . . to provide a forum to express, communicate, and resolve" issues of concern, such issues must be identified "unanimously" by the tribes (*id.* at Art. II). The Constitution also specifies twice—in Article IV, § 2, and Article XIII—that the Constitution itself does not in any way abridge the sovereign rights and authority of the member tribes.

How tribes are represented within the commission has changed over time. Originally, the commission was made up of five members, each of whom was elected by the set of tribes that had signed one of the five original treaties (Ott 2011). But these treaty groupings do not reflect natural, cultural, or habitat divisions; they were the product of the territorial governor's efforts to simplify negotiations (Harmon 2000). In 1984, the tribes amended the NWIFC Constitution to increase the number of commissioners to eight and to elect both a commissioner and an alternate from the tribes located in eight Puget Sound treaty drainage/watershed areas (NWIFC Const., Art. V, § II). The election procedures and terms of office are set by each of these eight watershed areas, and if the tribes in an area cannot agree, the tribes may each select their own commissioners (*id.*). Today, each of the 20 member tribes elects its own commissioner (Hollowed 2021).

The conditions that generated this coordination and independence began with the precontact history of the tribes. The Native inhabitants of Puget Sound lived in numerous autonomous villages and spoke several mutually unintelligible languages; many were strangers and even enemies to each other (Harmon 2000). In the words of Professor Alexandra Harmon, "What they shared was a system of communicating and conducting relationship[s] with outsiders—a system that drew them all into a region-wide social network" (*id.* at 6). Marriage outside one's village could enhance economic and social standing, and villages participated in each other's ceremonial functions and intervillage potlatches, practices that increased both intravillage diversity and local pride (*id.*). While some ethnographers treated them as one cultural group, others have said that "no American Indian population included a more diverse assortment of peoples" (*id.* at 8).

This diversity plus interconnection continued in the treaty era. Territorial governor Isaac Ingalls Stevens, who negotiated the five treaties reserving tribal fishing rights, had long tried and failed to get Puget Sound's Indigenous inhabitants to select Head Chiefs to represent the villages (Harmon 2000:79). Although Stevens appointed only a few Head Chiefs in the five treaties negotiated between 1854 and 1856, the treaties themselves list dozens of autonomous bands as parties, and the actual tribal signers outnumbered the bands (*id.* at 85). The treaties, however, created only eight reservations, and little effort was made to move Indigenous people onto them. In fact, the Indigenous negotiators were unified on little except preserving their off-reservation fishing rights and their rights to work and trade outside the designated reservations (*id.*).

The treaties also created more differences. The many Indigenous treaty signers who did not settle on the few reservations were left largely landless and received little support from federal Indian agents. In litigation over treaty fishing in the 1960s and 1970s, the state claimed these groups were no longer tribes entitled to treaty rights (*United States v. Washington*, 384 F. Supp. 312, 339 [W.D. Wash. 1974]). When the tribes petitioned the Bureau of Indian Affairs for recognition, some other Washington tribes initially opposed their claims, fearing that recognition would infringe on their rights or sovereignty (American Indian Policy Review Commission 1976). Further, because the treaties recognized the rights of all the tribes to fish at their "usual and accustomed places" and because the key resource, salmon, travels in its lifetime through many different places, disputes over how that resource should be divided among tribes became endemic.

Perhaps because of this, NWIFC does not take on some roles occupied by other major intertribal fisheries commissions. In contrast with NWIFC, for example, the Columbia River Inter-Tribal Fish Commission (CRITFC) declares that its mission is to "ensure a unified voice" in tribal management of fishery resources (https://www.critfc.org/). In addition to technical assistance, CRITFC helps enforce fishing regulations, with police cross-deputized by its four member tribes, the state of Oregon, and some counties in Washington (Columbia River Inter-Tribal Police Force). Similarly, the Great Lakes Indian Fish and Wildlife Commission (discussed further below) and the Chippewa Ottawa Resource Authority, which regulates fishing rights in Lake Michigan, provide the enforcement services for off-reservation tribal fishing and hunting rights in their coverage areas. Delegating such classically sovereign functions to NWIFC would likely be a very hard sell to its member tribes.

NWIFC embodies the member tribes' commitment to the idea that "by unity of action, we can best accomplish these things, not only for the benefit of our own people, but for all of the people of the Pacific Northwest" (NWIFC Const., preamble). And like the traditional potlatches at which diverse Puget Sound peoples gathered to exchange resources and form bonds while maintaining tribal independence, NWIFC has allowed unity of action despite the member tribes' differences. It provides a forum where its members can meet, share information, coordinate goals, and in many cases, resolve differences without going to court. With dozens of scientists on staff, NWIFC can provide cutting-edge information and technical assistance on the state of the fisheries and the Puget Sound habitat generally. These scientists provide expert testimony in ongoing litigation over treaty rights, draft and secure buy-in to model tribal fishing regulations, and provide valuable information to tribes, the state, the federal government, and the public. Further, NWIFC provides identification cards for tribal individuals, allowing them to prove their rights to fish under tribal, rather than state, law. Rather than replacing tribal efforts to manage the fisheries, NWIFC complements them. John Hollowed, the longtime legal and policy advisor of the commission, estimates that from three or four biologists working for the tribes when the Boldt Decision was issued, the tribes now employ over 500 people supporting the fisheries and their habitat (Hollowed 2021).

NWIFC contributes high-quality information and coordination to protect the wildlife and habitat of the Puget Sound watershed. For example, NWIFC provides coded wire tagging of millions of fish annually before their release from tribal hatcheries; genetic and statistical consulting for tribes; and pathology services to prevent, monitor, and address disease. Each year, NWIFC helps tribes coordinate with other managers of the fish resource to set fishing seasons and catch limits. Protecting habitat for all wildlife is an increasingly important part of its work, and NWIFC helps tribes address habitat destruction and climate change. This work reaches not just fish and shellfish, but includes protection of land animals and the water itself.

As state and federal officials come and go, NWIFC is "here to stay" (Hollowed 2021), a constant in the ongoing negotiations among groups interested in the resources of the Puget Sound. Its personnel, too, are remarkably stable: Billy Frank Jr., a Nisqually man whose resistance was key in establishing modern fishing rights, chaired NWIFC from 1981 to his death in 2014; legal and policy advisor John Hollowed has worked for NWIFC since 1987; and other key staff have served with NWIFC for decades. In the words of Frank, NWIFC is a clear embodiment of the reality that "[t]he tribes are here to stay as co-managers of the natural resources in western Washington. . . . We are confident that by working together—all of us—we can reach our common goals" (Northwest Treaty Tribes n.d.).

Great Lakes Indian Fish and Wildlife Commission

The Great Lakes Indian Fish and Wildlife Commission (GLIFWC) is an intertribal organization of 11 tribes with territories in Wisconsin, Michigan, and

Minnesota.[2] It has over 70 employees. GLIFWC was modeled on the success of NWIFC and has a similar history. The distinct interests and legal and cultural histories of the peoples and wildlife it serves, however, have led to important differences between the two organizations.

GLIFWC formed in 1984. At that time, courts had affirmed that tribes in Wisconsin and Michigan retained their treaty rights to hunt, fish, and gather resources in their ceded lands but had made clear that states could regulate tribal members in these rights, if necessary, for conservation or health and safety (*Lac Courte Oreilles Band of Lake Superior Chippewa Indians v. Voigt*, 700 F.2d 341, 364 [7th Cir. 1983]). One of the predecessors to GLIFWC, the Great Lakes Indian Fisheries Commission, had formed in 1982 as the brainchild of Henry Buffalo Jr., the newly graduated tribal attorney for the Red Cliff Band of Lake Superior Chippewa (Busiahn 2011). Buffalo approached biologist Tom Busiahn, who was working on developing fishing regulations for Red Cliff, and suggested they bring all six Lake Superior Chippewa tribes together to enhance their regulatory capacity (*id.*). In attending meetings between tribal leaders and the state in his law school summers, Buffalo had learned that the hard part "wasn't the affirmation of the right. The greatest challenge . . . was in re-emerging as governments with responsibilities over the resources" (Buffalo 2011). Many different organizations were making decisions about the resource, and the tribes needed "a capacity to reach into those organizations . . . take their position amongst those governments and be recognized as a part of the management of those governments" (*id.*).

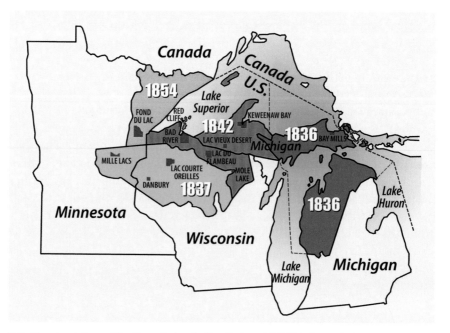

7.2. Member tribes of the Great Lakes Indian Fish and Wildlife Commission.
COURTESY OF THE GREAT LAKES INDIAN FISH COMMISSION

In 1983, the US Court of Appeals for the Seventh Circuit affirmed that the parties to the 1837 and 1842 treaties with the Lake Superior Chippewa Indians retained inland hunting and fishing rights in Wisconsin (*Lac Courte Oreilles Band of Lake Superior Chippewa Indians v. Voigt*, 700 F.2d 341 [7th Cir. 1983]). The court remanded, however, for trial on whether or how the state could regulate that right (*id.*). Lac Courte Oreilles Chairman Gordon Thayer recognized the importance of the victory, but worried that if tribal citizens kept getting cited for violating state law, that would generate a series of legal opinions that would jeopardize the hope of tribal regulation (Nesper and Schlender 2007). He organized a meeting of the other tribal parties to the treaties to request that they delegate their authority to a unified Voigt Intertribal Task Force to develop tribal hunting and fishing regulations and enforce them under negotiated agreements with the state.

The next year, the commission and the task force joined forces to become GLIFWC. The GLIFWC Constitution is modeled on NWIFC's, with a nearly identical preamble and a similar governance structure (GLIFWC Const., Art. IX[1][F]), but there are significant differences between the two. Where NWIFC's statement of purpose emphasizes the supplemental nature of its services, GLIFWC's announces a collective sense of tradition and responsibility:

> The Great Lakes Indian Fish and Wildlife Commission was begun in recognition of the traditional pursuits of the Native American people and the deep abiding respect for the circle of life in which our fellow creatures have played an essential life-giving role. As governments who have inherited the responsibilities for protection of our fish, wildlife, and plants we are burdened with the inability to effectively carry out tasks as protectors and managers. This is especially true now that the state and federal courts have recognized our traditional claims. We have never intended to abandon our responsibilities. The purpose of this agency is to ensure effective self-regulation and intertribal comanagement in support of the sovereignty of its member tribes. (*Id.* at Art. II.1–2)

While GLIFWC's Constitution declares that it does not abridge rights vested in the member tribes, it also makes clear that GLIFWC exercises "delegated sovereign authority on behalf of its member tribes" to accept funds, and has the power to "formulate a broad natural resources management program" in its areas of concern (*id.* at Art. IX.1[F], IX.2, X.1[B]). Consistent with these powers, GLIFWC officers act as the enforcement arm for its member tribes' off-reservation activities.

GLIFWC's member tribes are more similar to each other than are NWIFC's. Although GLIFWC includes 11 tribes across three states, all of its members are Ojibwe and share a common language and cultural traditions. In 1991, GLIFWC amended its mission statement to pledge to include Anishinaabe culture and tradition in its work (Nesper and Schlender 2007). It does this partly by incorporating traditional hunting seasons and practices and by collecting and disseminating information on traditional plants and their uses (*id.* at 289, 294). Although geography and history created differences among the tribes, and they continue to have different approaches to resource management, this common bond likely has increased willingness to delegate sovereign authority to the commission. GLIFWC is also careful to respect the limitations of its role. It does not monitor or regulate resource use on individual reservations, where tribal authority is exclusive (McCammon-Soltis 2021). It proposes elements of model regulations and creates the forum for tribal representatives to agree on them, but it is for tribes themselves to adopt them (*id.*). With respect to its member tribes, its work is governed by the idea that the tribes are the dog and GLIFWC the tail (*id.*).

From the beginning, GLIFWC has worked to monitor, preserve, and enhance a wide variety of animal and plant resources important to Anishinaabe people. This reflects both the diversity in Anishinaabe subsistence practices and the treaties themselves, which reserve rights to harvest all forms of animal and plant life the people utilized at the time the treaty was signed (*Lac Courte Oreilles Band of Lake Superior Chippewa v. Wisconsin*, 775 F. Supp. 321, 323 [W.D. Wis., 1991]). Wild rice (*Zizania palustris*; *manoomin*, "the good berry"), for example, is an important traditional staple. Since 1984, GLIFWC has conducted annual surveys of wild rice beds and worked to enhance existing beds and establish new ones. GLIFWC also has studied and worked to preserve access to other traditional plant life, including sugar maple (*Acer saccharum*), balsam fir (*Abies balsamea*) boughs, birch (*Betula* spp.) bark, and American ginseng (*Panax quinquefolius*). GLIFWC tracks and manages harvests of deer (*Odocoileus virginianus*), black bear (*Ursus americanus*), fisher (*Martes pennant*), bobcat (*Lynx rufus*), and other native terrestrial animals. GLIFWC also reports on environmental threats, including assessing the impact of logging on understory plants, how local mines degrade water quality, and the effects of invasive species like lamprey and purple loosestrife, and it has helped tribes develop federally approved migratory bird hunting regulations.

GLIFWC has made a tangible difference in legal and political disputes over these resources. Between 1983 and 1989, as litigation over state regulation continued, GLIFWC's Voigt Intertribal Task Force negotiated multiple interim agreements, worked with tribes to create model regulations, and provided monitoring and enforcement to prove that tribes could protect the resource (*Lac Courte Oreilles Band of Lake Superior Chippewa v. Wisconsin*, 707 F. Supp. 1034, 1050–52 [D. Wis., 1989]). In her 1989 decision, Judge Barbara Crabb declared that "[w]hat the parties in this case have done to give practical effect to plaintiffs' judicially recognized treaty rights is a remarkable story," and she highlighted GLIFWC's role in this story (*id.* at 1052, 1054). Because the tribes had shown they could protect the resource, she found, so long as the tribes maintained regulations meeting certain conditions, "the state cannot regulate tribal members' off-reservation fishing" (*id.* at 1056). She reached a similar decision regarding hunting and collecting natural resources in 1991, holding that the tribes could harvest an equal share of the treaty resources largely free from state regulation (*Lac Courte Oreilles Band of Lake Superior Chippewa v. Wisconsin*, 775 F. Supp. 321, 323 [W.D. Wis. 1991]). That year, the federal government released a report based in part on GLIFWC data proving that treaty fishing had not hurt fish populations (Nesper and Schlender 2007). After the report, Wisconsin chose not to appeal Crabb's decision (*id.*).

GLIFWC played a significant role in a more recent tribal victory: securing the right to hunt deer at night through "shining"—spotlighting with bright lights. Judge Crabb had prohibited this traditional practice in 1991, and the tribes had not appealed (*Lac Courte Oreilles Band of Lake Superior Chippewa Indians v. Wisconsin*, 769 F.3d 543, 544 [7th Cir. 2014]). At that time, the state prohibited all night hunting outside reservations, so there were no data on its safety (*id.* at 546). Over a decade later, the tribes sought to reopen the judgment after the state had authorized employees and contractors to engage in night hunting to control the deer population and address chronic wasting disease (*id.*). Crabb refused, but the court of appeals reversed, relying significantly on GLIFWC's work. Minnesota and Michigan had both safely allowed tribal night hunting for over a decade "managed by the same organization, the Great Lakes Indian Fish and Wildlife Organization [*sic*]," as would manage it in Wisconsin (*id.* at 549). The proposed regulations, moreover, were modeled on those in Minnesota and Michigan, but stricter, and hunters would receive the same GLIFWC firearm safety training as they did in Michigan (*id.*). Because of its intertribal nature, GLIFWC could show an established track record of safe night hunting, which was key to the tribes' victory in court.

GLIFWC data and publications have also reduced backlash against tribal treaty rights. In the 1980s, spearfishers on Wisconsin lakes faced abuse from crowds so virulent that the federal courts declared it a civil rights violation (*Lac du Flambeau of Lake Superior Chippewa Indians v. Stop Treaty Abuse–Wisconsin, Inc.*, 843 F. Supp. 1284 [W.D. Wis. 1994]). One of the key claims of the opponents was that tribes were wasting and destroying the resource, thereby threatening the way of life of non-Indigenous rural northern Wisconsin residents (Grossman 2017). Through its careful monitoring, however, GLIFWC was able to show that treaty fishing did not affect the resource and that the decline in tourism in northern Wisconsin was actually the result of the racist protests and other changes rather than due to the spearfishers themselves (*id.* at 218). More recently, by providing statistics to help tribes battle mining operations that harm water quality and fish populations, GLIFWC has helped tribes build bridges with their former enemies. Degradation from mines threatens rural non-Indians as well as Ojibwe; the interests that led them to protest at spearfishing lakes in the 1980s now have led them to recognize the tribes as important allies (*id.* at 228–30, 268).

GLIFWC has achieved Henry Buffalo's vision: tribes now exercise responsibility over natural resources and are recognized as necessary voices in managing them. In doing so, it has contributed to a resurgence of tribal people's connections to Anishinaabe culture and their tribal governments. Finally, it has enhanced the diversity and health of natural resources across the northern Great Lakes region.

InterTribal Buffalo Council

The InterTribal Buffalo Council was founded to restore buffalo to tribal ecosystems. While groups like NWIFC and GLIFWC focus on wildlife populations outside reservations, ITBC largely supports on-reservation herds. And while other organizations primarily support tribal regulation and monitoring of wildlife populations, ITBC works to provide tribes with access to buffalo and to influence state and federal regulations limiting that access. But like the other groups profiled here, ITBC's mission is to protect the connection of Indigenous people to wildlife of tremendous cultural importance. And like them, ITBC shows the power of intertribal action in achieving tribal and ecological goals.

The near destruction of the American buffalo at the end of the 19th century paralleled the near destruction of the Plains Indians. For many Plains tribes, buffalo played the same role that salmon did for the Indigenous peoples of the Pacific Northwest, shaping important aspects of their economies, cultures, and religions. But by the end of the 19th century the buffalo population had gone from 60 million to about a thousand; most were on private livestock ranches, and a handful were in national parks (Burton 2000). In the words of Fred DuBray, a founder and longtime president of ITBC: "When they destroyed the buffalo herds, they were destroying our culture. . . . if they can't be saved, then we can't either" (Zontek 2007:80).

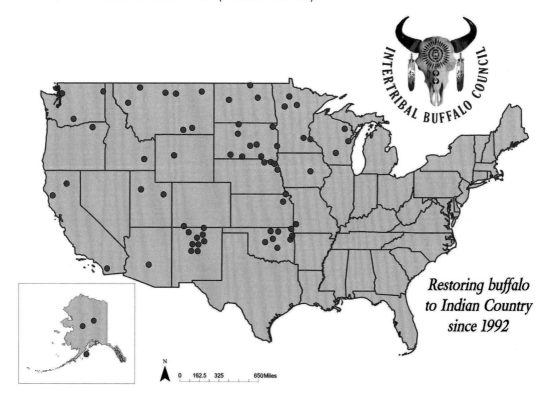

7.3. Locations of the member tribes of the InterTribal Buffalo Council.
COURTESY OF THE INTERTRIBAL BUFFALO COUNCIL

The tribes that founded ITBC first met through the Native American Fish and Wildlife Society. Several tribes had small buffalo herds, and they recognized that "[i]ndividually our goals may be different, but as a whole our goal is to get the buffalo back to the Indians" (*id.* at 80, quoting Ernie Robinson, Northern Cheyenne). In 1991, 19 tribes gathered in the sacred Black Hills of South Dakota to agree to collective action. They secured initial funding from the US Congress, and in 1992 met in Albuquerque, New Mexico, to form the Intertribal Bison Cooperative as a 501(c)(3) (Zontek 2007; ITBC n.d., Our History). In 2010, the organization secured recognition as a section 17 corporation under the Indian Reorganization Act and changed its name to the InterTribal Buffalo Council (Abold 2021).

Today, ITBC's membership includes 69 tribes ranging across 19 states (ITBC n.d., ITBC Today). Tribes join ITBC through resolution. The group is governed by a nine-member board of directors, with each tribe voting to elect officers and to elect the representative from their region of the country. ITBC supports tribes by distributing federal funds to tribal buffalo programs through competitive grants, providing tribes with technical assistance, and securing distribution of surplus bison from national parks. Further, the organization acts as a collective voice to promote the restoration and respectful use of bison with a range of federal and state government agencies and the public (Abold 2021).

From the beginning, ITBC has sought to restore buffalo in a distinctly Indigenous way. Its founders rejected both treating buffalo as livestock subservient to humans and treating them as wildlife fully separate from humans. Instead, ITBC sought to restore the buffalo as wild animals in relationship with Indigenous people. ITBC's mission is to "restor[e] buffalo to Indian lands in a manner which promotes cultural enhancement, spiritual revitalization, ecological restoration and economic development compatible with tribal beliefs and practices" (ITBC 2017). Consistent with this mission, ITBC promotes buffalo on "free ranges as large as political geographic limitations allow," avoids genetic manipulation or the division of family units, and recommends only the management and supplemental feeding necessary to keep the herds healthy and sustainable (Zontek 2007).

ITBC's efforts to distribute surplus buffalo from national parks to tribes quickly involved it in controversy. Because most buffalo in private herds have genetically mixed with cattle, one of the few sources of genetically undiluted buffalo is the descendants of the few hundred that escaped slaughter at Yellowstone National Park. Protected from poaching, the Yellowstone herds grew to over 3,000 by the 1990s. But buffalo are migratory, and in winter they regularly leave the park for lower grounds. Although these lands are mostly federally owned, private ranchers often graze their cattle there at low fees. Montana ranching interests feared that contact with buffalo would infect their cattle with *Brucella abortus*, which can cause abortions and sterility in cattle, bison, and elk, and fever in humans (*Fund for Animals, Inc. v. Lujan*, 794 F. Supp. 1015 [D. Mont. 1991]). The stakes were even higher than the potential loss of diseased animals, because exposure might lead the USDA Animal and Plant Health Inspection

Service to rescind Montana's certification as a brucellosis-free state, requiring expensive testing before any cattle could be shipped from the state. Montana authorized hunters to shoot any buffalo that left the park's grounds and later reached a management agreement with the federal government (*Greater Yellowstone Coalition v. Babbitt*, 952 F. Supp. 1435 [D. Mont. 1996]).

Conservation groups challenged these practices without success. In 1991, US District Court judge John Lovell upheld Montana's plan to authorize buffalo hunts, holding that "Montana has an absolute right under its police powers, in protecting the health, safety, and welfare of its inhabitants, to remove by reasonable means, possibly infected trespassing federal bison which migrate into Montana" (*Fund for Animals*, 794 F. Supp. at 1020). The court accepted Montana's allegation that brucellosis could travel from buffalo to cattle, although studies showed—and a later federal court decision accepted (*Fund for Animals v. Clark*, 27 F. Supp.2d 8, 14 [D.D.C. 1998])—that this was extremely unlikely. In 1996, Lovell also rejected a challenge to federal approval of the state-federal interim plan (*Greater Yellowstone Coalition v. Babbitt*, 952 F. Supp. 1435 [D. Mont. 1996]).

Soon after its founding, ITBC had begun trying to prevent the killing of buffalo that left the park grounds. In 1994, it proposed that animals be quarantined, tested, and transferred to tribes if healthy, offering facilities at two reservations for the project (Carlson 2020). After the winter of 1996–1997, when 1,100 buffalo were killed under the federal-state plan, ITBC challenged the plan in court, but it met with no more success than the conservation groups (*Intertribal Bison Co-op. v. Babbitt*, 25 F. Supp.2d 1135 [D. Mont. 1998]). ITBC kept working, achieving a quarantine program at Yellowstone, although the program could not accommodate all the buffalo, and many were still slaughtered. In 2009, ITBC finally won a seat as part of the coalition creating the Yellowstone Interagency Bison Management Plan (French 2018). In 2014, the Fort Peck Assiniboine and Sioux Tribes completed a quarantine facility on their Reservation with ITBC funding, but it remained empty for years because of continued resistance from Montana and the Animal and Plant Health Inspection Service (ITBC 2017). Finally, the first animals were sent to the facility, and 40 were shipped in 2020 from Fort Peck to tribes across the country, although the facility remains far under capacity, and many animals continue to be slaughtered (Carlson 2020).

ITBC's other work with the federal government has had unequivocal success. Although begun in litigation, ITBC now has a lasting and positive relationship with the National Park Service. ITBC now works with the National Park Service as an "integral partner" in developing the Bison Stewardship Strategy (Symstad et al. 2019:8). Perhaps influenced by ITBC advocacy, moreover, the National Park Service has "shifted away from managing bison from [a] livestock-based perspective toward a wildlife stewardship approach" (*id.* at vi). One of the pillars of this approach is supporting tribal buffalo culture (*id.*). ITBC has also helped elevate recognition of buffalo nationally, helping to lead a coalition that resulted in a 2016 declaration that the bison is the national mammal of the United States (Lee 2016).

Most striking is the transformed role of buffalo for tribal peoples. Enabled by ITBC grants, distribution of federal buffalo, and technical assistance, tribal buffalo holdings have multiplied from 1,500 head in 1992 to over 20,000 today. Although the herds are still dwarfed by the 400,000 buffalo kept as livestock, tribes now collectively hold four times as many wild buffalo as the federal government. And as tribes have restored buffalo to the landscape, their grazing patterns have restored the ecosystem, making room for native grasses, birds, and terrestrial animals that once joined them on the Plains (ITBC 2017).

Buffalo are healing not just the ecosystem but tribal people. ITBC works with the US Department of Agriculture to allow buffalo meat to be used in food programs for children and the elderly and has promoted buffalo meat on reservations as a way to address high diabetes and obesity rates (Zontek 2007). The ITBC network played a particular role in the COVID-19 shutdowns, as ITBC was able to direct tribal buffalo to tribal meat-processing plants to provide resources to commodity food programs (Abold 2021).

Some see other links between the recovery of the buffalo and the recovery of tribal people. In the words of Mike Faith, who managed the Standing Rock Sioux buffalo herd for 15 years and now serves as Chairman of the Tribe, "They were almost extinct. So were we as a people. . . . The buffalo are coming back strong and so are we as a community" (Springer 2019). By harnessing the power of collective tribal action, ITBC helped generate this rebirth.

ACCOMPLISHMENTS OF INTERTRIBAL WILDLIFE MANAGEMENT

As these case studies suggest, intertribal wildlife management organizations have played a crucial role for tribal people and the wildlife they cherish. This section focuses on three aspects of their impact: first, the ways they have enhanced preservation of wildlife and habitat; second, the ways they have furthered the development of distinctly Indigenous science; and third, the ways they have increased tribal sovereignty.

Wildlife and Habitat Improvement

Although most intertribal management organizations arose to facilitate tribal use of wildlife, all have become important forces in wildlife preservation. The Alaska Eskimo Whaling Commission, for example, negotiates annually with companies exploring for oil and gas off the Alaska coast to mitigate impacts on the bowhead whale population (Aiken 2011). Using results of its monitoring, scientific studies, and traditional knowledge, the commission has questioned the methodology and conclusions of federal agencies to advocate for more protection of whale populations (id. at 4–6).

Another example comes from the Columbia River Inter-Tribal Fish Commission. In 1995, CRITFC and its member tribes adopted a fish restoration plan called Wy-Kan-Ush-Mi-Wa-Kish (Spirit of the Salmon). Pursuant to this plan, they entered into the Columbia Basin Fish Accords with the Bonneville Power Administration, the US Army Corps of Engineers, and the US Bureau of Reclamation in 2008. The accords established coordinated and consistent funding for salmon habitat restoration in exchange for the tribes' promise not to litigate for breach of the Snake River dams over the next decade. A 2017 evaluation showed many accomplishments under the accords. Among other things, 397 barriers to fish passage had been removed; some fish populations had doubled; beavers, coho salmon, and Pacific lamprey had been reintroduced in a number of areas; and juvenile survival rates for wild Chinook, steelhead, and sockeye salmon had all increased (CRITFC 2017).

NWIFC, similarly, has used its position to protect water resources that affect the entire Puget Sound ecosystem. In litigation that went to the US Supreme Court, the tribes of western Washington won a ruling ordering the state to remove culverts that prevented anadromous fish from completing their life cycle and reaching tribal fishing grounds (*United States v. Washington*, 853 F.3d 946, 954 [9th Cir. 2017], *aff'd Washington v. United States*, 138 S. Ct. 1832 [2018]). NWIFC had supported the litigation since its filing in 2001, as a voice and negotiator for the tribes and through extensive reporting on the impact of the culverts on fish populations (Bougher 2001, Frank 2007). NWIFC also tracks many threats to Puget Sound watersheds, assists tribal members in seeking to address them, and negotiates with nontribal actors to do the same. In the words of NWIFC Chair Lorraine Loomis, "We know the status quo isn't working when it comes to salmon recovery. We know what the science says needs to be done. We know that we must move forward together to address habitat because it is the most important action we can take to recover salmon" (NWIFC 2020).

By winning the right to manage wildlife themselves, tribes have not supplanted state and federal management, they have augmented them. In 1983, for example, Wisconsin officials were conducting only about 12 spring fish counts annually at lakes throughout the state (Nesper and Schlender 2007:298). By 2001, the state and GLIFWC coordinated to undertake 60 such counts annually, a fivefold increase (*id.*). A 2007 evaluation reports that GLIFWC's work has helped produce what "may be the most tightly regulated and monitored walleye fishery in the world" (*id.* at 286).

Given the great differences between intertribal organizations and the wildlife they manage, a quantitative analysis of their impact is far beyond the scope of this chapter. But in numerous ways these organizations appear to have meaningfully enhanced wildlife protection and restoration.

Indigenous Science

Intertribal organizations also enhance the development of Indigenous science. "Indigenous science" here means something broader than the expansion and recognition of Traditional Ecological Knowledge, although it includes that too.

Indigenous science is what emerges when Indigenous people are directing the questions that science must address and influencing the sources of information used to answer them.

In part, intertribal organizations enhance Indigenous science simply through their collective power to hire scientists not enmeshed in existing bureaucracies. Henry Buffalo Jr. remarked on this in reflecting on GLIFWC's impact:

> Part of what we bring to the equation, is we stole from them some of their best younger minds and their biologists. These were the younger minds who had different theories than the guys who had been there for 30 years. Their theories were not given any daylight and they didn't want to wait around for 10 years or 15 years to get their theories up to the top. They became part of our staff and all of a sudden they were peers, they were colleagues, they were equals to the folks that were the head of the institutions in the states or provinces. And we began contributing with those new theories and developing science. (Buffalo 2011)

Because tribal authority over wildlife is still vulnerable to state or judicial abrogation, tribes have also had to monitor their catch more carefully. This has allowed them to reveal where state natural resource officials are "sidelining science for political concerns" (Nesper and Schlender 2007:294).

The products of scientific methods depend on the questions they are used to answer. Because tribes have different interests than non-Indigenous communities regarding wildlife, the science they generate will also be different. The impact of water quality on fish provides a good example. For many tribes, fish are not simply one food source among many; rather, eating fish regularly is a key cultural and subsistence practice. But most federal and state water-quality standards are based on the fish consumption rates of non-Indigenous people. Even when based on current Indigenous consumption rates, those rates may reflect consumption reduced by habitat degradation and suppression of cultural practices.

One innovative intertribal project in Maine sought to address this. For the Wabanaki Exposure Study, five Maine tribes partnered with the Environmental Protection Agency to commission a toxicologist and an anthropologist to determine the water-quality standards necessary to permit Indigenous people to safely consume fish as they would if traditional practices were not distorted or depressed (Harper and Ranco 2009:8). The study found that tribal sustenance practices included between 286 and 514 grams of fish per day, in contrast with the 32.4 grams per day Maine used in setting its water-quality standards (Dussias 2019:825). The EPA then relied on this study to reject Maine's water-quality standards on the grounds they did not sufficiently protect human safety (*id.* at 824).

Intertribal management organizations have also contributed to the growth and dissemination of Traditional Ecological Knowledge. Sometimes this is by actively seeking out traditional expertise. The Wabanaki Exposure Study report, for example, deliberately included discussions with Wabanaki elders and cultural experts and the vetting of academic literature by these cultural experts in developing its findings (Harper and Ranco 2009). Similarly, GLIFWC sought out knowledge from elders to publish two compendiums on the uses of over 300 plants traditionally used by Great Lakes Ojibwe (Nesper and Schlender 2007). Elsewhere, the dissemination of Traditional Ecological Knowledge occurs simply by creating settings in which non-Native scientific experts have access to the knowledge of the Indigenous people actually using the resource. By combining the knowledge of the tribal fishers across the Puget Sound watershed, for example, NWIFC learned that the state's data did not reflect differences between the different coastal areas in the sound and was able to create more accurate reports (Hollowed 2021). Similarly, the Alaska Eskimo Whaling Commission drew on the experience of Yupik and Inupiat whalers to correct the National Marine Fisheries Service on the migration patterns of bowhead whales (Aiken 2011:4–6). Ojibwe fishers informed GLIFWC staff that toxic mining wastes were contaminating an important Lake Superior spawning ground, leading to the creation of a multiagency effort to address the problem (McCammon-Soltis 2021).

In short, intertribal wildlife management has encouraged the development of science driven by the concerns, experiences, and traditional knowledge of Indigenous peoples.

Increased Sovereignty

Intertribal wildlife management organizations have also increased the sovereignty of their member tribes. This may seem surprising. Even organizations like NWIFC and ITBC, which do not assume classically sovereign functions, take on some roles (including monitoring, drafting model laws, and seeking funds) otherwise reserved for tribes themselves. But by taking on these tasks and coordinating communication with outside entities, these organizations have significantly increased the capacity and influence of their member tribes.

First, information is power. In litigation and negotiations with other governments, tribes too often have had to rely on data collected by others or counter false narratives built on no data at all. By using their collective resources to hire staff to collect and analyze data, intertribal management organizations can counter such imbalances and falsehoods. These data contributed to tribal victories in litigation in the Northwest and the Great Lakes, and empowered tribes to reshape policy debates to further their concerns. By facilitating information sharing among the member tribes, these organizations also maximize the knowledge of all members.

Second, relationships are influence. Resource management requires coordination with many government agencies and interest groups. Ongoing relationships create the trust and sharing of information necessary for that coordination. Although officials within the 20 tribes that comprise NWIFC, the 11 that comprise GLIFWC, or the 69 that comprise ITBC could form those relationships, it would take far more resources, and changes in leadership would make it necessary to build them again. By coordinating the tribal voices in a single, stable organization, intertribal associations can secure ongoing relationships and positions with many more entities than individual tribes can on their own.

Finally, unity is legitimacy. Tribal governments face serious questions as to their legitimacy, both from outside the tribe and from within. Challenges by nontribal entities to tribal authority and capacity are well known. Challenges from tribal citizens are less so. Just as tribes' governing capacity was distorted by generations of efforts to quash or co-opt their power, so was their citizens' respect for tribal authority. Having spent generations fighting state regulation of their fishing rights, citizens of Pacific Northwest and Great Lakes tribes were not eager to have any government, tribal or not, restrict them. Indeed, in the years before courts affirmed their treaty rights, "violating" (evading state wardens) had become a part of Ojibwe cultural identity in the Great Lakes region (Nesper 2002). As Henry Buffalo Jr. recalled, "tribal members didn't like" getting cited by GLIFWC wardens "any more than they liked getting cited by state officers" (Buffalo 2011).

To effectively manage wildlife, therefore, tribes must establish legitimacy in the minds of tribal people and outsiders. Dissension between tribes increases doubts about the legitimacy of tribal positions and authority, making them appear mercurial or self-interested. When, for example, the Sault Ste. Marie Tribe of Chippewa Indians, one of the five members of the Chippewa Ottawa Resource Authority, seemed to break from the other members on the renewal of their agreements with the state, it generated overblown attacks that the tribes were preparing to "end sportsfishing as we know it" (Green 2020). By coordinating a unified front, in contrast, intertribal organizations add legitimacy to the stances of individual tribal members.

Unity, of course, has its limits. Tribes cannot completely cede their decisions or voices to others. But the experience of the organizations profiled here provides no evidence that tribes have sacrificed their interests in favor of unity. Instead, the tribes have increased their power, influence, and legitimacy and have achieved their interests in protecting and using natural resources in the process.

CONCLUSION

Through collective tribal action, intertribal organizations have increased tribal peoples' capacity and authority in wildlife management. In so doing, they have served the interests of not just tribes, but the surrounding human and natural communities.

NOTES

1. The member tribes are the Lummi, Nooksack, Swinomish, Upper Skagit, Sauk-Suiattle, Stillaguamish, Tulalip, Muckleshoot, Puyallup, Nisqually, Squaxin Island, Skokomish, Suquamish, Port Gamble S'Klallam, Jamestown S'Klallam, Lower Elwha Klallam, Makah, Quileute, Quinault, and Hoh. NWIFC, About Us, https://nwifc.org/about-us/.

2. The member tribes are Misi-zaaga'iganiing (Mille Lacs), Nagaajiwanaang (Fond du Lac), Bikoganoogan St. Croix (Danbury), Gaa-miskwaabikaang (Red Cliff), Mashkiigong-ziibiing (Bad River), Ginoozhekaaning (Bay Mills), Waaswaaganing (Lac du Flambeau), Gete-gitigaaning (Lac Vieux Desert), Zaka'aaganing (Mole Lake/Sokaogon), Gakiiwe'onaning (Keweenaw Bay), and Odaawaa-zaaga'iganiing (Lac Courte Oreilles).

LITERATURE CITED

Abold, Arnell. 2021. Interview with Bethany Berger. 28 June.

Aiken, Johnny. 2011. Comment of Alaska Eskimo Whaling Commission to National Marine Fisheries Service on Incidental Harassment Application for Shell Offshore Inc. in the Chukchi Sea. 7 December. In possession of author

American Indian Policy Review Commission, Task Force Ten. 1976. Final Report to the American Indian Policy Review Commission: Report on Terminated and Non-Federally Recognized Indians. US Government Printing Office, Washington, DC.

Bougher, Bryan. 2001. Questions and Answers Regarding the Tribal Culvert Case. NWIFC Press Release, 16 January. https://nwifc.org/questions-and-answers-regarding-the-tribal-culvert-case/.

Buffalo, Henry, Jr. 2011. Untitled remarks. Pages 133–38 in Latish A. McRoy, editor. Minwaajimo—Telling a Good Story: Preserving Ojibwe Treaty Rights for the Past 25 Years. Great Lakes Indian Fish and Wildlife Commission, Odanah, Wisconsin. http://glifwc.org/minwaajimo/Speech/Henry%20Buffalo,%20Jr..pdf.

Burton, Lloyd. 2000. Wild Sacred Icon or Woolly Cow? Culture and the Legal Reconstruction of the American Bison. Political and Legal Anthropology Review 23:21–36.

Busiahn, Tom. 2011. Untitled remarks. In Latish A. McRoy, editor. Minwaajimo—Telling a Good Story: Preserving Ojibwe Treaty Rights for the Past 25 Years. Great Lakes Indian Fish and Wildlife Commission, Odanah, Wisconsin. http://glifwc.org/minwaajimo/Speech/Tom%20Busiahn.pdf.

Carlson, Ervin. 2020. Expand Quarantine for Buffalo beyond Yellowstone. Bozeman Daily. 12 December.

Columbia River Inter-Tribal Fish Commission [CRITFC]. 2017. Columbia Basin Fish Accords: Ten-Year Report, 2008–2017. https://www.critfc.org/blog/2018/08/14/fish-accords-10-year-summary/.

Dussias, Allison M. 2019. Water Quality and (In)equality: The Continuing Struggle to Protect Penobscot Sustenance Fishing in Maine. Connecticut Law Review 51:801–41.

Frank, Billy, Jr. 2007. Being Frank: Culvert Ruling Benefits Salmon, Everyone. NWIFC Blog, 4 September. https://nwifc.org/being-frank-culvert-ruling-benefits-salmon-everyone/.

French, Brett. 2018. Yellowstone's Quarantined Bison Won't Be Transferred to Tribes Anytime Soon. Billings Gazette, 11 November.

Green, Nick. 2020. Court Filing Could End Great Lakes Fishing as We Know It. Michigan United Conservation Clubs, 16 July. https://mucc.org/court-filing-could-end-great-lakes-fishing-as-we-know-it/.

Grossman, Zoltan. 2017. Unlikely Alliances: Native Nations and White Communities Join to Defend Rural Lands. University of Washington Press, Seattle.

Harmon, Alexandra. 2000. Indians in the Making: Ethnic Identities and Indian Relations around Puget Sound. University of California Press, Berkeley.

Harper, Barbara, and Darren Ranco. 2009. Wabanaki Traditional Cultural Lifeways Exposure Scenario. EPA, Region One. https://www.epa.gov/sites/production/files/2015-08/documents/ditca.pdf.

Hollowed, John. 2021. Interview with Bethany Berger. 7 June.

Ikuta, Hiroko. 2021. Political Strategies for the Historical Victory in Aboriginal Subsistence Whaling in the Alaskan Arctic: The International Whaling Commission Meeting in Brazil, 2018. Senri Ethnological Studies 104:209.

InterTribal Buffalo Council [ITBC]. 2017. Return of the Native: The 25 Year History of the InterTribal Buffalo Council. https://itbcbuffalonation.org/return-of-the-native-the-25-year-history-of-the-intertribal-buffalo-council/.

InterTribal Buffalo Council [ITBC]. n.d. ITBC Today. https://itbcbuffalonation.org/who-we-are/itbc-today/.

InterTribal Buffalo Council [ITBC]. n.d. Our History. https://itbcbuffalonation.org/who-we-are/history/.

Lee, Stephen. 2016. National Bison Day Marks America's New National Mammal. Capital Journal, 1 November.

Matheny, Keith. 2021. 10 Least Wanted Michigan Invaders. Detroit Free Press, 24 April.

McCammon-Soltis, Ann. 2021. Interview with Bethany Berger. 26 July.

Nesper, Larry. 2002. The Walleye War: The Struggle for Ojibwe Spearfishing and Treaty Rights. University of Nebraska Press, Lincoln.

Nesper, Larry, and James H. Schlender. 2007. The Politics of Cultural Revitalization and Intertribal Resource Management: The Great Lakes Indian Fish and Wildlife Commission and the States of Wisconsin, Michigan, and Minnesota. In David Rich Lewis and Michael Eugene Harkin, editors. Native Americans and the Environment: Perspectives on the Ecological Indian. University of Nebraska Press, Lincoln.

Northwest Indian Fisheries Commission [NWIFC]. 2020. State of Our Watersheds Report. https://nwifc.org/publications/state-of-our-watersheds/.

Northwest Treaty Tribes. n.d. About Us. https://nwtreatytribes.org/about-us/.

Office of the Assistant Secretary—Indian Affairs, US Department of the Interior. 2014. 2013 American Indian Population and Labor Force Report. US Government Printing Office, Washington, DC.

Ott, Jennifer. 2011. Northwest Indian Fisheries Commission. https://historylink.org/File/9786.

Springer, Patrick. 2019. Study Finds Clear Effects of Buffalo Slaughter. Bismarck Tribune, 14 October.

Symstad, Amy J., Brian W. Miller, Tanya M. Shenk, Nicole D. Athearn, and Michael C. Runge. 2019. A Draft Decision Framework for the National Park Service Interior Region 5 Bison Stewardship Strategy. National Park Service, Jamestown, North Dakota.

Wilkinson, Charles. 2018. "At Bears Ears We Can Hear the Voices of Our Ancestors": The First Native American National Monument. Arizona State Law Journal 50:317–33.

Wilkinson, Charles. 2020. The Belloni Decision. Environmental Law 50:331.

Zontek, Ken. 2007. Buffalo Nation: American Indian Efforts to Restore the Bison. Bison Books, University of Nebraska Press, Lincoln.

Chapter 8

State Regulation and Enforcing Usufructuary Treaty Rights

| GUY C. CHARLTON

INTRODUCTION

Despite the hard-fought community-led political and legal battles that led to the recognition and exercise of treaty-guaranteed hunting, fishing, and gathering rights for tribes, these rights remain under threat. Developmental pressures on habitat, climatic change that adversely effects flora and fauna, an increasingly conservative judiciary, and legal legerdemain by treaty right opponents—all have forced tribes to remain vigilant. In order to counter these threats both on and off their reservations, tribes have sought to protect their interests with less traditional legal arguments grounded in tribal property interests, historic and heritage preservation, religious claims, and tort claims. In this chapter, I consider some possible protective approaches to off-reservation hunting, fishing, and gathering rights.

OFF-RESERVATION TREATY RIGHTS

The extension of jurisdiction across the continent by the United States created a series of cultural and legal accommodations and boundaries between the Indigenous peoples and the American settlers. As this trend continued into the 19th century, the more accommodating nature of these relationships hardened, and the United States imposed its law and military force over previously independent tribes through treaties, war, coercion, removal, squatting, and population movements.

An important aspect of this history is treaty rights. Inheriting British practice, the United States entered into at least 366 treaties with various tribes across the continent prior to the practice ending in 1871.[1] These treaties, more often honored in the breach, guaranteed a range of self-determination, possessory, and usufructuary (hunting, fishing, and gathering) rights.[2] It is from this interaction that Native Americans retain, albeit in truncated form, usufructuary hunting, fishing, and gathering rights in a manner that would otherwise be prohibited by the applicable law of the state. These rights arise from the prior sovereignty and possession of territory by the tribe. They have been protected through treaties as tribal negotiators have sought to ensure the continued livelihoods and cultures of their groups in the face of American settler expansion. Generally, usufructuary rights reflect customary hunting, fishing, and gathering practices undertaken by the tribal members prior to the treaty agreement. Depending on the particular circumstances and treaty texts, these rights are called "usufructuary rights," "off-reservation rights," "reserved treaty rights," or "customary rights."

Legally, usufructuary rights have been characterized and analogized to profits à prendre, "access rights," or easements by the courts.[3] The rights are nonterritorial in the sense that they do not derive from and are independent of any present-day ownership interest in the land, but rather arise from historical occupation and use of particular lands and waters. They not only include the right to use resources for personal sustenance or religious purposes, but also may provide some insulation from governmental regulation, a right to a specific share of the harvested resource, and a right to preserve the resource from activities that might damage continued use.[4] Occasionally, the use rights can include commercial exploitation.

Courts in the United States have held that off-reservation hunting, fishing, and gathering rights are dependent on the tribe possessing the underlying Aboriginal title to a territory, or are reserved by way of federal treaty guarantees relating to a particular territory. Once the treaty right or occupancy and use are recognized as "Indian title" in judicial proceedings, usufructuary rights are included within the panoply of uses for a territory. The rights can be created by the treaty (e.g., where tribes have been relocated from their traditional territories or consolidated with other tribes in a single reservation), or the treaty may reserve a preexisting Aboriginal right based on occupancy.[5] Federal common law allows tribes to vindicate their Aboriginal rights, but the right to occupancy "need not be based on treaty, statute, or other formal Government action."[6] The content and scope of these various uses are in turn related to the particular culture, laws, and customs of the signatory or occupying tribe.[7] The rights are so bound up and such an intrinsic part of the "larger rights possessed by the Indians" resulting from tribal occupation and possession that the Supreme Court in the seminal case *United States v. Winans* compared their importance as equivalent to "the atmosphere they [the Indians] breathed."[8]

PRINCIPLES OF USUFRUCTUARY TREATY RIGHTS

Because tribes, the federal government, and the states have overlapping sovereignty within the United States, off-reservation usufructuary rights have been subsumed within the larger issues of Indian land title and treaty rights. Together, these areas have undergone numerous changes throughout US history, but six central doctrinal elements have been maintained since the first US Supreme Court decision involving Indian land, *Fletcher v. Peck*, was decided in 1810.[9]

1. *Tribes are independent entities that possess inherent sovereignty.* Native Americans' inherent sovereignty predates the founding of the United States and the colonial extension of British sovereignty across North America.[10] This sovereignty, while subject to complete extinguishment and regulation by Congress, nevertheless remains an independent source of authority over tribal members and land. It can also provide a basis for the replacement of state regulation with tribal regulation of off-reservation usufructuary activities.

2. *The federal government has plenary and exclusive authority over Indian tribes.* Justice Lewis F. Powell Jr. in *Montana v. Blackfeet Tribe* noted, "The Constitution vests the Federal Government with exclusive authority over relations with Indian tribes."[11] Where Congress clearly asserts its authority in contradiction of previous federal treaty and statutory undertakings, the actions are beyond the scope of judicial review in spite of the earlier treaties or federal fiduciary obligations. However, the plenary power over tribes and federal-tribal agreements must be exercised by Congress in a clear and unequivocal manner.[12] Of course, the nature of this "clear and plain" rule is subject to dispute. Some congressional action that may fall short of unambiguously abrogating or explicitly extinguishing reserved rights may survive judicial scrutiny. In the context of treaty-reserved usufructuary rights, the courts will not readily infer a congressional intent to repeal such rights.[13] Courts will construe successive treaties, agreements, and legislation in conjunction with the language, practice, and understandings of earlier agreements to determine whether any ambiguities exist that would prevent the extinguishment of tribal rights.[14] Nevertheless, if Congress does abrogate treaty-reserved rights, the federal government must compensate the tribe for the loss of those treaty rights.

3. *The power to regulate Indian tribes is completely federal; states are excluded from extending their jurisdiction and regulation to Indian tribes and land unless specifically authorized by Congress.* As Justice Sandra Day O'Connor, writing for the majority, stated in *Mille Lacs v. Minnesota*: "Although States have important interests in regulating wildlife and natural resources within their borders, this authority is shared with the Federal Government when the Federal Government exercises one of its enumerated constitutional powers, such as treaty making."[15] This doctrine is implied in the long-standing "policy of leaving Indians free from state jurisdiction and control[, which] is deeply rooted in this Nation's history."[16] As noted by Justice Neil Gorsuch in *McGirt v. Oklahoma*, this federal authority arises from an early recognition of tribes as "distinct political communities, having territorial boundaries,

within which their authority is exclusive," and they are subject to no state authority.[17]

4. *Tribes hold a unique legal status, and the federal government has a trust responsibility to steward tribal interests.* The unique position of tribes, arising from their prior occupation and possession of territory, coupled with the inapplicability of Anglo-American legal and constitutional categories to their existence, has given rise to distinctive approaches to statutory construction involving Native Americans, as well as to federal fiduciary and trust obligations. The statutory canons of treaty and statutory construction applicable to Indian law are more solicitous of tribal interests and "are rooted in the unique trust relationship between the United States and the Indians."[18] The concomitant fiduciary obligations interpose federal authority between the tribes and the states and provide a legally enforceable standard for federal action.[19]

5. *An individual state continues to have the authority to regulate off-reservation usufructuary activities for health, safety, and conservation purposes provided it does not discriminate against tribal harvesters.*[20] The general standard for permissible regulation is that the state may neither use its police power in a manner that qualifies the federal treaty rights by subordinating its practice to some other state policy or objective (such as tourism), nor can it regulate treaty rights in order to promote the welfare of the state's other citizens.[21]

   State authority is grounded in the state's sovereignty over its natural resources and people and the sovereign prerogative this inherent authority represents within the US constitutional structure. Thus, while the courts have narrowly defined the states' ability to regulate off-reservation hunting, fishing, and gathering activities, the courts have continued to hold that the state has legitimate interests in the area. They continue to apply structural considerations, the inherent right of states to control their natural resources under the public trust doctrine, and in some instances the equal footing doctrine to federal actions that may affect the state's authority.[22]

   Regulations are permissible when they are nondiscriminatory against tribal harvesters and when they are reasonably necessary to prevent or ameliorate a "substantial risk" to the public health, safety, or the environment.[23] In order to meet this test, the state first must show it needs to regulate a particular resource because there is a public health, safety, or conservation need. This requires the state to demonstrate that a substantial detriment or hazard to the environment, public health, or safety exists or is imminent.[24] The state must then demonstrate that the proposed regulation is necessary to prevent or improve a particular hazard.[25] Finally, the state must show that the regulation is the least restrictive alternative available to accomplish its purposes.[26]

6. *The existence of the treaty right and the extent of state regulation are dependent on the particular treaty language by which the off-reservation rights have been reserved.* The treaty text is understood in light of subsequent legal and developmental changes that may impact how and where the rights are exercised, whether the rights are exclusive, and how much of the resource is allocated for Indian harvest.[27] However, these rights are generally not exclusive and are shared in common with all citizens of the United States. They are also limited in other ways. For example, in the 1855 Treaty of Point No Point, the Klamath Tribe's hunting and gathering of roots and berries were limited to "open and unclaimed lands," language that the Supreme Court interpreted as ensuring "no exclusivity would be possible on lands open to non-Indians."[28] In addition, the 1855 treaty resource allocation is limited by the circumscribed territory and shared usages and the subsistence livelihood carried on by the Tribe in 1855.[29] Due to this treaty language, the tribes are limited to an allocation of 50% of the harvest, and the area of available land for hunting and gathering has been reduced by development and changes in tenure across the ceded territory, limiting the exercise of the right across private land. In contrast, the usufructuary rights in the 1868 treaty at issue in *Herrera v. Wyoming* were limited to territory not "occupied" and as such were extinguished as the land was developed and occupied since 1868.[30]

THE PROTECTION OF HUNTING, FISHING, AND GATHERING RIGHTS

Despite treaty assurances and clear legal authority, the legal recognition of Aboriginal and treaty rights outside of reservation boundaries has been a relatively recent phenomenon and has included contentious political and doctrinal disputes involving the extent to which the rights continue to exist and the extent to which the state can regulate the rights. The historic inability of tribal members to exercise their usufructuary rights outside the reservation (as well as within the exterior boundaries of the reservation through the early part of the 20th century) has been due to the vigorous extension of state jurisdiction based on the public trust and equal footing doctrines, as well as state conservation efforts to protect resources for non-Indian exploitation. State efforts were often given currency by the federal government's indifference toward its fiduciary and treaty obligations or the federal government's implementation of destructive assimilation policies.

Implied Right of Environmental Protection

The review of these causes of action demonstrates a logic that emphasizes the notion that treaty negotiations were done by real persons with real intentions and requirements, which underpin the treaty text. Too often, despite the protection canons of treaty interpretation, legal decision-makers impose a European perspective on Indigenous political organization, decision-making, rhetoric, and legal understandings, and they fail to consider Native Americans' agency and the assumptions, values, and aspirations that Indigenous groups brought to treaty negotiations. For example, a major basis for the Winters doctrine, which holds that water rights sufficient to support Native American populations are reserved by the treaties establishing the reservations, is that it is difficult to believe that a tribe would negotiate for or acquiesce in the creation of a reservation if the reserved territory could not sustain the community.[31] (See chapter 6 for more discussion of Indian water rights.)

Similarly, absent some express language in the agreement, there is an implied material term that the parties would make efforts to conserve the reserved flora and fauna in a manner that would ensure future subsistence and other uses by the tribe, even if the parties at the time did not foresee that the resources in question would be under threat.[32] This is evident in the fair share doctrine, which reserved 50% of the fishery to Native American fishers in the Pacific Northwest.[33] While the courts have avoided this issue of an implied right of habitat protection when confronted with it in other circumstances, it is evident that the conceptual basis for avoiding the implication that there is an implied right to habitat protection embedded in the treaty agreement is strained.[34] Thus, efforts that seriously degrade a resource, including land development and clearance, pipelines, mining, and hydrologic works, would fall within a range of activities proscribed by the agreement and a protected treaty right. Put another way, it is unlikely that treaties only reserve an extant right to take whatever flora and fauna remain in the area traditionally used by the tribe. Rather, a tribe's off-reservation rights entitlement likely will give rise to a cause of action for degradation of the reserved flora and fauna that formerly inhabited the territory in which the tribe traditionally carried on its activities.

Nuisance

One area of law that has been used to prevent infringement of off-reservation rights, either directly or by preventing degradation of the environment in which the resources survive, has been nuisance. Nuisance is found in both the common and statutory law across a number of states. There can be either a private nuisance or a public nuisance. As scientific, technological, and factual issues change, nuisance law has generally followed. Thus, nuisance can shift from private to public as circumstances change. The pollution of a stream that only affects riparian owners and their private interests in the stream is a private nuisance, but it becomes a public nuisance when it kills any fish, the ownership of which is enjoyed by the public. Tribes have often resorted to state law nuisance claims against those who are undertaking actions damaging to their rights.

The use of nuisance claims to address impacts within the exterior boundary of the reservation is a matter of proving a substantial inference with the enjoyment and use of reservation property, and to this extent, it is an unremarkable extension of state nuisance law. This is evident in *Skokomish Indian Tribe v. United States* where the tribe brought a nuisance and trespass challenge against the United States, the city of Tacoma, and Tacoma Public Utilities. The tribe alleged harm caused by the Cushman Hydroelectric Project, a city-owned water project, for damage to reservation waters, associated fish, and other fauna.[35] The claim, while dismissed on statute-of-limitations grounds, was clearly cognizable under state nuisance law.

One difficulty that a plaintiff tribe may have in a state law nuisance claim for off-reservation flora and fauna is the conceptualization of flora and fauna as "property" of the tribe. Nuisance claims do not rely on damage to an individual's property, but require interference with a plaintiff's use and enjoyment of their property.[36] The nonrecognition of tribal property interests in flora and fauna and the assertion of state jurisdiction over treaty-reserved off-reservation rights have been justified under the public trust doctrine. The public trust doctrine presumes that state ownership and regulation of wildlife *for the benefit of all state citizens* is a fundamental aspect of state sovereignty.[37] Given the tripartite relationship among the state, tribal, and national governments, a treaty relationship recognizes a corresponding tribal sovereign property interest (treaty servitude) in the reserved flora or fauna, or a conservation easement in the flora or fauna. This interest coexists with public trust notions of property asserted by the state and is not incompatible with state regulatory interests. The source of the treaty servitude is affirmed by the pretreaty territorial sovereignty that tribes exercised across their lands, which was incorporated through the treaty process into US federal law. This property interest provides legal rights against both the state and third parties in the event the resource has been damaged.[38] While nuisance law in general has only a limited foreseeability requirement, it is likely that this servitude or easement may protect only habitat degradation that was foreseeable and covered by any express treaty language. Case law regarding these types of servitude has skirted finding this type of interest, but underlying policy reasons in resource-scarce habitats give traction to these types of argument.

The second nuisance argument is that usufructuary rights, like water rights, are treaty guaranteed property interests or, at minimum, "quasi-property" rights. These interests are analogous to interests in water allocations and uses.[39] The justification for this form of interest is that the federal government has provided constitutional protection against the governmental taking of reserved treaty rights for public purposes without compensation. In addition, the various usufructuary allocation issues under treaties have been dealt with in a manner similar to a "legitimate expectation" estoppel argument by the courts.

Finally, whether or not the rights are characterized as "property," it is evident that a treaty-protected off-reservation usufructuary "use of property" qualifies as a property interest protectable under 42 USC § 1982.[40] Under this statute the Supreme Court has broadly construed what constitutes a property interest in order to effectuate the underlying policies.[41] As the Seventh Circuit Court of Appeals noted in *Lac Courte Oreilles Band of Lake Superior Chippewa Indians v. Voigt*, "treaty-recognized rights of use depend neither on title nor right of permanent occupancy; rather, they are similar to a profit [à] prendre."[42] A profit à prendre, in turn, is "an easement that confers the right to enter and remove timber, minerals, oil, gas, game, or other substances from land in the possession of another," and as such is a property interest cognizable under nuisance law.[43] These property rights have been recognized under state law and can be vindicated through nuisance actions.[44]

Public nuisance as a cause of action has certain advantages over private nuisance in that the plaintiff tribe does not need to show a property interest. Public nuisances are typically broad in scope and can encompass any unreasonable interference with a right common to the general public. A public nuisance, unlike a private nuisance, does not necessarily involve an interference with the use and enjoyment of land nor an invasion of another's interest in the private use and enjoyment of land, but encompasses any unreasonable interference with a right common to the general public. Thus, an action for public nuisance may be brought even though neither the plaintiff nor the defendant acts in the exercise of private property rights.[45]

The authoritative *Restatement (Second) of Torts*, which is often referred to by the courts, sets out factors to be considered in deciding whether an interference is unreasonable as it relates to public nuisance: (1) whether the conduct involves a significant interference with the public health, the public safety, the public peace, the public comfort, or the public convenience; (2) whether the conduct is proscribed by a statute, ordinance, or administrative regulation; or (3) whether the conduct is of a continuing nature or has produced a permanent or long-lasting effect.[46] The plaintiff must show a special injury in order to bring an action in public nuisance; otherwise, such actions are left to public authorities.[47]

The use of public nuisance to protect Indigenous off-reservation interests can be undertaken under state or federal common law. In some states, the claims have narrowed over the past decades for tribal plaintiffs, but there remain opportunities to proceed under this legal theory. In *Hopi Tribe v. Arizona Snowbowl Resort Ltd. Partnership*, the Supreme Court of Arizona held that environmental damage to public land with religious, cultural, or emotional significance to the plaintiff is not a special injury for public nuisance purposes.[48] The majority held that public nuisance and the special injury necessary to bring a public nuisance claim had to involve a property or pecuniary interest. This decision was contrary to other approaches to public nuisance, which have acknowledged that a nuisance may interfere with a variety of nonproprietary rights or interests.[49] Public nuisance law has focused on the invasion of an interest more generally and the type of harm or damage occasioned by the alleged tort-feasor, including such things as noxious odors, aesthetic offensiveness, and a threat to health caused by the discharge of dangerous substances. Moreover, the action has generally adapted to changes in circumstances, including the degradation of natural resources. This is evident from other cases, such as *In re Exxon Valdez*, *Native Village of Kivalina v. ExxonMobil Corp.*, and *Nevada v. United States*.[50]

There is a potential state public nuisance action arising out of the trust doctrine. The trust doctrine holds that the ownership of wildlife is vested in the state, which manages the stock of resources as the trustee for present use and future generations. As such, the state and the larger public have a distinct public interest in wildlife. Given that the state possesses the power to regulate, conserve, and protect wildlife resources as part of its regulatory police power, there exists an opportunity for tribes with treaty-guaranteed off-reservation hunting, fishing, and gathering rights to seek protection of the resource under the public nuisance theory.[51]

In federal common law there are potentially additional opportunities to pursue public nuisance claims. In *Michigan v. United States Army Corps of Engineers*, the court of appeals dismissed a claim against the Army Corps of Engineers for injunctive relief based on public nuisance due to the corps' failure to completely eliminate the risk that Asian carp would invade the Great Lakes through the Chicago ship and sanitary canals. However, the court noted that the basic federal structure creates a need for a federal common law "where there is an overriding federal interest in the need for a uniform rule of decision or where the controversy touches basic interests of federalism."[52] Within this structure, the federal common law of public nuisance can provide remedies to environmental degradation and pollution. In *Michigan* the court agreed with the plaintiff states and the tribe that the invasion of exotic carp into the Great Lakes would constitute a public nuisance. The army corps argued that public nuisance only reaches traditional sources of nuisance. However, the court rejected the corps' contention: "It would be arbitrary to conclude that this type of action extends to the harm caused by industrial pollution but not to the environmental and economic destruction caused by the introduction of an invasive, nonnative organism into a new ecosystem (assuming that the states have correctly forecast the depletion of the Great Lakes fishery and the corresponding damage to the multi-billion-dollar sports fishing industry). *Public nuisance* traditionally has been understood to cover a tremendous range of subjects."[53]

Given the broad scope of the potential subjects covered by public nuisance mentioned by the court, it is arguable that federal common law may become an area of future efforts to protect treaty-reserved usufructuary rights. First, any federal cause of action is generally limited to entities bound up in the federal structure of the US republic. This sovereign aspect of Indian tribes was underscored by the US Supreme Court in the *Inyo County* opinion where the Court determined that the Paiute-Shoshone Indian Tribe "does not qualify as a 'person' who may sue under 42 USC § 1983."[54] As such, tribes are precluded from bringing actions in their sovereign capacity to vindicate their sovereign rights, which include treaty-reserved usufructuary rights under 42 USC § 1983.[55] Second, it is evident that there needs to be a standard rule about any implied restrictions on state actions relating generally to off-reservation usufructuary rights. While the rights are reserved and delimited by the treaty text and implications of the text, once the right is reserved, the permissible scope of state regulation of the right has been and needs to be determined by federal law. From this perspective, a tribe may be able to avail itself of the federal common law of public nuisance for treaty rights.

CONCLUSION

The centralizing and totalizing claims of the unitary US federal and state authority have not sat comfortably with the alternative pluralistic notions of law and authority advocated by Indigenous groups. This is particularly salient in hunting, fishing, and gathering disputes because, for Indigenous groups, these issues are often about asserting their "sovereignty rights at the 'grass roots' level."[56] Although some tribes have been successful in effectuating the recognition and exercise of their treaty rights, there remain significant challenges to the availability of natural resources in the present and the future. These problems are compounded by the poor "fit" that legal remedies (as in nuisance and public nuisance) provide tribes should they wish to take legal action to protect off-reservation flora and fauna and by narrow judicial decision-making. Nevertheless, for Native American tribes to obtain full recognition of their sovereign treaty rights, it will be necessary for the larger society, the state, and the federal judiciary to reevaluate how rights in general are understood and enforced, and how treaty rights fit within the federal constitutional structure.

NOTES

1. US National Archives, American Indian Treaties: Catalog Links, https://www.archives.gov/research/native-americans/treaties/catalog-links. There are additional treaties signed by individual states. New York, for example, entered into 92 treaties, cessions, and deeds of transfer with the various tribes in the state, most recently in 1997. University of Buffalo, Indians and New York State: Treaties by Tribe, https://research.lib.buffalo.edu/indians-and-ny-state/treaties-by-tribe.

2. A "usufruct" is a legal right accorded to a person or party that confers the temporary right to use and derive income or benefit from someone else's property.

3. *Corpus Juris Secundum*, vol. 28A, "Easements" at § 9. *Black's Law Dictionary* defines a usufruct as "[a] right to use another's property for a time without diminishing or damaging it, although the property might naturally deteriorate over time." *Black's Law Dictionary*, 7th ed. (St. Paul, MN: West Group, 1999).

4. Michael C. Blumm, "Native Fishing Rights and Environmental Protection in North America and New Zealand: A Comparative Analysis of Profits à Prendre and Habitat Servitudes," 8 Wis. Int'l L. J. 2 (1989–1990).

5. United States v. Michigan, 653 F.2d 277, 279 (6th Cir.), *cert. denied*, 454 US 1124 (1981).

6. County of Oneida v. Oneida Indian Nation, 470 US 226, at 236 (1985).

7. Mitchel v. United States, 34 US (9 Pet.) 711, at 746 (1835).

8. United States v. Winans, 198 US 371, at 381 (1905).

9. Fletcher v. Peck, 10 US (6 Cranch.) 87 (1810).

10. United States v. Wheeler, 435 US 313, at 322–23 (1978).

11. Montana v. Blackfeet Tribe, 471 US 759, at 764 (1985); Oneida Indian Nation v. County of Oneida, 414 US 661, at 670 (1974).

12. McGirt v. Oklahoma, 140 S. Ct. 2452, at 2462–63 (2020).

13. Menominee Tribe of Indians v. United States, 391 US 404, at 413 (1968).

14. United States v. Dion, 476 US 734, at 740 (1986).

15. Minnesota et al. v. Mille Lacs Band of Chippewa Indians et al., 526 US 172, at 204 (1999).

16. Rice v. Olson, 324 US 786, at 789 (1945).

17. McGirt v. Oklahoma, 140 S. Ct. 2452, at 2476–77 (2020), citing Worcester v. Georgia, 31 US 515 (6 Pet.), at 557 (1832).

18. Oneida County v. Oneida Indian Nation, 470 US 226, at 247 (1985).

19. These Indian tribes are the wards of the nation. They are communities dependent on the United States. Dependent largely for their daily food. Dependent for their political rights. They owe no allegiance to the States, and receive from them no protection. Because of the local ill feeling, the people of the States where they are found are often their deadliest enemies. From their very weakness and helplessness, so largely due to the course of dealing of the Federal Government with them and the treaties in which it has been promised, there arises the duty of protection, and with it the power. This has always been recognized by the Executive and by Congress, and by this court, whenever the question has arisen. (United States v. Kagama, 118 US 375, at 383–84 [1886])

United States v. Mitchell, 463 US 206, at 225 (1983) (citations omitted): "Our construction of these statutes and regulations [relating to federal management of forests owned by the tribe or allottees within the reservation] is reinforced by the undisputed existence of a general trust relationship between the United States and the Indian people. This Court has previously emphasized 'the distinctive obligation of trust incumbent upon the Government in its dealings with these dependent and sometimes exploited people.'"

20. I owe this typology to William C. Canby Jr., *American Indian Law in a Nutshell* (St. Paul, MN: West Group, 1998), 1–3.

21. Lac Courte Oreilles Band of Lake Superior Chippewa Indians v. Voigt, 668 F. Supp. 1233, at 1237–38 (1987); Washington v. Washington State Commercial Passenger Fishing Vessel Assn., 443 US 658 (1979); Antoine v. Washington, 420 US 194 (1975).

22. South Dakota v. Gregg Bourland, 508 US 679 (1993); Montana v. United States, 450 US 544 (1981).

23. Lac Courte Oreilles Band of Lake Superior Chippewa Indians v. Voigt, 668 F. Supp. 1233, at 1241–42 (W.D. Wis. 1987).

24. Minnesota v. Mille Lacs Band of Chippewa Indians, 952 F. Supp. 1362, at 1369 (D. Minn. 1997).

25. Id.

26. Id.

27. Oregon Department of Fish and Wildlife v. Klamath Tribe, 473 US 753 (1985).

28. Article 4, Treaty of Point No Point, 1855, State of Washington, Governor's Office of Indian Affairs https://goia.wa.gov/tribal-government/treaty-point-no-point-1855; Oregon Department of Fish and Wildlife v. Klamath Tribe, at 767.

29. Oregon Department of Fish and Wildlife v. Klamath Tribe, at 767.

30. Herrera v. Wyoming, 139 S. Ct. 1686 (2019).

31. Winters v. United States, 207 US 564 (1908); Arizona v. San Carlos Apache Tribe of Arizona, 463 US 545 (1983).

32. Lac Courte Oreilles Band of Lake Superior Chippewa Indians v. Wisconsin, 653 F. Supp. 1420, at 1426, 1434 (1987). This notion is evident in Judge James Doyle's discussion of the extent of the Chippewas' usufructuary interest. "The effect is this: On lands privately owned, the Chippewa have lost their reserved usufructuary rights. However, if

at a given time the Chippewa can show, in a lawsuit if necessary, that this diminution in their usufructuary rights is preventing them from enjoying a modest living, appropriate measures must be taken for Chippewa activity on privately owned lands to permit the Chippewa to enjoy a modest living" (*id.* at 1432).

33. Washington v. Washington State Commercial Passenger Fishing Vessel Ass'n, 443 US 658 (1979).

34. United States v. Washington, 759 F.2d 1353 (1985).

35. Skokomish Indian Tribe v. United States, 410 F.3d 506 (2004).

36. Quechan Indian Tribe v. United States, 535 F. Supp. 2d 1072 (2008).

37. Because of the necessarily incidental relationship between the ownership and regulation of wildlife and state sovereignty, wildlife regulation was considered beyond the reach of federal power under the 1789 Constitution in the 19th and early 20th centuries. Michael C. Blumm and Lucus Ritchie, "The Pioneer Spirit and the Public Trust: The American Rule of Capture and State Ownership of Wildlife," 35 Environmental Rev. 673 (2005); Christina Wood, "The Tribal Property Right to Wildlife Capital (Part I): Applying Principles of Sovereignty to Protect Imperiled Wildlife Populations," 37 Idaho L. Rev. 1, 62 (2000). The public trust doctrine as a source of authority to regulate wildlife differs from the state's police power in that it places an affirmative duty on the state to protect wildlife for the benefit of future generations. The state must protect the corpus of the wildlife within its boundaries. The state's general police power for the health, safety, and welfare of residents has no similar obligation.

38. Mary Christina Wood, "The Tribal Property Right to Wildlife Capital (Part II): Asserting a Sovereign Servitude to Protect Habitat of Imperiled Species," 25 Vt. L. Rev. 355 (2001); Michael C. Blumm and Brett M. Swift, "The Indian Treaty Piscary Profit and Habitat Protection in the Pacific Northwest: A Property Rights Approach," 69 U. Col. L. Rev. 407 (1998).

39. Morton v. Ruiz, 415 US 199 (1974).

40. Enacted as part of the Civil Rights Act of 1871, 42 USC § 1982 guarantees to all citizens of the United States "the same right . . . as is enjoyed by white citizens thereof to inherit, purchase, lease, sell, hold and convey real and personal property."

41. Memphis v. Greene, 451 US 100 (1981).

42. Lac Courte Oreilles Band of Lake Superior Chippewa Indians v. Voigt, 700 F.2d 341, at 352 (1983); Lac du Flambeau Band of Lake Superior Chippewa Indians v. Stop Treaty Abuse–Wisconsin, Inc., 991 F.2d 1249 (7th Cir. 1993).

43. Restatement (Third) of Property: Servitudes § 1.2(2) (1998); Lobato v. Taylor, 71 P.3d 938 (2002); Douglaston Manor v. Bahrakis, 89 N.Y.2d 472 (1997); Connerty v. Metropolitan Dist. Com., 398 Mass. 140 (1986).

44. The district court in Lac du Flambeau Band of Lake Superior Chippewa Indians v. Stop Treaty Abuse–Wisconsin, Inc. (759 F. Supp. 1339) held that the off-reservation interests reserved by the 1837 and 1842 treaties between the Chippewas and the United States constituted a property interest under Wisconsin law (Wis. Stat. § 840.01). This statute defines property as "estates in, powers . . . over, and all present and future rights to, title to, or interests in real property, including without limitation by enumeration, security interests and liens on land, easements, profits, rights of appointees under land, rights under covenants running with the land, powers of termination and homestead rights; the interest may be such as was formerly designated legal or equitable; the interest may be surface, subsurface, suprasurface, riparian or littoral; but 'interest' does not include interests held only as a member of the public nor does it include licenses."

45. 58 Am. Jur. 2d Nuisances § 31, at 592 (2002).

46. Restatement (Second) of Torts § 821B(2).

47. Eric T. Freyfogle, Michael C. Blumm, and Blake Hudson, *Natural Resources Law: Private Rights and the Public Interest* (Minneapolis, MN: West, 2015), 152.

48. Hopi Tribe v. Ariz. Snowbowl Resort Ltd. P'ship, 245 Ariz. 397 (2018).

49. Richard A. Epstein, "Nuisance Law: Corrective Justice and Its Utilitarian Constraints," 8 J. Legal Stud. 49, at 50 (1979).

50. *In re* Exxon Valdez, 104 F.3d 1196, 1198 (9th Cir. 1997); Weirton Area Water Bd. v. 3M Co., 2020 US Dist. (LEXIS 244141); Native Village of Kivalina v. ExxonMobil Corp., 696 F.3d 849, 855 (9th Cir. 2012); Nevada v. United States, 2019 US Dist. (LEXIS 181403); Miotke v. Spokane, 101 Wn.2d 307 (1984).

51. Matthew Russo, "Note: Productive Public Nuisance: How Private Individuals Can Use Public Nuisance to Achieve Environmental Objectives," 18 U. Ill. L. Rev. 1969, at 1989–90 (1969).

52. Michigan v. United States Army Corps of Engineers, 667 F.3d 765, at 771 (2011). See also Michigan v. United States Army Corps of Engineers, 758 F.3d 892 (2014). The Grand Traverse Band of Ottawa and Chippewa Indians was an intervenor-plaintiff in this suit.

53. Michigan v. United States Army Corps of Engineers, 667 F.3d 765, at 771.

54. Inyo County, California v. Paiute-Shoshone Indians of the Bishop Cmty. of the Bishop Colony, 538 US 701 (2003).

55. "[W]hether a sovereign entity may be considered a 'person' depends on the specific rights that it is asserting." Keweenaw Bay Indian Cmty. v. Rising, 569 F.3d 589, at 595 (6th Cir. 2009). See also Skokomish Indian Tribe v. United States, 410 F.3d 506 (2005).

56. Anthony G. Gulig and Sidney L. Harring, "Symposium: 'An Indian Cannot Get a Morsel of Pork . . .': A Retrospective on Crow Dog, Lone Wolf, Blackbird, Tribal Sovereignty, Indian Land, and Writing Indian Legal History," 38 Tulsa L. Rev. 87, at 102 (2002).

Chapter 9

Tribal Perspectives on the Endangered Species Act

| STEVEN ALBERT *and* RILEY PLUMER

INTRODUCTION

The 574 federally recognized Indian tribes and Alaska Native villages in the United States are recognized as "separate sovereigns pre-existing the Constitution."[1] While tribes have become "domestic dependent nations," they continue to "exercise inherent sovereign authority."[2]

The total land base under tribal jurisdiction in the continental United States is more than 70 million acres, about as much land as is under the jurisdiction of the National Park Service and the US Fish and Wildlife Service combined. Alaska Native corporations control an additional 17 million acres. The largest reservation, the Navajo Nation, covers 14 million acres in three states and is approximately the size of West Virginia—larger than 10 states. Ecosystems across many tribal lands remain largely intact, biologically robust, and with as high or higher rates of biodiversity than surrounding areas—ideal habitat for rare, threatened, or endangered species.

Since passage of the Endangered Species Act (ESA) in 1973, there has been considerable debate over the respective roles of the federal government and tribes concerning its implementation. Many tribes view the federal government's attempts to manage any species on tribal land as an affront to tribal sovereignty.[3] Since the 1990s, however, a new dialogue has occurred between tribes and the US Fish and Wildlife Service that has led to better cooperation between these entities and may ultimately be more effective at protecting endangered species.

TRIBAL SOVEREIGNTY AND TRUST RESPONSIBILITY

The foundation of the federal-tribal relationship is built on several principles that have been articulated through court decisions and the directives of the executive branch. Two of the most important are tribal sovereignty and federal trust responsibility. Like many aspects of the federal-tribal relationship, interactions of the concepts and practices of these precepts are often complex and occasionally contradictory: tribal sovereignty leads toward greater autonomy, while trust responsibility tends toward more federal responsibility.

The inherent sovereignty of Indian tribes, exemplified by the treaty process, has long been recognized by the US government and is explicitly incorporated in the US Constitution, which establishes a sovereign-to-sovereign relationship between the federal government and Indian tribes.[4] While often honored more in the breach than in fact over the past 250 years—numerous atrocities committed by the government and settlers in the 18th and 19th centuries, the parceling out of Indian lands under the Dawes Act in the late 19th and early 20th centuries, and the unilateral termination of many tribes by the US government—this principle remains a fundamental basis of the federal-tribal relationship.[5] The notion that tribes are the source of their own laws relating to their lands and their members has remained intact, and this authority is embedded in the US federal system and has been repeatedly reemphasized. The Nixon administration recognized and reinvigorated tribal sovereignty to achieve many significant beneficial legal and policy outcomes for Native Americans. In 1970, President Richard Nixon stated: "[T]he goal of any new national policy toward the Indian people [must be] to strengthen the Indian's sense of autonomy without threatening th[e] sense of community. We must assure the Indian that he can assume control of his own life without being separated involuntarily from the tribal group. And we must make it clear that Indians can become independent of Federal control without being cut off from Federal concern and Federal support."[6]

In the context of natural resource management, tribal sovereignty has been strengthened with the issuance of several executive branch directives. Tribes and the lands they administer are not necessarily subject to the same public laws that govern other public or private lands in the United States. It has been legally well established that tribes may regulate hunting and fishing on their lands, rights conferred either by treaties or by the fact of Aboriginal occupancy.[7] (See chapters 6, 7, and 8 of this volume for much more on this topic.) However, the courts have also generally ruled that the federal government maintains some power over governmental relations with tribes that affect these rights. For example, Congress has the power to abrogate treaties—including those with tribes—extinguishing any rights that may have been reserved by the tribal negotiators, and it also has the power to make certain categories of laws affecting tribal rights.[8]

The courts have held that Congress must clearly express its intent to extinguish Native American rights guaranteed by treaties.[9] In many instances, on-reservation and off-reservation (on both public and private lands) hunting and fishing rights continue to be used by tribes under the respective treaties.[10] (See chapters 7 and 8 of this volume.) In fact, many tribes have relatively small reservations but large areas in which they maintain off-reservation treaty hunting and fishing rights and have an active stake in management. This concept was affirmed by the Supreme Court in *United States v. Winans*, which upheld that concurrent with the establishment of a reservation, tribes retain all rights that were not explicitly given up.[11]

While recognizing tribal sovereignty, the United States is still legally required to act as a fiduciary, administrator, and caretaker for Indian interests, including the protection of the health, welfare, and land and water resources of Indian people. Tribal land and resources are held in trust by the federal government, a policy known as the government's "trust responsibility." The US Supreme Court has long acknowledged "the undisputed existence of a general trust relationship between the United States and the Indian people," which informs its interpretation of specific statutes.[12] Through this trust relationship, the United States "has charged itself with moral obligations of the highest responsibility and trust," and it "should therefore be judged by the most exacting fiduciary standards."[13]

In managing trust resources, or assisting tribes to do so, the federal government must act for the exclusive benefit of tribes and ensure that tribal trust or other lands are protected and used for the purposes for which they were intended: the physical, economic, social, and spiritual well-being of tribal members.[14] Ironically, this stringent obligation on the federal government to protect tribal interests has also led in practice to some of the most restrictive regulations on tribes concerning development, a situation that often leads to conflicts with the practice of tribal sovereignty.[15]

EXECUTIVE BRANCH DIRECTIVES

In matters of natural resource or wildlife law, several executive branch administrative directives bear directly on the relationship of the US Fish and Wildlife Service and other Interior Department agencies to tribes:

- *Secretarial Order 3175 (1993)* and *Interior Departmental Manual (512 DM 2)*. These two directives require all Interior Department agencies to identify potential effects from their activities on Indian trust resources and to have meaningful consultation with tribes when department activities directly or indirectly affect tribal resources.

- *Presidential Memorandum : Government-to-Government Relations with Native American Tribal Governments (29 April 1994)*. This memorandum reminds all executive branch departments and agencies of the government-to-government relationship between tribes and the United States, requiring them to consult with tribal governments to "the greatest extent practicable" prior to taking actions that affect tribal governments and to assess the impact of federal activities on tribal trust resources.

- *Native American Policy of the US Fish and Wildlife Service (1994)*. This policy reiterates the government-to-government relationship, establishes a framework for joint projects and formal agreements, directs the service to assist tribes in identifying federal and nonfederal funding sources for wildlife management activities, and provides a framework for the service to give technical assistance to tribes, when requested.

- *Secretarial Order 3206: American Indian Tribal Rights, Federal-Tribal Trust Responsibilities, and the Endangered Species Act (1997).* This order reacknowledges the trust and treaty responsibilities of the US government; instructs federal agencies to be sensitive to Indian cultures, religions, and spirituality, recognizing that the basis for these often relies on the use of natural resources; and instructs the service to support tribal measures that preclude the need for federal laws governing the conservation of natural resources. The order also reminds agencies in the Interior Department that Indian lands are not subject to the same controls or regulations as federal public lands. Perhaps most important, it reaffirms that tribes themselves are the appropriate governmental entities to manage their lands. The order strives to harmonize tribal concerns about the ESA with federal mandates to enforce it and makes allowances for tribes to develop their own conservation plans for listed species. If a tribe develops a viable conservation plan, the service is directed to defer to it. Further, because listing tribal lands as critical habitat can greatly impede tribal economic development, the order provides that the federal government will not designate critical habitat on tribal lands unless doing so is essential to conserve a listed species. The order also requires that tribes do not bear a disproportionate burden of conserving endangered species on tribal land.

- *Executive Order No. 13084: Consultation and Coordination with Indian Tribal Governments (1998).* This order instructs all executive branch agencies to establish a process whereby elected officials and other representatives of Indian tribal governments can provide meaningful and timely input in the development of regulatory policies on matters that significantly or uniquely affect their communities. It also instructs agencies to consider any application by a tribal government for a waiver of statutory or regulatory requirements with a general view of increasing opportunities for flexible policy approaches. This opportunity for administrative flexibility has the potential to play a key role in how the service implements endangered species recovery on tribal land.

- *Native American Policy of the US Fish and Wildlife Service (2016).*[16] The purpose of this policy is to "carry out the United States' trust responsibility to Indian tribes by establishing a framework on which to base [the service's] continued interactions with federally-recognized tribes and Alaska Native corporations." The policy recognizes the sovereignty of federally recognized tribes; states that the service will work on a government-to-government basis with tribal governments; and includes guidance on co-management, access to and use of cultural resources, capacity development, law enforcement, and education.

UNIQUE STATUS OF TRIBES WITH REGARD TO NATURAL RESOURCE MANAGEMENT AND THE ESA

In 1973, Congress passed the Endangered Species Act with the broad purpose of protecting endangered and threatened species.[17] Under the act, virtually all dealings with endangered species, including taking, possession, transportation, and sale, were prohibited, except in extremely narrow circumstances.[18] The ESA's implementing regulations provide that "[e]ach Federal agency shall review its actions at the earliest possible time to determine whether any action may affect listed species or critical habitat[s]."[19] The Ninth Circuit has elaborated:

"The purpose of consultation is to obtain the expert opinion of wildlife agencies to determine whether the action is likely to jeopardize a listed species or adversely modify its critical habitat and if so, to identify reasonable and prudent alternatives that will avoid the action's unfavorable impacts. The consultation requirement reflects a conscious decision by Congress to give Endangered species priority over the 'primary missions' of federal agencies."[20]

Courts have interpreted "agency action" triggering the section 7 consultation requirement to mean that such consultation is necessary as long as the federal action agency has "'some discretion' to take action for the benefit of a protected species."[21] Though the intent of these regulations is to protect tribal resources, a side effect can be an excessive bureaucracy that slows even the most benign projects.

Pervasive federal influence has made many tribes wary of the potential impacts of the ESA, since most feel that they have been excellent land stewards and consequently have many rare and endangered species on their lands. Most tribes prefer to keep large areas of their lands intact and free from development so tribal lands in general have a higher proportion of rangelands, forests, or de facto wilderness and thus higher biodiversity than surrounding lands.[22] In practice, this sometimes means that tribal lands have the potential to act as a safe haven for species that are driven off surrounding private land as it gets developed. Tribes feel penalized for this good stewardship when restrictions are placed on development activities that they view as their right. While most tribes cherish and rely on their protected and undeveloped lands, they do not want to be restricted from pursuing a level of economic development that other communities across the country would take for granted.

Perhaps the most noteworthy example of federal legal restrictions impacting tribal uses that were previously undertaken in a sustainable manner is the decades-long dispute over the congressional prohibition of the hunting of eagles and the use of eagle parts under the Bald and Golden Eagle Protection Act of 1940.[23] Though tribes have taken relatively small numbers of eagles for ceremonial uses for millennia, it was only when the bald eagle (*Haliaeetus leucocephalus*) became endangered from non-Indian activities, especially the widespread use of the pesticide DDT, that it became a conflict.

Does the service have the authority to enforce the ESA on tribal land? There have been few cases specifically testing this question. The ESA does not mention tribes, and some court cases have upheld the concept that unless tribal treaty and other rights are specifically abrogated by an act of Congress or other legislation, they remain in force.[24] In the case that came the closest to testing this question of the enforceability of the ESA when it conflicts with ceremonial use, *United States v. Dion*, Dwight Dion Sr., a member of the Yankton Sioux Tribe, was convicted of taking of a bald eagle.[25] The statute under which the case was prosecuted, however, was the Bald and Golden Eagle Protection Act, not the ESA. The circuit court ruled that the eagle was taken for commercial use and was thus prohibited; the ESA question was left unanswered.

In 2019, in *United States v. Turtle*, a federal district court of Florida declined to dismiss an indictment against a member of the Seminole Tribe for selling American alligator eggs in violation of the Lacey Act, which prohibits trade in wildlife that has been illegally taken, possessed, transported, or sold.[26] While the court concluded that the ESA did not abrogate the usufructuary rights of a tribal member to sell American alligator eggs gathered from a reservation, it found that the government may "enforce reasonable and necessary conservation measures against members of the Seminole Tribe," and the Department of the Interior has adopted a rule including such a measure.[27]

Interestingly, not all tribes have fought against applying the ESA on their lands. In many cases, the distinction between trust land and off-reservation land is key. Tribes generally want to maintain all rights to manage their own trust land, including decisions about endangered species, while for off-reservation hunting and fishing rights, tribes fight for the protection of resources over which they have less day-to-day control and which have benefited from implementation of the ESA. This is especially true in regard to protection of fisheries. In the Pacific Northwest and in the Great Lakes states, for example, off-reservation treaty fishing rights for salmon and other species are often protected by mandatory conservation measures that are backed with the strong arm of the ESA.

Conflict has also arisen over administrative aspects of the ESA. When a species has been declared endangered, the law mandates the development of a recovery plan. Previous recovery efforts have not always utilized tribal input to the greatest extent possible. While some tribes have been included at the level of stakeholders or interested parties, their participation, comments, or suggestions were accorded only the same weight as, for example, large private landowners in the region. The Tulalip Tribes of the Northwest have charged that they were largely ignored by the US Fish and Wildlife Service during the section 7 consultation of a major habitat conservation plan that directly affected them. Several tribes in the Southwest were shocked to find that critical habitat for the Mexican spotted owl (*Strix occidentalis lucida*) had been designated on tribal land without prior consultation. Critical habitat for the Rio Grande silvery minnow (*Hybognathus amarus*) was declared on Pueblo land in New Mexico over the objections of tribal leaders.

Species recovery efforts can also greatly affect tribal water rights, particularly in the desert Southwest, where water is scarce and the number of users is growing rapidly. Tribes along the Rio Grande are already involved in issues surrounding the Rio Grande silvery minnow and the southwestern willow flycatcher (*Empidonax traillii extimus*), a riparian-dependent migratory bird. While tribes are supportive of protection for these species, they are also wary of shouldering a large share of the burden of recovery for fear that this will lead to restrictions on their use of water. (See chapter 6 for much more about tribal water rights.)

In nearly all instances, Indian water rights are senior to those of all other users, generally dating back at least to the date of the establishment of a tribe's reservation. If a tribe's occupancy of an area predates the establishment of the reservation, tribes may exert a "time immemorial" water claim. Tribal water rights are generally referred to as "federal reserved" or "Indian reserved" water rights, meaning that when Indian reservations were created, although water rights were not always specifically addressed, federal law recognizes that it was clearly the intent to include them. Any establishment of an Indian reservation without rights to water would have been ridiculously unfair. This concept is referred to as the "Winters doctrine" and is one of the cornerstones of Indian water law.[28] This doctrine has been affirmed to apply to both surface water and groundwater.[29]

In many cases, the water rights in a basin or a river have not yet been adjudicated, but water development has gone on apace. When the water rights are finally determined, it's likely that tribes will have rights senior to all other users. The implications for the continued use of water by non-Indian parties in many instances are profound. Even in cases where the water rights have been adjudicated, tribes have not always made full use of their water rights—yet. These rights are not subject to forfeiture due to nonuse and retain their senior priority; they may be exercised in the future. This could become problematic when, for example, a watercourse is already fully appropriated and further water use has been deemed to jeopardize a listed species. In the tribes' view, the ESA should not—and does not—reprioritize tribal water rights, and the suggestion that tribes should be precluded from exercising their reserved water rights in order to protect an endangered species ignores the causes of habitat deterioration that led to the species' decline.

Possibilities for Federal-Tribal Cooperation

The service and many tribes have expressed a willingness to work collaboratively on ESA issues. Doors have opened for tribal participation on a broader level among many federal agencies, and tribes have rapidly developed sophisticated management programs (as many of the chapters in this volume attest).[30]

- On Washington's Olympic Peninsula, the Quinault Nation and the service have worked to protect the marbled murrelet, northern spotted owl, snowy plover, and bull trout. Citing Secretarial Order 3206, they have developed management plans that include a mandate to uphold Quinault's treaty rights, a commitment to protect Indian wildlife assets on federal land, and a recognition that tribal priorities take precedence over those of the general public. Conservation actions have included the creation of more than 4,000 acres of easements.

- The Navajo Nation has its own endangered species list, largely but not entirely overlapping with the federal list, and manages additional species that the Nation asserts are likely to be in jeopardy in the foreseeable future. Development such as coal leasing is a significant source of jobs and revenue for the Nation, but it has taken the lead in balancing these elements. Resource development must comply with tribal procedural and environmental requirements, and those seeking to develop tribal land must undergo multistep biological evaluations done by the Navajo Nation, including an assessment of potential effects.[31]

- The Lower Brule and Cheyenne River Sioux wildlife management programs were instrumental in the recovery of the black-footed ferret (see chapter 21), a species that was thought to be extinct before it was rediscovered on tribal land.

- The Nez Perce Tribe has taken an active lead in the recovery of the Rocky Mountain wolf (*Canis lupus*).

- The White Mountain Apache Tribe is involved in efforts to recover the Apache trout and the Mexican wolf. Its wolf program was the recipient of the service's 2018 Recovery Champion Award for their outstanding management.

- The Pyramid Lake Paiute Tribe has made significant contributions to the recovery and restoration of the Lahontan cutthroat trout (*Oncorhynchus clarkii henshawi*) and the cui-ui (*Chasmistes cujus*).

Overall, the US Fish and Wildlife Service and Indigenous peoples are currently involved in over 100 service-tribal partnerships to restore and recover endangered and threatened species (S. Rinkevich, pers. comm. 2022). Central to this approach is the service's use of some of its "administrative flexibility" (a phrase taken from some of the executive branch directives) to work with tribes to develop mutually satisfactory solutions to previously seemingly intransigent issues. In 2021, the Wisconsin Natural Resources Board had to reduce the number of allocated permits for hunting the recovered gray wolf when tribes exercised their rights to 50% of the harvest in their ceded territory, approximately the northern one-third of the state.[32]

In summary, while the history of federal-tribal interaction over the ESA has been contentious and litigious, optimism exists over the possibilities for reshaping the relationship in an extralegal framework. Flexibility and openness on both sides are key to the success of this approach. The goal of the recovery process, of course, is not only higher populations of a particular species, but secure and improved habitat. Tribes are not advocating abandoning the ESA or nonparticipation in the recovery process. Instead, they are insisting on the flexibility to be able to perform these functions in a manner consistent with tribal goals and tribal sovereignty, and they are pressing for funding to carry out these mandates, consistent with the federal government's trust responsibility.

ACKNOWLEDGMENTS

Portions of this chapter were originally published by the *Buffalo Environmental Law Journal* and are reprinted here with permission. Research for this paper was by the Zuni Tribal Council, the Zuni Fish and Wildlife Department, supported and the Zuni Heritage and Historic Preservation Office. The manuscript benefited greatly from comments and input from Jim Cooney, Sylvia Cates, Charles Wilkinson, Norman Jojola, Bruce Finke, Les Ramirez, David Mikesic, John Nystedt, and Stuart Leon.

NOTES

1. Indian Entities Recognized and Eligible to Receive Services from the United States Bureau of Indian Affairs, 86 Fed. Reg. 7554 (29 January 2021); Michigan v. Bay Mills Indian Cmty., 572 US 782, 788 (2014) (quoting Santa Clara Pueblo v. Martinez, 436 US 49, 58 [1978]).

2. Michigan v. Bay Mills Indian Cmty., at 788 (quoting Okla. Tax Comm'n v. Citizen Band Potawatomi Tribe of Okla., 498 US 505, 509 [1991]) (quoting in turn Cherokee Nation v. Georgia, 30 US [5 Pet.] 1, 17 [1831]).

3. Ronnie Lupe, Chairman of the White Mountain Apache Tribe, Address at the 20th Annual National Indian Timber Symposium I (13–17 May 1996) (on file with author).

4. US Const., Art. I, § 8 (authorizing Congress to regulate commerce "with foreign Nations, and among the several States, and with Indian Tribes"). See also Cherokee Nation v. Georgia, 30 US (5 Pet.) 1, 18 (1831) ("The objects, to which the power of regulating commerce might be directed, are divided into three distinct class—foreign nations, the several states, and Indian tribes. When forming this article, the convention considered them as entirely distinct"); Cotton Petroleum v. New Mexico, 490 US 163, 191–92 (1989) ("[T]he Commerce Clause draws a clear distinction between 'States' and 'Indian Tribes.' ").

5. Brian Czech, *American Indians and Wildlife Conservation*, 23 Wildlife Society Bulletin 568–73 (1995). The Dawes Act had two primary components: (1) converting trust land belonging to Indian tribes into fee land and allotting the land to individual Indians, and (2) opening of "surplus" land on Indian reservations to non-Indians. See Michigan v. Bay Mills Indian Ctmy., 572 US 782, 811 (2014) (Sotomayor, J., concurring). The selling of lands to non-Indians was part of the policy of forced assimilation of Indigenous people. Nell Jessup Newton, ed., Cohen's Handbook of Federal Indian Law, § 1.04, at 75 (2012) ("Under the federal allotment and assimilation policy, the Native American was to become another lost race in the American melting pot. He was to abandon his nomadic lifestyle, leave tribal land, own his own farm, and compete for material goods."). Due to the Dawes Act, "there are millions of acres of non-Indian fee land located within the contiguous borders of Indian tribes" (Plains Commerce Bank v. Long Family Land and Cattle Co., 554 US 316, 328 [2008]). From 1954 to 1968, Congress terminated the federal recognition of over 100 Indian tribes. Charles F. Wilkinson and Eric R. Biggs, *The Evolution of the Termination Policy*, 3 Am. Indian L. Rev. 139, 151 (1977).

6. Richard Nixon, Special Message to the Congress on Indian Affairs, 8 July 1970, https://www.ncai.org/attachments/Consultation_IJaOfGZqlYSuxpPUqoSSWIaNTkE JEPXxKLzLcaOikifwWhGOLSA_12%20Nixon%20Self%20Determination%20Policy.pdf.

7. See Menominee Tribe of Indians v. United States, 391 US 404 (1968); New Mexico v. Mescalero Apache Tribe, 462 US 324 (1983); United States v. Adair, 723 F.2d 1394 (9th

Cir. 1983), *cert denied*, 467 US 1252 (1984); Washington v. Wash. State Commercial Passenger Fishing Vessel Ass'n., 443 US 658 (1979).

8. See, e.g., United States v. Dion, 436 US 734, 738 (1986) ("Indians enjoy exclusive treaty rights to hunt and fish on lands reserved to them, unless such rights were clearly relinquished by treaty or have been modified by Congress.").

9. Menominee Tribe v. United States, 391 US 404, 413 (1968).

10. Herrera v. Wyoming, 139 S. Ct. 1686, 1696 (2019) ("If Congress seeks to abrogate treaty rights, 'it must clearly express its intent to do so.'") (quoting Minnesota v. Mille Lacs Band of Chippewa Indians, 526 US 172, 202 [1999]).

11. United States v. Winans, 198 US 371 (1905); Bruce Davies, Treaty Rights and the Endangered Species Act (n.d.) (unpublished paper on file with authors).

12. United States v. Mitchell, 463 US 206, 225 (1983).

13. Seminole Nation v. United States, 316 US 286, 297 (1942).

14. United States v. White Mountain Apache Tribe, 537 US 465, 475 (2003) ("Elementary trust law, after all, confirms the commonsense assumption that a fiduciary actually administering trust property may not allow it to fall into ruin on his watch.").

15. Gary Morishima, *Indian Tribes and Endangered Species*, in Proceedings of the 21st National Indian Timber Symposium, 1–6 June 1997 (on file with author).

16. The service's 2016 policy replaced the 1994 Native American policy.

17. 16 USC § 1531(b).

18. *Id*.

19. 50 CFR § 402.14(a).

20. Nat. Res. Def. Council v. Jewell, 749 F.3d 776, 779 (9th Cir. 2014) (citations omitted).

21. See *id*. In Karuk Tribe of California v. United States Forest Service, 681 F.3d 1006 (9th Cir. 2012), the court determined whether the Forest Service's approval of notices of intent to conduct mining activities in the Klamath National Forest was "agency action" that "may affect" threatened coho salmon and thus should trigger consultation. The court held that authorization of a notice of intent did not constitute an agency action sufficient to trigger an ESA assessment. *Id*. at 1017.

22. C. Corrigan, H. Bingham, Y. Shia, E. Lewisa, A. Chauvenet, and N. Kingston, Quantifying the Contribution to Biodiversity Conservation of Protected Areas Governed by Indigenous Peoples and Local Communities, Biological Conservation 227 (2018):403–12; R. Schuster, R. R. Germain, J. R. Bennett, N. J. Reo, D. L. Secord, and P. Arcese, Biodiversity on Indigenous Lands Equals That in Protected Areas, BioRxIV:10.1101/321935.

23. 16 USC §§ 668-668c.

24. Sylvia Cates, Endangered Species and Tribal Lands: Approaches to Addressing Endangered Species Issues on Indian Lands and Meeting Tribal Resource Management Priorities (1999) (unpublished paper on file with authors); Charles Wilkinson, Symposium: The Role of Bilateralism in Fulfilling the Federal-Tribal Relationship: The Tribal Rights–Endangered Species Secretarial Order, 72 Wash. L. Rev. 1063–1107 (1997).

25. 476 US 734 (1986).

26. 365 F. Supp. 3d 1242 (M.D. Fla. 2019).

27. *Id*. at 1247–50.

28. Winters v. United States, 207 US 564 (1908).

29. *In re* General Adjudication of All Rights to Use Water in the Gila River System and Source, 989 P.2d 739 (1999), *cert. denied*, 120 S. Ct. 2705 (2000).

30. D. Kraniak, *Conserving Endangered Species in Indian Country: The Success and Struggles of Joint Secretarial Order 3206 Nineteen Years On*, 26 Colo. Natural Resources, Energy and Environmental L. Rev. 332–41 (2015).

31. *Id*.

32. https://www.jsonline.com/story/sports/outdoors/2021/02/22/state-licensed-hunters-can-kill-119-wolves-wisconsin-february-hunt/4543149001/.

Chapter 10

Sinixt Hunting
A Test of Tribal Sovereignty

E. RICHARD HART *and* CODY DESAUTEL

THE SINIXT PEOPLE

The Sinixt (Sn̓ʕaýckstx, pronounced *sin-eeeye-ch-kiss-ta* and said by some to mean "people of the place of the bull trout"), or the Lakes Tribe, lived along the Columbia River in what is now northeastern Washington state and southeastern British Columbia. Their land ranged from Kettle Falls in the southern portion of their territory to the "big bend" in the north, which is north of Revelstoke, BC. The present-day Sinixt is a constituent tribe within the Confederated Tribes of the Colville Reservation. The Colville confederacy consists of 12 distinct tribes, whose combined traditional territories encompass approximately 39 million acres, extending from northeastern Oregon, across much of eastern Washington state, and into southeastern British Columbia.

While descendants of the Sinixt were documented as living on both sides of the US-Canada border at the time, the Canadian federal government nonetheless declared them extinct in 1956. To correct this historic wrong, the Confederated Tribes of the Colville Reservation brought forward a test hunting case aimed at reestablishing recognition of the Sinixt as an "Aboriginal People of Canada" under the Canadian Constitution. The Sinixt have been found to have a constitutionally protected right to hunt by British Columbia courts (*Regina v. Richard Lee Desautel*), including a decision from the Supreme Court of Canada.

There were many legal issues at question in that case that are unique to tribal hunting rights and management. One of the most prominent was, could a US-based tribe hold Aboriginal hunting rights in Canada as an "Aboriginal People of Canada" under the Canadian Constitution? This raises another question: does a tribe lose Aboriginal rights on one side of an international border regardless of its historical use and occupation of that territory dating back thousands of years? Now that the lower court decision has been affirmed by the Supreme Court of Canada, how can tribal co-management and use be implemented?

To answer those questions, two important aspects of the Sinixt history and present-day tribal government must be understood. First, who were the Sinixt people, where did they come from, what aspects of their lives defined their culture, and how did they come to exist on the Colville Reservation today? Second, how can the present-day tribal government and management direction, which honors the culture and traditions of the Sinixt people, be incorporated into the co-management of wildlife in their traditional territory in British Columbia?

The history of the Sinixt people has been both passed down through oral traditions and documented by anthropologists and other scholars. That history is similar to those of many other tribes across North America, where movements around their traditional territory were influenced by early settlers and disease. The 1846 Treaty of Oregon established the international border, and both Canadian and US Indian policy was geared toward removal and assimilation. A review of that history below demonstrates the importance of hunting to both the culture and subsistence of the Sinixt people.

Archaeological investigations, anthropological studies, and historical documentary accounts demonstrate that hunting was very important to the Sinixt when they first came into contact with Europeans (1811–1846). Because of the nature of their territory, game was especially important to their subsistence, and big game played a central role in their survival, especially after the Tribe's depopulation as a result of European diseases.

A HISTORY OF SINIXT HUNTING

The unique habitat of the Sinixt contributed greatly to the importance of game to the people. The portion of the country utilized by the Sinixt, located along the Arrow Lakes and the Columbia River, required specialized summer and winter lodges, which were different even from those of closely related Indigenous communities. Various types of seasonal lodges and shelters allowed the Sinixt to travel throughout their territory with the seasons, pursuing game and other traditional subsistence foods as they became available.[1]

Because of the Tribe's location, hunting was of more importance to the Sinixt than to neighboring Indigenous communities. Different game was found in different locations of their territory during the changing seasons.[2] As anthropologists Dorothy Kennedy and Randy Bouchard pointed out, "Culturally, the Lakes [Sinixt] Indians differed in several respects from their southern neighbours, the Colville [Skoyelpi] . . . , adopting instead lifeways more suitable to their Arrow Lakes/Slocan Lake habitat. The Lakes people were more mobile, were canoe oriented rather than horse or foot oriented, placed a greater emphasis on hunting than fishing or root-gathering, and formed smaller village groups."[3]

James Teit observed: "The Lake tribe hunted deer, caribou, goat, and bear. Deer were not so plentiful as in the territories of the other tribes. Elk and sheep were very scarce, and antelope were not found. On the other hand, caribou, goat, and bear were more plentiful than in the countries of the other tribes. Moose, it seems, were occasionally met with."[4]

Verne F. Ray also described Sinixt hunting. He said that deer was "the most important game animal":

> Of secondary importance, but extensively utilized, were the caribou, elk, moose and the brown bear. Grizzly bears were also frequently taken as game animals. The antelope was not found in this region. Both mountain goat and mountain sheep were numerous in the mountainous northern regions and were taken in considerable numbers for food purposes and for their hides and horns. Beaver were plentiful in the numerous streams and were taken in considerable numbers. Rabbits were also of economic importance.... The principal areas for deer hunting were the hilly regions between the Columbia River and Pend Oreille River in the southern part of Lakes territory, and the extensive and more mountainous areas of the headwaters of the Kettle River. Elk were quite scarce but sometimes wandered into Lakes territory from the more frequented areas of northern Idaho and northwestern Montana. Bears, black and brown, were found sporadically distributed throughout the territory but were hunted particularly in the regions of lakes and in the mountain meadows where food was to be found, particularly berries. The grizzly bear was found in the mountain regions above the lakes; caribou in the plains around the lakes. Mountain sheep and mountain goat frequented all of the craggy areas of the mountains.[5]

Teit described Sinixt manufacture of woven rabbit skin blankets and goat hair blankets woven on a loom.[6]

Sinixt hunted game in communal hunts with the use of drive lanes, and they regularly hunted near water or ice in winter. Sinixt hunters used highly trained dogs to hunt, and women sometimes joined the men on communal hunting expeditions. Winter hunting trips were usually composed of from three to six hunters, and in the autumn during big game migration, snares were placed on the migration routes. Lone individuals, especially after the onset of European diseases, also hunted big game.[7]

Ray provided detailed descriptions of the communal methods used by the Sinixt for hunting game. In autumn, Sinixt hunters constructed drive lanes and drove deer over cliffs, harvesting the meat at the base of the cliffs. Groups of hunters also drove deer into the water or onto ice in winter. Sinixt also used dogs to hunt, sometimes driving deer into a lake or river where a Sinixt hunter could dispatch the deer using bow and arrow.[8]

Lone individuals hunted for bear. Ray said, "Bear meat was prized above that of the deer and the hides were more highly valued, since they made excellent robes." A different type of snare was used to capture bear. When a bear was hibernating, hunters would sometimes crawl into the den to dispatch it. Beaver were snared or taken with a deadfall. They were not taken with club or spear, nor were their dams destroyed. Gaff hooks and nets were also unknown. The musk was utilized as a perfume.[9]

Bouchard and Kennedy reported that "Caribou and the occasional elk and moose were stalked by individual Sinixt hunters."[10] Sinixt hunters had an unusual method of hunting elk.

Elk and moose, but not deer or rabbits, were called with a tubular whistle made from the stem of the elderberry or rhubarb plant. The same animals were called by a whistle made of a leaf.[11]

The Sinixt hunted for all the big game that was found in their territory. Anthropologist Verne F. Ray concluded, "there was practically no part of the territory that was not deer and elk hunting territory."[12] Archaeologist Gordon W. Mohs reported that the Sinixt, who had lived in their Aboriginal territory for 2,000 to 3,000 years, hunted elk both individually and communally, sharing the meat throughout the community.[13] Other anthropologists also commented on the importance to the Sinixt of elk at the time of Aboriginal subsistence practices.[14]

The traders at the Hudson's Bay Company (HBC) post at Fort Colvile, where the Lakes (Sinixt) people traded after 1825, prized the elk, among other fur-bearing animals, both for the fur and for the meat.[15] The same traders at Fort Colvile had special use for the elk skins, for which they traded with the Sinixt, sometimes using them to bale furs in order to properly ship them.[16] In 1830, HBC official John Work reported on the animals that the Sinixt hunted and the territories in which they were found:

> The Lake Indians inhabited the Columbia from or above the Athabasca portage to the White goat river or little Dalls [Dalles] not far above Kettle falls, and the small streams that fall into it. This is generally a rugged hilly country, covered with woods. Black tail & common long tail chivexeau [mule deer], Rein deer, White mountain goats & some Elk with several smaller animals are found here. The river in the summer season abounds with salmon & trout, Sturgeon and other fish of a smaller size are also found in it but the latter are little sought after by the Natives. Beaver are still pretty numerous in this part of the country.[17]

Table 10.1, taken from John Work, shows what trade was concluded with the Sinixt in just the short period between 1827 and 1828, but it nevertheless gives a good indication of the extent of Sinixt hunting in the early 19th century.[18] Hudson's Bay Company traders at Fort Colvile regarded the Sinixt as the best hunters in the region. Great quantities of animal pelts were traded by the Sinixt to HBC employees at Fort Colvile from the 1820s through the 1840s. As a result of this trade, game was somewhat reduced in Sinixt territory during the period.

The Sinixt had a regular "seasonal round" during which they harvested the foods necessary for their annual subsistence. They gathered berries and other plants, dug roots, hunted, fished, and gathered medicinal plants and necessary nonorganic materials. The Sinixt people knew each plant that they used, whether for food or other uses. They knew when each plant matured to

TABLE 10.1. Sinixt Trade with Fort Colvile in 1827 and 1828

ITEM	1827	1828
Beaver skins	281	336
Beaver tails	8	6
Black bears	5	3
Brown bears	2	5
Castoreum (pairs)[a]	60	108
Chivreu[b]	22	8
Colts	0	1
Ducks	1	0
Elk skins	5	0
Fishers	13	9
Foxes	1	5
Fresh beaver	2	5
Fresh venison (lbs.)	235	94
Grizzly bears	5	1
Leggins (Indian leather)	2	0
Martens	72	157
Minks	13	33
Musquash [muskrat]	76	188
Otters	15	28
Pack saddles	0	1
Reindeer [caribou?]	1	0
Reindeer skins	2	2
Roots and berries (kegs)	0.75	2
Skin pants	1	
Swans	1	0
Wolverines	0	1

[a] This probably refers to pairs of castor sacs from beavers.
[b] A "chevreau" is a kid goat, so this probably refers to a mountain goat.

the stage needed in order for them to be able to use it, and they knew where each of the species was located within their territory. Each year they migrated seasonally in order to acquire plants, roots, berries, fish, and especially game. In *The Geography of Memory: Recovering Stories of a Landscape's First People*, British Columbia author Eileen Delehanty Pearkes reported, "The early 19th century territory of the Sinixt encountered by early explorers has been described as being rich with mammals. The Sinixt followed seasonal rounds, pursuing deer, mountain sheep and mountain goat in spring; deer, sheep, elk and bear in fall; deer and caribou in mid-winter and sheep again in spring."[19]

Each of the activities came at different times of the year as the food resources became available.[20] Fish, especially salmon (Chinook, coho, and sockeye), provided the Sinixt with much of their annual subsistence needs. Each year in June Sinixt would travel down the Arrow Lakes and Columbia River to the important fishery at Kettle Falls where an annual Chinook salmon migration took place, "the second largest salmon fishery on the Columbia River System."[21] Some salmon migrated upstream from Kettle Falls, and most of those fish spawned in Sinixt territory, especially below Bonnington Falls on the Kootenay River and on its tributary the Slocan River.[22]

Hunting was often carried out in association with these other subsistence activities, especially on gathering expeditions. Berries used by the Sinixt included serviceberries, soapberries, and especially huckleberries.[23] Edible roots also were important to the Sinixt people, including tiger lily, mariposa lily, blue camas, yellowbell, and "Indian potatoes."[24] Other aboveground plants used by the Sinixt included balsamroot, cow parsnips, and "Indian celery."[25]

The Sinixt have long had a unique and essential focus on hunting for subsistence as well as cultural and spiritual well-being. Even prior to the time of first contact with Europeans, European diseases caused a dramatic depopulation (perhaps as much as 95%) of the Tribe, which increased the necessity of individuals' hunting activities. First contact led to trade with the Hudson's Bay Company and other fur companies. The Sinixt traded pelts for guns, ammunition, and metal traps, which increased the ability of individuals to hunt and decreased the necessity for communal hunts.

After the Treaty of Oregon in 1846, the Sinixt dealt with both British and US authorities. The importance of Sinixt hunting continued through the end of the 19th century, through the 20th century, and into the 21st century.

CONFLICTS WITH WHITE SETTLERS AND THE IMPACTS TO SINIXT HUNTING

As mentioned above, the Sinixt used highly trained dogs to assist them in hunting deer. Sinixt hunting with dogs was reported on a number of occasions by trappers and traders who came to their territory. Dogs would catch the scent of deer, run them down, and force the deer into one of the lakes or rivers.[26] While trading for skins with the Sinixt in 1823, John Work noted that a small band had "a number of small dogs."[27] In 1825 HBC trapper-trader Samuel Black shot a deer at the edge of a lake, but a Sinixt claimed the deer because he said his dogs had driven it to the water.[28] In 1831 William Kittson complained that deer meat brought into Fort Colvile by the Sinixt was of poor quality because dogs had been used to run the deer down.[29]

In 1859 Paul Kane described Sinixt hunting and provided the following account:

> We fell in with the Indian chief of the lakes, and procured some bear's and deer's meat from him, of which he seemed to possess a plentiful supply. A small species of dog was tied to the bushes near his lodge, to prevent them from hunting on their own accord, and driving away all the deer. The chief told me that when disposed to hunt with them, he had only to

find a fresh deer-track, set his dogs on it and lie down to sleep, as they never fail to find the deer, and turn them back to the place where they had left him lying. We saw some of these dogs, apparently on the track of some deer, full twelve or fifteen miles from the chief's lodge.[30]

Especially after the Tribe's depopulation from European diseases and the depletion of game as a result of the fur companies, the highly trained dogs were especially important to the Sinixt.[31] Yet the Hudson's Bay Company decided to slaughter 90% of the Tribe's dogs at Fort Colvile in 1839 because the HBC's chief trader found them annoying.[32]

Many traditions are maintained today among Sinixt hunters, who continue to use their traditional hunting territory in what are now the province of British Columbia and Washington state. Hunters say they should take only one bullet when deer hunting. A hunter should never wound a deer instead of killing it with one shot. Nor should a hunter's bullet destroy or taint any of the meat on the animal. Traditions are also followed to instruct young hunters in marksmanship and hunting techniques. In the 21st century, the Sinixt continue to hunt in their Aboriginal territory in British Columbia. Sinixt hunting in the portion of their Aboriginal territory now in British Columbia remains integral and essential to Sinixt culture to the present day.

Some traditional activities, however, are no longer possible in the portion of Sinixt territory in British Columbia since the construction of dams along the Columbia River blocked the salmon migrations in Sinixt territory, and reservoirs now inundate many locations previously used by the Sinixt. The establishment of towns and cities has also created obstacles. Nevertheless, in addition to hunting, the Sinixt continue to engage in other important subsistence practices, such as the gathering of huckleberries, the digging of roots, and the collecting of plants for medicinal purposes. Hunting is associated with traditional medicinal practices and other religious, cultural, and spiritual activities.

Considerable testimony from Sinixt tribal members have documented Sinixt hunting in the portion of their Aboriginal territory now in British Columbia. The following accounts are representative of this available documentary material.

In 1954, Sinixt leader Peter Lemery testified before the US Indian Claims Commission. He explained the reasons the Sinixt people had for hunting in the winter: "Deer and elk can be more readily tracked and found and with a method of group hunting . . . they could more effectively carry on this operation of hunting."[33] He continued: "The Lakes area was one in which there was practically no part of the territory that was not deer and elk hunting territory. All of this area is relatively rugged and relatively productive of game, but they were inclined to use the more accessible regions, of course, and when they were hunting deer by the drive method, they would go fairly far up the mountain defiles in order to surround the game and would drive them down so as to have them in more open territory before they actually attempted to dispatch them."[34]

In 1976 author Ruth Lakin interviewed Sinixt elder Mary Marchand, then over 70 years of age. In the interview, Mary Marchand said, "I miss the days when we had so much fish and other animals. . . . Besides the deer, we had elk, bear, cougar, beaver, muskrat, mart[e]n, and rabbits."

Marchand said she had spent "many years" with her grandmother Nancy, Chief Aurapakan's wife, also a Sinixt Indian, and she recalled stories her grandmother told about how the Sinixts lived in their traditional territory. The materials obtained through hunting were necessary to their survival: "Winters were spent around the Falls and near-by Nancy Creek. Winter quarters were snug and warm, nestled in holes about 6 feet deep and covered teepee-style with green peeled bark of cedar. The covering was banked with dirt to keep out drafts. A fire in the center of the dirt floor warmed the dwelling. Smoke escaped through a hole at the top of the teepee. Blankets were made of animal skins and the floor was covered with tule mats. [I] remember as a child seeing many of these holes where underground homes were once located."[35]

Marchand's grandmother told her that animal hides were used to make the winter pit houses: "They dig down, then they make their winter quarters. They put cedar bark, then dirt, then canvas on top of that. When they didn't have canvas, they put hides 'round . . . the upper part."[36]

Lucy Swan, who was living on the Colville Reservation in 1976, reported traveling to Canada to get huckleberries and described the traditional method of tanning hides.[37] She said, "There was always lots of game."[38]

In 1986, Father Pat Conroy, a Jesuit priest associated with St. Mary's Mission on the Colville Reservation, set out to document the history and culture of local Sinixt people by interviewing surviving elders in the region. A decade after Ruth Lakin had interviewed Mary Marchand, Father Conroy also conducted an extensive interview with her at her home in the Sinixt Kelly Hill community, adjacent to Kettle Falls in the southern end of Sinixt Aboriginal territory. Then 81 years old, Marchand described the Sinixt people:

> Lake Indians, the Tribe of the Lake Indians, the boundary was at that island there at Kettle Falls, that island belonged to the Lake Indians. That's where they spent their summer, well year 'round. And about that, well that's the boundary. And they owned that, from there clear up to the end of the Arrowhead Lake. I've seen it, I went around the end of the Arrowhead Lake, I've seen it. I went around the [lake] and I've seen where, it's about a 300-mile lake, that's where the Columbia comes from. . . . And our boundary was at the end of that lake and that's where these Indians here, in the summer they go up there to hunt and pick berries—[of] course there was everything up there, everything plenty. And that's where they get fish, deer, all kinds of meat, like they had sheep and elk, moose, all kinds [of] bear, grizzly, any kind they want and plenty of it.[39]

During the same year, Father Conroy conducted an interview with Eva Orr, another Sinixt elder, who also lived in the Sinixt community at Kelly Hill. Orr reported that her family continued to travel into a portion of Sinixt territory located in British Columbia. She said the family traveled to Nelson and Fernie to get huckleberries every summer: "Like my Dad said, he went up there when his Great-Grandfather was up there, and he was saying how he stayed on the edge of the lake, then he would see the wolves, then he would rush away from them. . . . We asked him if he liked it up there. He said it was a good country, a person would never starve there in them days, because like he said there was a lot of lush [sic], there was deer, elk, all kinds of stuff up there, you know, that they survived on."[40]

William Barr described going up and down the river to the Arrow Lakes from Kettle Falls with his grandmother. He said they hunted and fished as they went. Barr also described the pressures Sinixt people faced in Canada. He said his mother's first husband was shot and killed by whites.[41]

Charles Quintasket also described hunting around the Arrow Lakes. He said the men were "always hunting, dry[ing] meat for the winters, and there was all different species there to hunt, many species in fact. They had moose, caribou, deer, all kinds of different Blue Grouse, Ruffled Grouse, and of course they had to have the [predators] to go along with these game, such as the coyotes, the wolves. Then they had also Badger, [Lynx] Cat, Mart[e]n, [of] course those are all [fur-bearing], the Mart[e]n, which they would trap. So it was a very busy life for the Native population."[42]

RECLAIMING AN ABORIGINAL RIGHT

In the early 21st century, other Sinixt people provided extensive testimony about hunting activities in their Aboriginal territory in British Columbia. Edward and Joseph Marchand described family stories about "herding caribou," saying that the hunters "would imitate wolves and scare the caribou to the river and . . . be waiting in their canoes."[43] Rosalie Bragg-Gabriel recalled hunting for big game.[44] Elsie M. Picard reported that the women made hook rugs using string made from the backs and legs of deer.[45] Philip Grunlose said his father trapped for beaver.[46] Alice M. Stewart pointed out that the Arrow Lakes people dressed in animal furs because of the weather in the mountains and that they hunted for caribou.[47]

The Sinixt also recalled difficulties they faced as whites filled up their country. Carmilita Theresa Campbell, born in 1900, was the granddaughter of Sinixt Chief Edward. She said the reason the Sinixt came south to Colville was because they were being persecuted.[48] Martin Louie Sr. recalled a more recent problem affecting hunting: a Sinixt man was prevented from bringing deer or caribou meat across the border. When that began to happen, he said, it affected the lives of the people.[49]

Many Sinixt people described other subsistence activities in addition to hunting, which were associated with their continuing trips to British Columbia. They collected berries, including huckleberries, serviceberries, wild cherries, and Oregon grape.[50] Sinixt from the Colville Reservation also dug for roots, including camas, black camas, and bitterroot.[51] Noel Campbell reported that he continued to be drawn to his traditional territory in Canada: "I still travel to Canada, I have traveled as far as Revelstoke, Castlegar, and Arrowhead and have been there many times for sentimental reasons, because of the stories I have heard about that area."[52]

When Sinixt visit the British Columbia portion of their territory, they not only hunt, collect berries, dig roots, andengage in other subsistence activities, they also engage in ceremonial and religious activities. Rosalie Bragg-Gabriel described in detail her grandfather's Spirit Quest in Canada and the ensuing ceremony.[53] Other Sinixt described Chinook (Winter) Dances on the northern side of the international boundary. Veronica M. Orr, born in 1945, recalled that in the past the Canadian government tried to prohibit traditional religious activities (and US officials made the same effort). She said: "In Canada the government did stop us from having our dances years ago but that didn't stop the people they still had the dances. I remember I was about six years old going up with my grandmother and it was all done in secret."[54]

Orr, Sheilah Hall-Cleveland, Lewis Butch Lemery, and Rosalie Bragg-Gabriel described Chinook Dances held in British Columbia.[55] Other Sinixt traditional practices carried out in British Columbia by Sinixt included sweats, funerals, and traditional medicinal practices. Elsie M. Picard, born in 1932, said, "The adults would go up to Revelstoke and Castlegar for funerals and for sweats. In Canada people would use different kinds of rocks and would only take sweats at certain times of the year. The sweats were our rituals and celebrations."[56]

Considering all the ties the Sinixt have to their Aboriginal territory, Lewis Butch Lemery concluded that in addition to old photographs and documents, "We have an oral history. . . .

When it comes to the land we lived on, the land we walked on, then we are part of the land. Spiritually we are tied to the land. Our ties are to the Upper/Lower Arrow Lakes and Slocan Valley, British Columbia, Canada, and to the Columbia River all the way up into Canada. Our grandparents knew them, our generation heard about them, and we still talk about them, it left an impression and stories that will stay with us forever."[57]

In her expert report prepared for the *Desautel* hunting case, Andrea Laforet provided ample evidence to demonstrate that hunting remains important to people of Sinixt descent for food as well as for social and ceremonial purposes.[58]

Pryce learned that many of the Sinixt have lived on the Colville Reservation for generations, even though the Canadian government found them to be extinct in 1956. She cited "heavy silver and gold mining" in British Columbia as a cause for "much social disruption since the 1850s." She concluded that although Sinixt moved to the Colville Reservation, that did not change their Aboriginal territory, even though the Canadian government seemed largely ignorant of their history: "Despite their obscurity in Canada and the scattered documentation of their presence in the area, both archival and published materials show that the Sinixt Interior Salish resided along the Columbia River,

Arrow Lakes, Slocan Valley, and parts of Kootenay Lake in what is now called the West Kootenays of Canada."[59] Pryce was careful to point out that the Sinixt retained ties to their traditional territory and that their "contact with the area since then has been limited but consistent."[60]

Pryce reviewed some of the documentary evidence showing Sinixt presence in their Aboriginal territory throughout the 19th century. She noted that the Sinixt "were considered the best beaver trappers in the Fort Colvile region," depended more on hunting than did neighboring Plateau Indigenous communities, and continued to cross the international boundary for traditional subsistence purposes.[61]

In *The Geography of Memory*, Eileen Delehanty Pearkes recounted the 10,000 years of archaeological material in Sinixt territory and provided a summary of Sinixt history. She noted how the Hudson's Bay Company forts influenced the Sinixt in their hunting practices, and when the posts closed, the Sinixt really had no choice but to move to the Colville Reservation across the border: "Without a reserve yet in Canada, the majority of Lakes Indians appear to have focused cultural activities south of the 49th parallel during the last few decades of the 19th century. However, many of them continued to return to their traditional hunting grounds in the north during the warmer months." Settlers in the Arrow Lakes area reported "Colville" coming north for hunting and fishing. These were Sinixt, but they were called Colville because they came from the Colville Reservation. Settlers brought pressure to have the Sinixt returned to the United States.[62]

On the US side of the border, the grandson of Alexander Christian, one of James Teit's informants, published two books about his Sinixt heritage. Lawney L. Reyes was the son of Mary Christian, the daughter of Alexander Christian, who was also known by the Salish term for White Grizzly Bear (rendered in Salish as "PIC Ah Kelowna" by Reyes). Reyes described the Sinixt territory and traditional subsistence practices that he learned from his mother: "During the late fall, hunters went out alone or in teams to hunt deer, elk, moose, bear, and caribou in the mountains. This was the best time to hunt large game."[63] Alexander Christian received the name White Grizzly Bear when he refused to kill a white grizzly bear, which was rare but native to the upper Sinixt territory.[64]

Much of Reyes's book is devoted to describing the knowledge of Sinixt culture that he learned from his mother, who in turn learned it from her father. It is a remarkable tribute to the cultural tenacity of the Sinixt. Through this oral tradition, Reyes learned folktales of the Sinixt, subsistence practices, the extent of the Sinixt traditional territory, and Sinixt history.[65]

Hunting for game throughout their Aboriginal territory, including that portion now in British Columbia, remains integral to Sinixt culture to the present day. Hunting is important as a source of food and is associated with many social and ceremonial activities of the Sinixt. As we have shown in this chapter, this has been well documented in professional, government, and popular publications in North America over the past half century.

MANAGING WILDLIFE ACROSS BORDERS

Today, the Sinixt use a combination of Traditional Ecological Knowledge and modern technology to manage game in their territory and continue to hunt for subsistence and cultural and spiritual purposes throughout their Aboriginal territory. Natural resources are managed on the Colville Reservation under the guidance of an integrated resource management plan. This plan uses a whole system management approach, viewing all resources (natural, social, cultural, and economic) as being interrelated in such a manner that management actions directed at one resource also affects others. The tribal Wildlife Department's overall goal is to ensure that wildlife populations are maintained on the Colville Indian Reservation (including the North Half) and within the Tribe's "usual and accustomed areas" in order to meet the cultural and subsistence needs of the members by maintaining healthy, self-sustaining game and nongame populations. The Wildlife Department also contributes toward population recovery of federally threatened and endangered species and other priority species that occur within the boundaries of the Colville Indian Reservation. To accomplish these goals, a fish and wildlife program staff of over 150 employees is responsible for coordination with co-management agencies and a variety of operational tasks.

To the extent possible, hunting regulations are developed to provide maximum opportunity for tribal members. These include extended seasons and limited restrictions on harvest, which allow members to hunt, fish, and gather as they would have historically. In many situations, young hunters are taught the traditional ways, which ensures they hunt ethically and only harvest what they need for subsistence. Hunters are mindful of the appropriate seasons to harvest animals, and they regularly consider how their actions impact herd dynamics. For example, when populations are declining, hunters will make a conscious effort to avoid harvesting females. In other situations, hunters will shift focus to species that are not in decline, such as hunting elk while putting a temporary moratorium on hunting deer. When hunters take this type of responsibility on themselves, it creates a self-regulating management system. However, when management action is needed to protect or help recover a population, regulations give the wildlife program an opportunity to make changes quickly.

Population tracking is a critical component of big game management and informs much of the decision-making process when developing hunting regulations. Aerial surveys allow the wildlife program to monitor game populations over time and react to changing herd populations or herd dynamics. These flights happen on a one- to three-year basis, depending on predicted wildlife migration and comparisons of herd trends by geographic area. They allow biologists to identify changes in populations, determine what potential causes might be, and if needed make management changes to ameliorate the decline. Causes could include predation, hunting pressure, disease, fire, forest management, or access. Trends can be both positive and negative, and identifying and documenting the change agent is important to informing management decisions in real time.

In addition to tracking existing species, the wildlife program evaluates for reintroduction species that used to occur on the landscape. A number of species of importance to the constituent tribes that now reside on the Reservation were extirpated due to overhunting or disease from domestic animals when non-Indian immigrants settled in the area. Two of those species have since been reintroduced to their historic ranges. Bighorn sheep were reintroduced in the Hellgate Game Reserve starting in 2008, with subsequent releases in 2009 and 2010. This population has grown to over 200 animals with an annual hunt with a limited number of tags. Pronghorn antelope were reintroduced to the Reservation in 2016. A subsequent release happened in 2017 to augment the population. That population has spread across much of the available habitat, including an off-reservation group on the south side of the Columbia River.

In addition to reintroduction efforts, the wildlife program has used targeted management to dramatically increase the number of elk and moose on the Reservation. In 1977, a small herd of elk were released to augment the existing population in what became the Hellgate Game Reserve. The reserve was created to provide a sanctuary for the augmented elk population and provide a source population to expand to other suitable habitat around the Reservation. Within five years, the herd had grown to a huntable population. Since that augmentation, elk have expanded in population numbers and distribution to occupy almost all the available habitat on the Reservation. A similar situation, though less targeted, occurred with moose populations. An active forest management program that included targets for early seral structure provided ample foraging opportunities for moose. With a distributed mix of forage, hiding, and thermal cover scattered across the landscape, moose quickly took advantage of the available habitat. Currently, moose have a limited hunting season for tribal members, who must draw a lottery tag. However, the moose population and number of tags allocated have grown substantially in the 21st century.

Wildlife populations have expanded under an integrated approach to resource management that includes considerations of all natural resources and how the health and availability of those resources affects the natural, cultural, and social components of life. This holistic, integrated approach varies from many wildlife management agencies, but it has been successful for the Colville Reservation and its tribal members. Through co-management agreements with other fish and wildlife managers, the Colville Tribes share this approach, in hopes that populations outside the Reservation will see similar improvements in total numbers, structure, and distribution.

The management approach used by the Colville Tribes is unique when compared to other state and federal agencies. Most agencies have a mission focused on one aspect of resource management and accordingly have staffs that focus on that distinct resource. For example, many national forests have management plans that consider wildlife habitat, but forest restoration projects are geared toward forest health and wildfire resilience. While those are important considerations, planning typically does not take a long-term approach to defining wildlife habitat components and where those exist on the landscape spatially and through time. Similarly, the US Fish and Wildlife Service manages areas primarily for the benefit of wildlife, but it may create an increased risk of habitat loss to catastrophic wildfire because forest health and fuels are lesser considerations in their management decisions. By using an integrated approach that maximizes benefits to as many resources as possible, the Tribes have attained healthy wildlife populations that meet their subsistence needs. This integrated approach considers disturbance, climate change, and succession through time to ensure that the benefits are maintained for the current members and the generations to come.

WINNING THE LEGAL BATTLE FOR CO-MANAGEMENT

The co-management discussed above was defined through negotiation and legal battles. Hunting rights on the former North Half of the Reservation were affirmed in the *Antoine v. Washington* case. In that case, a tribal member was cited by a state game officer for hunting deer during a closed season. Mr. Antoine was found guilty in the superior court and the Washington State Supreme Court, but the case was overturned in the US Supreme Court. Prior to the US Supreme Court decision, the state had maintained it had jurisdiction over the North Half. Since that decision in 1975, the relationship between the state and the Tribes has changed. Subsequent agreements were reached between Washington state and the Tribes that consider these retained rights and the Tribes' right to have decision-making authority in the management of those resources.

This same situation has been litigated in the Canadian courts. A Sinixt descendant, Richard Lee Desautel, harvested a cow elk near Castlegar, British Columbia, on 1 October 2010. Desautel reported the harvest to British Columbia conservation officers, who issued Desautel an appearance notice for several days later. Desautel was charged with hunting without a license and hunting big game while not being a resident. A quote from the introduction of the provincial court decision summarizes the situation well. The judge stated, "At issue in this case is the very existence of an aboriginal right to hunt in British Columbia by the descendants of the Sinixt, a people well known to have lived, travelled, hunted, fished and gathered in these parts for thousands of years."[66] The legal argument raised by the province was that Desautel was not exercising an Aboriginal right because no right had ever existed in Canada. The Crown further argued that the Sinixt hunting right in Canada did not survive the Canadian assertion of sovereignty. Counter to Desautel's claim that Sinixt had continually used and maintained rights in their traditional Canadian territory, the Crown contended that the Sinixt voluntarily had left Canada and did not maintain continuity with the precontact group. Because of this migration south and the people's absence from Canada, the Crown argued that no rights remained north of the international border.

The provincial court found compelling evidence that the Sinixt had existed for thousands of years in the area claimed by Desautel. The court ruled that Desautel was exercising an Aboriginal right, and the violations he was charged with, according to the Canada Wildlife Act, unjustifiably infringed on that right. The court ruled the charges were not applicable, and Desautel was acquitted of all charges. The decision was upheld in the Supreme Court of British Columbia, the Court of Appeal for British Columbia, and the Supreme Court of Canada.

Now that the Canadian Supreme Court has upheld the lower court's decision, the Colville Reservation will engage with regional wildlife managers to develop co-management agreements similar to those used in the United States. This work will likely include a review of wildlife populations, an assessment of habitat goals and objectives, and a review of harvest management. In addition, engagement in other aspects of resource management will be reviewed to determine where opportunities exist to collaborate with provincial staff. By utilizing capacity within the Colville Tribes' Natural Resource Division in combination with local wildlife staff and organizations, we can find management solutions that will restore declining populations in the West Kootenay region.

NOTES

1. Teit, "Okanago," 198, 226–27; Ray, "Cultural Relations," 135; Kennedy and Bouchard, "Northern Okanagan," 242–43.
2. Teit, "Okanogon," 219, 242; Ray, "Aboriginal Economy," 136; Ray, "Testimony," 96.
3. Bouchard and Kennedy, "Lakes Indian," 29; Kennedy and Bouchard, "Northern Okanagan," 239–41.
4. Teit, "Okanogon," 242.
5. Ray, "Aboriginal Economy," 136. Ray, "Testimony," 96, said, "there was practically no part of the territory that was not deer and elk hunting territory."
6. Teit, " Okanogon," 219.
7. Ray, "Aboriginal Economy," 136–41.
8. Ray, "Aboriginal Economy," 136–39.
9. Ray, "Aboriginal Economy," 139–41, quote at 141.
10. Kennedy and Bouchard. "Northern Okanagan," 241.
11. Ray, "Aboriginal Economy," 142.
12. Ray, "Testimony," 96.
13. Mohs, "Archaeological Investigations," 26, 53; Mohs, "Prehistoric Settlement Patterns," 15, 26.
14. Ray, "Aboriginal Economy," 136. See also Bouchard and Kennedy, "Indian Land Use," 103; Bouchard and Kennedy, "Lakes Indian," 29, 34, 101; Bouchard and Kennedy, "First Nations' Ethnography," 267; Bouchard and Kennedy, "First Nations' Aboriginal Interest," B-1.
15. Bouchard and Kennedy, "Indian Land Use," 103.
16. Chance, "Fort Colvile," 6. Heron, Fort Colvile Journal, B.45/a/1, reported on specific requests to traders to acquire elk skins for use at Fort Colvile.
17. Work, "Some Information," B.45/e/3, fols. 1–11.
18. Work, "Some Information," B.45/e/3, fols. 1–11.
19. Pearkes, *Geography of Memory*, 83.
20. Eldridge, "Vallican Archeological," 5–6.
21. Ray, "Aboriginal Economy," 127–28, 132; Kennedy and Bouchard, "Northern

Okanagan," 241; Freisinger, "Phase I. Report," 4; Layman, *Native River*, 148; Ray, *Sanpoil and Nespelem*, 115–16.

22. Ray, "Aboriginal Economy," 127–28, 132; Ray, *Sanpoil and Nespelem*, 115–16; Freisinger, "Phase I. Report," 4; Kennedy and Bouchard, "Northern Okanagan," 241. Layman, *Native River*, 148, was one of several authors citing traditional narratives of the Sinixt relating to the Kettle Falls fishery.

23. Teit, "Okanagon," 209, 237; Ray, "Native Village and Groupings," 126; Bouchard and Kennedy, "Lakes Indian," 46; Eldridge, "Vallican Archeological Site," 5–6.

24. Teit, "Okanagon," 209; Bouchard and Kennedy, "Lakes Indian," 47; Douglas, *Journal*, 250; De Smet, *Oregon Missions*, 216–17.

25. Bouchard and Kennedy, "Lakes Indian," 47; Douglas, *Journal*, 250; Mohs, "Archaeological Investigations," 38.

26. Ray, "Aboriginal Economy," 137–38.

27. Work, "Journal," 40–42.

28. Bouchard and Kennedy, "Lakes Indian," 30.

29. Bouchard and Kennedy, "Indian Land Use," 104.

30. Kane, *Wanderings*, 229. A number of contemporary authors reported that the Sinixt hunted with dogs, including Reyes, *White Grizzly Bear's Legacy*, 6.

31. In an email to Richard Hart, 14 January 2021, Eileen Delehanty Pearkes wrote of a non-Indian family who recalled a Sinixt named Alex Christian using trained dogs to hunt deer.

32. Chance, "Influences," 96.

33. Lemery, "Testimony," 130–31.

34. Lemery, "Testimony," 132–33.

35. Lakin, *Kettle River Country*, 5–6.

36. Mary Marchand interview by Father Pat Conroy with Sheila Cleveland at Kelly Hill on 21 April 1986 (hereafter Marchand interview). Joanne Signor was present. Some spelling and punctuation in all of the interview transcripts have been adjusted.

37. Lucy Swan interview by Joanne Signor, 29 November–27 December 1977, transcript, 9–10, 18.

38. Lakin, *Kettle River Country*, 5.

39. Marchand interview.

40. Eva Orr interview by Father Pat Conroy and Sheila Cleveland on 29 July 1986 in Omak, Washington.

41. William Barr interview by Father Pat Conroy with Sheila Cleveland in Okanogan on 29 July 1986.

42. Charles L. Quintasket interview by Father Pat Conroy with Sheila Cleveland in Omak, Washington, 9 June 1986. Quintasket also described bear hunting and the traditional reason that Sinixt do not like to take the last food on a platter.

43. Marchand and Marchand, "Affidavit." They also described the traditional methods of constructing canoes.

44. Bragg-Gabriel, "Affidavit."

45. Picard, "Affidavit."

46. Grunlose, "Affidavit."

47. Stewart, "Affidavit."

48. Carmilita Theresa Campbell interview by Father Pat Conroy with Sheila Cleveland on 21 April 1986.

49. Martin Louie Sr. interview by Father Pat Conroy with Sheila Cleveland at his home on the Columbia River, 23 September 1986.

50. Picard, "Affidavit"; Bragg-Gabriel, "Affidavit."

51. Picard, "Affidavit"; Orr, "Affidavit"; Boyd, "Affidavit"; Campbell, "Affidavit"; Fry, "Affidavit"; Hall-Cleveland, "Affidavit"; Seymour, "Affidavit"; Steele-Caru, "Affidavit"; Stewart, "Affidavit."

52. Campbell, "Affidavit."
53. Bragg-Gabriel, "Affidavit."
54. Orr, "Affidavit."
55. Orr, "Affidavit"; Hall-Cleveland, "Affidavit"; Bragg-Gabriel, "Affidavit"; Lemery, "Affidavit."
56. Picard, "Affidavit." See also Boyd, "Affidavit"; Steele-Caru, "Affidavit."
57. Lemery, "Affidavit."
58. Laforet, "Sinixt (Lakes) Familial Connections."
59. Pryce, *Keeping the Lakes' Way*, 7–8.
60. Pryce, *Keeping the Lakes' Way*, 8.
61. Pryce, *Keeping the Lakes' Way*, 25, 29, 37.
62. Pearkes, *Geography of Memory*, 16–17, quote at 16.
63. Reyes, *White Grizzly Bear's Legacy*, xi–xii, 5–6, quote at 17. He also described the Sinixt use of dogs to hunt deer.
64. Reyes, *White Grizzly Bear's Legacy*, 32.
65. Reyes, *White Grizzly Bear's Legacy*, for instance, 19–21, 53, 59–71, 84. A great-grandson of Baptiste Christian, Jim Boyd, is also well known in North America. He is a seven-time Native American Music Awards winner who now lives on the Colville Reservation. His 13th full-length recording, released in 2010, is called *Voices from the Lakes* and references the people's strong ties to the portion of their Aboriginal territory in British Columbia.
66. Regina v. Richard Lee Desautel, BCPC 84, 1 (2017).

LITERATURE CITED

Bouchard, Randall T., and Dorothy I. D. Kennedy. "First Nations' Aboriginal Interest and Traditional Use in the Waneta Hydroelectric Expansion Project Area: A Summary and Analysis of Known and Available Background Information." Rev. 11/205. Report prepared at the request of Waneta Expansion Power Corporation as Reference Information for Project Area First Nations, 20 August 2004.

———. "First Nations' Ethnography and Ethnohistory in British Columbia's Lower Kootenay/Columbia Hydropower Region." Report prepared for the Columbia Power Corporation, Castlegar, BC, Canada, August 2000.

———. "Indian Land Use and Occupancy in the Franklin D. Roosevelt Lake Area of Washington State." Report prepared for the Colville Confederated Tribes, Washington, and the US Bureau of Reclamation, Seattle, Washington, 1984.

———. "Lakes Indian Ethnography and History." Report prepared for the BC Heritage Conservation Branch, Victoria, British Columbia, August 1985.

Boyd, Mathew J., Sr. "Affidavit." In the Matter of the Province of British Columbia Evidence Act, R.S.B.C. 1996, Chap. 124, and a Statutory Declaration Pursuant to Section 69 Thereof. Nespelem, Washington, 29 September 2009.

Bragg-Gabriel, Rosalie. "Affidavit." In the Matter of the Province of British Columbia Evidence Act, R.S.B.C. 1996, Chap. 124, and a Statutory Declaration Pursuant to Section 69 Thereof. Nespelem, Washington, 7 May 2009.

Campbell, Noel. "Affidavit." In the Matter of the Province of British Columbia Evidence Act, R.S.B.C. 1996, Chap. 124, and a Statutory Declaration Pursuant to Section 69 Thereof. Nespelem, Washington, 22 September 2009.

Chance, David H. "Fort Colvile: The Structure of a Hudson's Bay Company Post, 1825 to 1871 and After." Unpublished paper from the Department of Sociology/Anthropology, University of Idaho, Moscow, 1972.

———. "Influences of the Hudson's Bay Company on the Native Cultures of the Colville District." *Northwest Anthropological Research Notes* 7(1, pt. 2) (1973):2.

De Smet, Pierre-Jean. *Oregon Missions and Travels over the Rocky Mountains in 1845–46.* 1847; rpt., Fairfield, Washington: Ye Galleon Press, 1978.

Douglas, David. *Journal Kept by David Douglas during His travels in North America.* London: William Wesley and Son, 1914.

Eldridge, Morley. "Vallican Archeological Site (DjQj 1): A Synthesis and Management Report." Submitted to the Heritage Conservation Branch, 19 April 1984.

Freisinger, Michael A. "Phase I. Report of the Boundary Archaeological Survey, September 25, 1978–June 1, 1979." Permit No. 1978-27. In possession of author.

Fry, Frank R. "Affidavit." In the Matter of the Province of British Columbia Evidence Act, R.S.B.C. 1996, Chap. 124, and a Statutory Declaration Pursuant to Section 69 Thereof. Nespelem, Washington, 16 July 2009.

Grunlose, Philip. "Affidavit." In the Matter of the Province of British Columbia Evidence Act, R.S.B.C. 1996, Chap. 124, and a Statutory Declaration Pursuant to Section 69 Thereof. Nespelem, Washington, 14 September 2009.

Hall-Cleveland, Sheilah. "Affidavit." In the Matter of the Province of British Columbia Evidence Act, R.S.B.C. 1996, Chap. 124, and a Statutory Declaration Pursuant to Section 69 Thereof. Nespelem, Washington, 16 July 2009.

Heron, Francis. Fort Colvile Journal, 1830–1831 [copied by William Kittson between 13 April and 23 May 1830]. Hudson's Bay Company Archives, Provincial Archives of Manitoba, Winnipeg. B.45/a/1.

Kane, Paul. *Wanderings of an Artist among the Indians of North America.* 1859; rpt., Mineola, New York: Dover, 1996.

Kennedy, Dorothy I. D., and Randall T. Bouchard. "Northern Okanagan, Lakes, and Colville." In *Handbook of North American Indians*, vol. 12: *Plateau*. Washington, DC: Smithsonian Institution, 1998.

Laforet, Andrea. "Sinixt (Lakes) Familial Connections to British Columbia." Report prepared by Andrea Laforet Consulting, 12 February 2015, for the *Desautel* hunting case. In author's possession.

Lakin, Ruth. *Kettle River Country.* Colville, Washington: Statesman Examiner, 1976.

Layman, William D. *Native River: The Columbia Remembered.* Pullman: Washington State University Press, 2002.

Lemery, Lewis Butch. "Affidavit." In the Matter of the Province of British Columbia Evidence Act, R.S.B.C. 1996, Chap. 124, and a Statutory Declaration Pursuant to Section 69 Thereof. Nespelem, Washington, 4 September 2009.

Lemery, Peter. "Testimony." In *The Confederated Tribes of the Colville Reservation, et al. v. The United States of America.* Indian Claims Commission, Docket No. 181, Washington, DC, 26 October 1954.

Marchand, Edward, and Joseph Marchand. "Affidavit." In the Matter of the Province of British Columbia Evidence Act, R.S.B.C. 1996, Chap. 124, and a Statutory Declaration Pursuant to Section 69 Thereof. Nespelem, Washington, 16 July 2009.

Mohs, Gordon W. "Archaeological Investigations at the Vallican Site (DjQj 1), Slocan Valley, Southeastern British Columbia." 1982. Unpublished manuscript in author's possession.

———. "Prehistoric Settlement Patterns in the Columbia/Lakes Region of Southeastern British Columbia and Northeastern Washington." December 1982. In possession of author.

Orr, Veronica M. "Affidavit." In the Matter of the Province of British Columbia Evidence Act, R.S.B.C. 1996, Chap. 124, and a Statutory Declaration Pursuant to Section 69 Thereof. Nespelem, Washington, 16 July 2009.

Pearkes, Eileen Delehanty. *The Geography of Memory: Recovering Stories of a Landscape's First People.* Nelson, British Columbia: Kutenai House Press, 2002.

Picard, Elsie M. "Affidavit." In the Matter of the Province of British Columbia Evidence Act, R.S.B.C. 1996, Chap. 124, and a Statutory Declaration Pursuant to Section 69 Thereof. Nespelem, Washington, 15 September 2009.

Pryce, Paula. *"Keeping the Lakes' Way": Reburial and the Re-Creation of a Moral World among an Invisible People.* Toronto: University of Toronto Press, 1999.

Ray, Verne F. "Aboriginal Economy and Polity of the Lakes (Senijextee) Indians" [1947]. Journal of Northwest Anthropology 50(2) (2016):145–66. https://www.edpearkes.com/wp-content/uploads/2022/04/Verne-Ray-Sinxit-Economy-Polity.pdf.

———. "Cultural Relations in the Plateau of Northwestern America." Publications of the Frederick Webb Hodge Anniversary Publication Fund 3. Los Angeles, California, 1939.

———. "Native Village and Groupings of the Columbia Basin." *Pacific Northwest Quarterly* 27(2) (April 1936):126.

———. *The Sanpoil and Nespelem: Salishan Peoples of Northeastern Washington.* 1933; rpt., New Haven, Connecticut: Human Relations Area Files, 1954.

———. "Testimony." In *The Confederated Tribes of the Colville Reservation, et al. v. The United States of America.* Indian Claims Commission, Docket No. 181. Washington, DC, 26 October 1954.

Reyes, Lawney L. *White Grizzly Bear's Legacy: Learning to Be Indian.* Seattle: University of Washington Press, 2002.

Seymour, Elsie A. "Affidavit." In the Matter of the Province of British Columbia Evidence Act, R.S.B.C. 1996, Chap. 124, and a Statutory Declaration Pursuant to Section 69 Thereof. Nespelem, Washington, 22 September 2009.

Steele-Caru, Vivian. "Affidavit." In the Matter of the Province of British Columbia Evidence Act, R.S.B.C. 1996, Chap. 124, and a Statutory Declaration Pursuant to Section 69 Thereof. Nespelem, Washington, 16 July 2009.

Stewart, Alice M. "Affidavit." In the Matter of the Province of British Columbia Evidence Act, R.S.B.C. 1996, Chap. 124, and a Statutory Declaration Pursuant to Section 69 Thereof. Nespelem, Washington, 22 September 2009.

Teit, James A. "The Okanagon." In *Forty-Fifth Annual Report of the Bureau of American Ethnology, 1927–1928,* edited by Franz Boas. Washington, DC: US Government Printing Office, 1930.

Work, John. "Journal." 18 July–28 October 1823, York Factory to Spokane House. Typescript A/B/40/W89.1A in Provincial Archives of British Columbia, Victoria.

———. "Some Information Relative to Colvile District." April 1830. Hudson's Bay Company Archives, Provincial Archives of Manitoba, Winnipeg. B.45/e/3, fols. 1–11.

Chapter 11

Poems by **NILA NORTHSUN**

we always knew

as natives close to
mother earth
she takes care of us
if
we take care of her
you don't despoil the water
plunder her greenery
desecrate the air
humankind was moving
in the wrong direction
we need to protect our
biodiversity
we pray to the 4-leggeds
winged and those
in the water
to the sun the moon
earth
and four directions
we always knew.

wetlands

in the dry desert of nevada
an irrigation system was
created to help farmers farm
in the desert
that's a crazy idea
water from lake tahoe
and the sierra mountains
trickled then tumbled into
the truckee river
channeled into ditches
with water rights
and concrete
the land became
an oasis
the runoff
the wetlands
perfect for
the migratory bird path
for bird watching
and
hunting.

Part III

Resource Use, Protection, and Management

The diversity of the hundreds of tribal fish and wildlife programs cannot be adequately represented in one volume of work; nevertheless, we believe that the selected examples in this part provide an introduction to the array of approaches being undertaken by tribes. From Alaska to the Southwest, from the Rocky Mountains to the Southeast, we highlight some of the unique work being conducted by tribes in the 21st century.

Chapter 12

The Indigenous Sentinels Network
Community-Based Monitoring to Enhance Food Security

LAUREN M. DIVINE, BRUCE W. ROBSON, CHRISTOPHER C. TRAN, PAUL I. MELOVIDOV, *and* AARON P. LESTENKOF

INTRODUCTION

The Pribilof Islands are a five-island archipelago in the eastern Bering Sea, Alaska, and provide vital breeding and feeding habitat for more than half of the world's population of *laaqudan* (Unangam Tunuu [Aleut], northern fur seal, *Callorhinus ursinus*) and foraging and resting areas for the federally endangered western population of *qawan* (Unangam Tunuu [Aleut], Steller sea lion, *Eumetopias jubatus*). The precontact users of the Pribilof Islands created one of the world's most specialized and successful maritime hunter-gatherer traditions, lasting from roughly 4000 BP to the time of Russian contact in 1741 CE (Veniaminov 1984). The Indigenous peoples of the area, known among themselves by regional synonyms, were called by their historical colonizers "Aleuts." Prehistoric culture across the Aleutian and Pribilof Islands was based almost entirely on the utilization of marine resources, including hunting local sea mammals, fishing and foraging in coastal and offshore waters, and hunting birds on land and at sea.

Today, there are two permanent Alaska Native communities in the Pribilof Islands: St. Paul (2021 population of 477, > 80% Alaska Native) and St. George (2021 population of 97, > 90% Alaska Native). The Indigenous peoples collectively sharing the time-honored bond of living together on St. George and St. Paul prefer to be called Unangax̂ (singular noun, adjective), Unangan (plural collective), or Unangas. These communities rely heavily on access to marine resources for sustenance; for physical, emotional, and spiritual health; and to maintain the continuity of their customary and traditional ways of life.

Even after the Russian occupation of the islands and enslavement of Unangan for the purpose of harvesting *laaqudan* pelts on the Pribilof Islands (Black 1983, Torrey 1983), the predominance of marine mammals in Unangax̂ subsistence diets, despite the availability of Russian foodstuffs, demonstrates the maintenance of cultural traditions during a time of intense colonialism. Since the late 20th century, St. Paul has observed rapid and dramatic changes in the distribution and abundance of both *laaqudan* and *qawan*. Historically, a majority (75%) of the worldwide population of *laaqudan* have bred on the Pribilof Islands (Gentry 1998). The Pribilof *laaqudan* herd has declined ~70% since the 1970s, with a majority of *laaqudan* breeding on St. Paul (Short et al. 2021). Since 1998, pup production on St. Paul has declined by 57.7% (Towell et al. 2018). *Laaqudan* birth and survival rate decline factors currently under investigation include climate change (Francis et al. 1998, Hare and Mantua 2000); competition with commercial fisheries (McHuron et al. 2020, Short et al. 2021); predation (Springer et al. 2003, Springer et al. 2008, Newman et al. 2008); and emigration (Muto et al. 2019, Short et al. 2021). Understanding the factors influencing *laaqudan* population dynamics in order to promote population recovery is a high priority for resource managers at the local and regional levels. At the local level, it is vital in the context of ensuring food security.

In recorded history, *qawan* were abundant in the Bering Sea and bred in large numbers on the Pribilof Islands (Trites and Larkin 1996). However, populations of *qawan* on the islands have declined to exceedingly low levels, and the sole remaining breeding rookery, located at Walrus Island, is currently in danger of extinction. In the 1870s, approximately 10,000–12,000 sea lions were distributed at breeding rookeries on the Pribilof Islands (Elliott 1880). By 1916, these rookeries had been largely extirpated due to hunting and culling (Loughlin et al. 1984). Since the mid-20th century, *qawan* pup production on Walrus Island has declined by over 98%, from 2,866 pups born in 1960 to 48 in 2015 (Fritz et al. 2015). Subsistence takes of *qawan* on St. Paul have also declined. The cause of the decline of *qawan* on the Pribilof Islands remains poorly understood, though the most severe threats to the recovery of the western Alaska stock of *qawan* are environmental variability, competition with fisheries, killer whale predation, and toxins in the marine environment (NMFS 2008).

The decline of keystone subsistence resources such as *laaqudan* and *qawan*, in conjunction with increasing environmental variability and the unpredictability of other subsistence resources, has created an urgent need for ongoing resource monitoring. The communities of St. Paul and St. George, technical and scientific advisors, agency partners, and local volunteers together have developed a regional community-based coastal and ocean monitoring program that has enabled community members to collect reliable local data to inform decisions that affect these key species. The first iteration of the monitoring program, called the Sentinel Database, started in 2004, using personal digital assistant devices. The database was taken online in 2008 and renamed BeringWatch. BeringWatch data collection was rooted in both Traditional Ecological Knowledge and Western science, recognizing the diversity of qualitative and quantitative data and that both are valuable in decision-making, especially in a rapidly changing environment. In 2018, in response to a growing need for practical, streamlined, Indigenous-driven monitoring programs across

Alaska and Canada, the BeringWatch program evolved into the Indigenous Sentinels Network (ISN). ISN has broadened its reach in several Bering Sea and Aleutian Islands communities, including Akutan, False Pass, King Cove, and Unalaska.

In this chapter, we explore how the data collection and dissemination of components of ISN have been successful from a food security perspective. We use marine mammal harvest and winter sea duck and gull monitoring on St. Paul Island, Alaska, as case studies to demonstrate how ISN addresses a collective need for long-term subsistence resource monitoring for communities, agency and academic partners, and resource managers, while providing consistency in data collection.

METHODS

The Indigenous Sentinels Network Framework

ISN fills a niche for communities seeking to implement a monitoring program for their key environmental resources without high start-up costs or significant monetary investments in software or other infrastructure. The focus of ISN is on real-time ecological monitoring by local community members who are traditional knowledge-holders. ISN is taxonomically broad in scope and provides a flexible and customizable framework that enables local "sentinels" to monitor ecological phenomena and anomalies through standardized and repeatable surveys. Each participating community develops and customizes the structure of its monitoring program by identifying focal species or phenomena of concern. ISN also assists other organizations to build the capability to monitor focal and novel (e.g., invasive) species and address community needs to maintain or improve long-term food security. ISN was designed with flexibility to include a customized, community-based program with web-based and mobile data collection tools, a handbook of standardized protocols, hands-on training materials, an embedded quality control system, and technical assistance. The database is password protected for secure long-term archiving, interpretation, and dissemination of information.

Marine Mammal Harvest Monitoring

The real-time marine mammal harvest monitoring methods (Zavadil et al. 2006) were established in 1999 under the BeringWatch program. The Ecosystem Conservation Office (ECO) collects harvest information on St. Paul during the federally regulated subadult male *laaqudan* harvest season, 23 June to 31 December. During each subsistence harvest, ECO staff monitor and record the number of *laaqudan* harvested, released, flipper-tagged, or entangled in marine debris and successfully disentangled. Data are recorded using a mobile app on a handheld device in the field and then uploaded to the BeringWatch/ISN database.

ECO uses similar standardized methods to collect *qawan* harvest data from hunters within 12–48 hours following a subsistence hunting trip. *Qawan* hunting is monitored daily throughout the year and supplemented with weekly marine mammal stranding surveys. Data collected include the sex and age class of successfully hunted *qawan*, the location of the take, the status of the animal (retrieved or struck and lost), and any other relevant information (e.g., biological samples). Consistent year-round shoreline monitoring also allows ECO to document changes in the frequency and causes of marine mammal strandings. Active participation by ECO staff during *laaqudan* subsistence harvests entails voluntary reporting of *qawan*; this reporting has remained very high (up to 100%) since 2001 (Lestenkof et al. 2018). All harvest monitoring data for *laaqudan* and *qawan* are shared with the federal co-managers of marine mammals on St. Paul, the National Oceanic and Atmospheric Administration (NOAA), and the tribal community of St. Paul through annually published subsistence harvest reports and oral reports at tribal meetings.

Winter *San* and *Slukan* (Sea Duck and Gull) Surveys

Prior to ECO beginning surveys of winter *san* (birds, including sea ducks) and *slukan* (gulls) in 2008, the only estimate of sea duck and gull populations on the Pribilof Islands were from one month in 1993 and one month in 1996 (Sowls 1997). Biological data are currently collected on seabird species occurring on the islands during the summer (breeding) months on both St. Paul and St. George by the US Fish and Wildlife Service (USFWS) Alaska Maritime National Wildlife Refuge (AMNWR). However, the islands are known overwintering sites for a number of sea duck species, including *kasimax̂* (common eider, *Somateria mollissima*), *saakum aliĝii* (king eider, *Somateria spectabilis*), *kaangadgiix̂/kaaxadgix̂* (harlequin duck, *Histrionicus histrionicus*), *aalngaaĝix̂* (long-tailed duck, *Clangula hyemalis*), white-winged scoter (*Melanitta deglandi*), red-breasted merganser (*Mergus serrator*; Sowls 1993, 1997), and multiple species of *slukan*/gulls (*Larus* spp.; Insley et al. 2017). Subsistence hunters on St. Paul are known to hunt birds year-round, with effort concentrated in the winter and spring months before *laaqudan* return to the islands in April and May annually (Tran et al. 2020). Thus, monitoring subsistence waterfowl and seabird population trends is important for the cultural continuity for future generations of subsistence activities for Pribilof Islands communities, which are currently at risk due to environmental and societal changes that impact access to traditional and customary ways of life (Young et al. 2014, Romano et al. 2019).

Sea duck and gull surveys were initiated in 2008 in collaboration with the AMNWR. Data collection focuses on the overwintering months from autumn (typically beginning around 1 October) to early spring (mid-April). Focal species include eiders (*Somateria* spp.), harlequin ducks, long-tailed ducks, white-winged scoters, Eurasian wigeons (*Mareca penelope*), and glaucous-winged gulls (*Larus glaucescens*). Here, we review nine years of data collection: 2008–2011 (previously published in Insley et al. 2017) and 2015–2021. Survey locations were initially chosen due to high concentrations of focal species and visibility, accessibility, and repeatability across surveys. The same sites were used across all survey years (Sowls 1997, Insley et al. 2017). Only one site was discontinued (i.e., the city landfill) during the survey period because access was difficult and the site had consistently low counts; thus it has been excluded from this analysis.

RESULTS

Qawan Subsistence Hunting Monitoring

The number of *qawan* harvested has remained consistent since 2001 with local hunters taking between 15 and 40 *qawan* per year (ECO, unpub. data). Subsistence hunters typically target small male *qawan* (pups, juveniles). Preference for small animals is also seen in subsistence harvesting of *laaqudan* and reindeer on St. Paul. *Qawan* age 3+ are rarely harvested because they are too large to retrieve and butcher along the shore, especially in inclement weather. Based on ECO data, from 2001 to 2020, 92% of the 366 *qawan* harvested were juvenile males, with very few adults or females killed (table 12.1).

TABLE 12.1. Age Class and Sex of Subsistence Harvested Qawan (Steller Sea Lions) Retrieved by Local Hunters on St. Paul Island, Alaska, 2001–2020 (percentage retrieved in parentheses)

SEX	AGE CLASS				
	ADULT	JUVENILE	PUP	UNKNOWN	TOTAL
Male	4 (1.1%)	338 (92.3%)	6 (1.6%)	0	348 (95.1%)
Female	1 (0.3%)	11 (3.0%)	0	1 (0.3%)	13 (3.6%)
Unknown	0	3 (0.8%)	0	2 (0.5%)	5 (1.4%)
Total	5 (1.4%)	352 (96.2%)	6 (1.6%)	3 (0.8%)	366 (100%)

Since 2001, annual retrieval rates have varied from 50% to 91% with an average of 68% across all years and no obvious trends over time. *Qawan* hunting is evaluated on an annual as well as a seasonal basis (table 12.2) due to differences in effort and retrieval success among areas and seasons. There is virtually no hunting in June to August because of the presence of *laaqudan* at *qawan* hunting locations and the desire of St. Paul subsistence hunters to improve retrieval rates to ensure nonwasteful use of *qawan*. Retrieval rates have been 70% or higher in 10 of the last 20 years. Hunting on St. Paul only occurs when the hunter is on land, but animals may be shot while the animal is swimming within 100 yards of the shoreline or after they have hauled out on land. Retrieval rates are higher when *qawan* are hunted on land (table 12.2).

Contributing factors to observed differences in hunting performance include the environmental conditions following a hunt, wind direction and strength and oceanic current dynamics, the sinking rate of an animal, the presence of scavengers, and accessibility to the animal when it washes ashore. Hunters typically see more *qawan* traveling along the shoreline in the water during the spring and fall. From late May to early July, *qawan* are primarily located onshore at rookeries for pupping and breeding. *Qawan* are not known to migrate, but individuals disperse throughout western Alaska outside of the breeding season.

TABLE 12.2. Subsistence Harvested Qawan (Steller Sea Lions) That Were Retrieved or Struck and Lost after Being Hunted while the Animal Was on Land, Swimming in the Water, or in an Unknown Location on St. Paul Island, Alaska, 2005–2020

YEAR	LAND		IN WATER		UNKNOWN	
	RETRIEVED	STRUCK/LOST	RETRIEVED	STRUCK/LOST	RETRIEVED	STRUCK/LOST
2005	7	0	11	2	2	0
2006	5	2	14	4	0	1
2007	13	1	9	11	0	0
2008	5	0	15	2	0	0
2009	1	0	17	8	0	0
2010	5	0	10	5	0	1
2011	7	0	17	2	0	6
2012	5	0	11	7	0	1
2013	1	0	20	4	6	3
2014	8	0	13	12	0	2
2015	4	0	13	7	0	0
2016	7	0	10	14	0	0
2017	5	1	12	12	0	0
2018	9	0	13	6	0	0
2019	6	0	9	9	0	0
2020	11	0	13	9	0	0
Subtotals	99 (96%)	4 (4%)	207 (64%)	114 (36%)	8 (36%)	14 (64%)
Total Take	Land (23%)		In Water (72%)		Unknown (5%)	

Biosampling

ECO has collected biosamples (blood, muscle, liver, blubber) of *laaqudan* and *qawan* since 1998 to assess body condition, diets, and contaminant loads (i.e., mercury, lead, organohalogen contaminants) of marine mammals used for subsistence foods. Detailed harvest data and chain of custody information for samples are critical for collaborating partners such as the Alaska Marine Mammal Tissue Archival Project (AMMTAP), an interagency project involved in the collection, archiving, and analysis of tissues from marine mammals from Alaska for retrospective research on contaminant levels and animal health (Kucklick et al. 2013, Reiner et al. 2016). For example, archived liver and blubber biosamples collected from 1987 to 2007 from *laaqudan* provided to AMMTAP were analyzed for persistent organic pollutants and vitamins (Kucklick et al. 2013). Ultimately, a suite of persistent organic pollutants (POPs) exponentially increased each sampling year, with only concentrations of polybrominated diphenyl ethers appearing to plateau over time (Kucklick et al. 2013). These results highlight the need to continue biosampling *laaqudan* because of increasing concentrations of current-use POPs and declining *laaqudan* populations in the northern Pacific (Kucklick et al. 2013).

ADDITIONAL MARINE MAMMAL SURVEYS

Rapid environmental shifts in the Arctic have impacted the ecology and phenology of Arctic marine mammals, such as occupation timing and habitat use of ice-dependent seals and cetaceans (Moore and Huntington 2008). St. Paul Sentinels conduct regular surveys of *laaqudan* at standardized and repeatable vantage points (observation locations) to determine timing of the arrival of adult males, timing of the departure of adult females and pups, and numbers of overwintering *laaqudan*. These surveys are designed to evaluate the timing of occupation for each age and sex class of *laaqudan* to understand the potential impacts of climate change on migration patterns and habitat use, and ultimately to address food security questions. In addition, traditional knowledge from northern and western Alaskan communities has documented how marine mammal migratory routes and timing have been affected by physical changes in Arctic marine environments, such as warmer water temperatures and changes in sea ice dynamics (Huntington et al. 2017).

Winter *San* and *Slukan* (Sea Duck and Gull) Monitoring

Sea duck abundance has remained consistent over time, but diversity has declined. Species such as Barrow's goldeneye (*Bucephala islandica*), common eider, and white-winged scoter were commonly seen in early survey years but have become rare or absent since 2018. In contrast, *saakum aliĝii* (king eider) have increased dramatically in abundance (weekly average of 21.6 ± 6.5 individuals from 2008 to 2011; 100.8 ± 34.4 from 2018 to 2021). The increase may be explained by niche space made available by the declining abundance of other sea duck species. The declining trend in sea ducks over the survey period was substantiated by St. Paul Sentinels, who noted that scoters were once counted in the hundreds at one vantage point but have not been seen in the same area since 2016. Observations and recollections by community members on the island also note a marked decrease in local scoter populations since 2011.

The decline in sea duck diversity may be attributable to trends in sea ice retreat in the southeastern Bering Sea (e.g., Hunt et al. 2018). Data obtained from the National Snow and Ice Data Center indicate that winter Bering Sea ice extent decreased an average of ~ 2% annually from 2008 to 2020 (Fetterer et al. 2017). St. Paul Sentinels, who also collect observations of sea ice, last observed sea ice around the island in 2017. Historically, winter sea ice has shown interannual variability in extent and thickness, but typically sea ice is at or just north of the Pribilof Islands. Community observers note that after 2010–2011, warmer winter air and ocean temperatures have been accompanied by a lack of sea ice around the Pribilof Islands and a decrease in predictability in the timing of the ice. For example, some have observed earlier arrival and longer periods of sea ice and noted a noncyclical pattern of sea ice over time (Lestenkof et al. 2013, Romano et al. 2019).

Previous studies have shown a negative correlation between winter sea ice coverage and bufflehead and common goldeneye lipid reserves, while long-tailed ducks (*aalngaaĝix̂*) were not similarly impacted (Schummer et al. 2012). This negative correlation supports Sentinels' observations of reduced populations for bufflehead and common goldeneye, while long-tailed duck populations remained stable. Intermediate ice coverage is known to provide ideal conditions for spectacled and king eider because it provides a platform for roosting. Heavy ice impedes foraging for benthic prey (e.g., clams), and sparse ice increases thermoregulation costs due to higher wave action (Christie et al. 2018, Oppel et al. 2009). Common eiders had higher breeding success during early ice retreat years, and it is predicted that the warming of the Arctic due to climate change may increase common eider populations in the future (Lechikoninen et al. 2006, Mehlum 2012).

Consistent with early data published by Insley et al. (2017), overwintering *slukan* continue to increase in abundance on St. Paul. This may be due to fish offal concentrations near the village area and harbor at the southern end of the island, providing a steady food source that can support high numbers of nonbreeding *slukan*. *Slukan* are considered to pose a threat to other seabird species because they prey on eggs and chicks (Lowney et al. 2018). If populations of *slukan* continue to increase, they may significantly impact the diversity and abundance of other species; however, if *slukan* can continue to be supported with fish offal, they may not be a threat to other seabirds utilizing the island (Insley et al. 2017). Further investigation into the linkages between fish offal and *slukan* population are warranted.

DISCUSSION

Alaska is experiencing some of the most severe impacts of coastal change observed in the United States, having warmed an average of 1.5°C since the mid-20th century, more than twice as fast as the continental United States (USGCRP 2014) Alaska's coasts are also home to over 80% of the state's residents, including the large majority of the 228 federally recognized Alaska Native tribes. These coasts also support many of the world's most economically important fisheries (Worm et al. 2009) and were home to 5 of the top 10 US fishing ports in value and landings in 2017 and 2018 (NMFS 2020).

The rapid pace of climate change has increased scientific uncertainty regarding the future productivity and resilience of the Bering Sea and other Arctic ecosystems. The value of the ISN approach and the proven competence of Indigenous communities to collect these data have been demonstrated through various BeringWatch/ISN projects conducted in the 21st century by Alaskan tribal organizations, such as the Aleut Community of St. Paul Island ECO, the St. George Traditional Council Kayumixtax Eco-Office, the Agdaagux Tribe of King Cove, and the villages of False Pass, Akutan, and Unalaska.

One of the strengths of ISN is its ability to scale from local- to landscape-level observations, providing an existing framework to build coordinated community-based monitoring programs customized to fit each community's needs. For example, the technical infrastructure necessary to support a series of new observation protocols for use in the interior of Alaska to monitor subsistence species in freshwater streams and rivers was implemented in the ISN database and mobile apps in partnership with the Tanana Chiefs Conference during a 2018 community-based monitoring project. The project was implemented to fill data gaps and improve monitoring regarding the major species of Pacific salmon and whitefish, which are important to Yukon River communities as subsistence foods.

We have been successful with previous efforts at building new partnerships with regional tribes and Native organizations and are currently planning to expand data collection procedures and protocols to include coastal vulnerability assessments, coastal erosion detection systems, and assessing real-time changes in inventories of indicator species. We look forward to building on existing partnerships and creating new partnerships as we expand the ISN network to new tribal communities. It is a proven, effective, and tribally driven environmental program that can enhance tribal resilience efforts throughout the Arctic.

ACKNOWLEDGMENTS

The authors say *txichin qaĝaasakuqing*, many thanks, to the subsistence hunters on St. Paul Island who voluntarily report their subsistence harvests of marine mammals and birds to ECO, contribute their Indigenous knowledge to provide context to scientific data collections, and make biosampling possible. We also thank the editors of this volume, Steven Albert and Serra Hoagland, for their leadership and support in completing this contribution; Aaron Poe and Leanna Heffner for support on early versions of this chapter and on broader ISN efforts; and the anonymous reviewers during submission. Data were collected by the Ecosystem Conservation Office, Aleut Community of St. Paul Island tribal government under NMFS Scientific Research Permit Nos. NA14NMF4390160, NA15NMF4390421, NA16NMF4390123, NA17NMF4390134, NA18NMF4390163, NA19NMF4390120, and NA20NMF4390060. Funding was also provided by the NOAA Regional Ocean Program through Grant No. NA13NOS4730097. Sea duck and gull data were partially funded through the Indian General Assistance Program (EPA) and Tribal Wildlife Grant (USFWS) No. U2-23-NA-1.

LITERATURE CITED

Black, L. 1983. Some problems in the interpretation of Aleut prehistory. Arctic Anthropology 20:49–78.

Christie, K. S., T. E. Hollmen, P. Flint, and D. Douglas. 2018. Non-linear effect of sea ice: spectacled eider survival declines at both extremes of the ice spectrum. Ecology and Evolution 8(23):11808–18. https://doi.org/10.1002/ece3.4637.

Elliott, H. W. 1880. Report on the Seal Islands of Alaska. Elliott's field notes transmitted by him to F. A. Walker, superintendent, 10th Census, 31 March. https://www2.census.gov/library/publications/decennial/1880/vol-08-news-press/1880v8-14.pdf.

Fetterer, F., K. Knowles, W. N. Meier, M. Savoie, and A. K. Windnagel. 2017. Sea Ice Index, version 3 [2008–2021]. National Snow and Ice Data Center, Boulder, Colorado. doi:https://doi.org/10.7265/N5K072F8. Accessed 23 April 2021.

Francis, R. C., S. R. Hare, A. B. Hollowed, and W. S. Wooster. 1998. Effects of interdecadal climate variability on the oceanic ecosystems of the NE Pacific. Fisheries Oceanography 7(1):1–21. https://doi.org/10.1046/j.1365-2419.1998.00052.x.

Fritz, L., K. Sweeney, R. Towell, and T. Gelatt. 2015. Results of Steller sea lion surveys in Alaska, June–July 2015. Memorandum to D. DeMaster, J. Bengtson, J. Balsiger, J. Kurland, and L. Rotterman, 28 December. AFSC, Marine Mammal Laboratory, NOAA, Seattle, Washington.

Gentry, R. L. 1998. Behavior and ecology of the northern fur seal. Princeton University Press, Princeton, New Jersey.

Hare, S. R., and N. J. Mantua. 2000. Empirical evidence for North Pacific regime shifts in 1977 and 1989. Progress in Oceanography 47(2–4):103–45. https://doi.org/10.1016/s0079-6611(00)00033-1.

Hunt, G. L., M. Renner, K. Kuletz, S. Salo, L. Eisner, P. Ressler, C. Ladd, and J. Santora. 2018. Timing of sea-ice retreat affects the distribution of seabirds and their prey in the southeastern Bering Sea. Marine Ecology Progress Series 593:209–30. https://doi.org/10.3354/meps12383.

Huntington H. P., L. T. Quakenbush, and M. Nelson. 2017. Evaluating the effects of climate change on Indigenous marine mammal hunting in northern and western Alaska using traditional knowledge. Frontiers in Marine Science 4(319):1–17. doi:10.3389/fmars.2017.00319.

Insley, S. J., P. I. Melovidov, D. J. Jones, B. W. Robson, P. A. Zavadil, and R. Paredes. 2017. Multi-year counts of sea ducks and gulls in the nearshore of the Pribilof Islands, Alaska. Northwestern Naturalist 98(3):215–27. https://doi.org/10.1898/nwn16-20.1.

Kucklick, J., J. Reiner, M. Schantz, J. Keller, J. Hoguet, C. Rimmer, T. Ragland, R. Pugh, A. Moors, J. Rhoderick, J. Ness, D. Peterson, and P. Becker. 2013. Persistent organic pollutants and vitamins in northern fur seals (*Callorhinus ursinus*) collected from St. Paul Island, Alaska, as part of the Alaska Marine Mammal Tissue Archival Project. NIST Report. http://dx.doi.org/10.6028/NIST.IR.7958.

Lechikoninen, A., M. Kilpi, and M. Öst. 2006. Winter climate affects subsequent breeding success of common eiders. Global Change Biology 12(7):1355–65. https://doi.org/10.1111/j.1365-2486.2006.01162.x.

Lestenkof, P. M., P. I. Melovidov, A. P. Lestenkof, and L. M. Divine. 2018. The subsistence harvest of Steller sea lions on St. Paul Island, Alaska, from 2005–2016. Available from Aleut Community of St. Paul Island, Tribal Government, Ecosystem Conservation Office, St. Paul Island, Alaska.

Lestenkof, P. M., P. A. Zavadil, S. M. Zacharof, and E. M. Melovidov. 2013. Subsistence harvest monitoring results from 1999 to 2010 and local and traditional knowledge interview results for St. Paul Island, Alaska.
https://2pj6gg2zevt02q7hkno9xf7f-wpengine.netdna-ssl.com/wp-content/uploads/2021/03/BSIERP-St-Paul-final-report_LTK-interviews-and-data.pdf.

Loughlin, T. R., D. J. Rugh, and C. H. Fiscus. 1984. Northern sea lion distribution and abundance, 1956–1980. Journal of Wildlife Management 48:729–40.

Lowney, M. S., S. F. Beckerman, S. C. Barras, and T. W. Seamans. 2018. Gulls. USDA Animal and Plant Health Inspection Service, Wildlife Damage Management Technical Series. https://www.aphis.usda.gov/wildlife_damage/reports/Wildlife%20Damage%20Management%20Technical%20Series/Gulls-WDM-Technical-Series.pdf.

McHuron, E. A., K. Luxa, N. A. Pelland, K. Holsman, R. Ream, T. Zeppelin, and J. T. Sterling. 2020. Practical application of a bioenergetic model to inform management of a declining fur seal population and their commercially important prey. Frontiers in Marine Science 7:1–17. https://doi.org/10.3389/fmars.2020.597973.

Mehlum, F. 2012. Effects of sea ice on breeding numbers and clutch size of a high Arctic population of the common eider, *Somateria mollissima*. Polar Science 6(1):143–53. https://doi.org/10.1016/j.polar.2012.03.004.

Moore, S. E., and H. Huntington. 2008. Arctic marine mammals and climate change: impacts and resilience. Ecological Applications 18:S157-65. doi:10.1890/06-0571.1

Muto, M. M., V. T. Helker, J. Delean, R. P. Angliss, P. L. Boveng, J. M. Breiwick, M. M. Brost, M. F. Cameron, P. J. Clapham, and S. P. Dahle. 2019. Alaska marine mammal stock assessments, 2019. NOAA Technical Memo NMFS-AFSC-404. National Technical Information Service, Springfield, Virginia. https://media.fisheries.noaa.gov/dam-migration/2019_sars_alaska_508.pdf.

National Marine Fisheries Service [NMFS]. 2008. Recovery plan for the Steller sea lion (*Eumetopias jubatus*). Revision. NMFS, Silver Spring, Maryland.

National Marine Fisheries Service [NMFS]. 2020. Fisheries of the United States, 2018. US Department of Commerce, NOAA Current Fishery Statistics No. 2018. https://www.fisheries.noaa.gov/national/commercial-fishing/fisheries-united-states-2018.

Newman, K., A. Springer, and C. Matkin. 2008. Killer whales at the Pribilof Islands: who are they and what are they doing. Journal of the Acoustical Society of America 124(4):2507. https://doi.org/10.1121/1.4782886.

Oppel, S., A. N. Powell, and D. L. Dickson. 2009. Using an algorithmic model to reveal individually variable movement decisions in a wintering sea duck. Journal of Animal Ecology 78(3):524–31. https://doi.org/10.1111/j.1365-2656.2008.01513.x.

Reiner, J. L., R. P. Becker, M. O. Gribble, J. M. Lynch, A. J. Moors, J. Ness, D. Peterson, R. S. Pugh, T. Ragland, C. Rimmer, J. Rhoderick, M. M. Schantz, J. Trevillian, and J. R. Kucklick. 2016. Organohalogen contaminants and vitamins in northern fur seals (*Callorhinus ursinus*) collected during subsistence hunts in Alaska. Archives of Environmental Contamination and Toxicology 70(1):96–105. https://doi.org/10.1007/s00244-015-0179-y.

Romano, M. D., L. M. Divine, and S. Merculief. 2019. The role of red-legged kittiwakes in the subsistence culture of the Pribilof Islands, Alaska. AMNWR 2019/08. US Fish and Wildlife Service Report, Homer, Alaska.

Schummer, M. L., A. Petrie, R. C. Bailey, and R. B. Badzinski. 2012. Factors affecting lipid reserves and foraging activity of buffleheads, common goldeneyes, and long-tailed ducks during winter at Lake Ontario. Condor 114(1):62–74. https://doi.org/10.1525/cond.2012.110050.

Short, J. W., H. J. Geiger, L. W. Fritz, and J. J. Warrenchuk. 2021. First-year survival of northern fur seals (*Callorhinus ursinus*) can be explained by pollock (*Gadus chalcogrammus*) catches in the eastern Bering Sea. Journal of Marine Science and Engineering 9:975. https://doi.org/10.3390/jmse9090975.

Sowls, A. L. 1993. Trip report: winter wildlife and oil contamination surveys, St. Paul Island, Alaska, 8–12 March 1993. AMNWR 93/15. US Fish and Wildlife Service Report, Homer, Alaska.

Sowls, A. L. 1997. Winter observations associated with the Citrus oil spill of February/March, 1996. AMNWR 97/23. US Fish and Wildlife Service Report, Homer, Alaska.

Springer, A. M., J. A. Estes, G. B. Van Vliet, T. M. Williams, D. F. Doak, E. M. Danner, K. A. Forney, and B. Pfister. 2003. Sequential megafaunal collapse in the north Pacific

Ocean: an ongoing legacy of industrial whaling? Proceedings of the National Academy of Sciences 100:12223–28.

Springer, A. M., J. A. Estes, G. B. Van Vliet, T. M. Williams, D. F. Doak, E. M. Danner, and B. Pfister. 2008. Mammal-eating killer whales, industrial whaling, and the sequential megafaunal collapse in the north Pacific Ocean: A reply to critics of Springer et al. 2003. Marine Mammal Science 24:414–42.

Torrey, B. 1983. Slaves of the harvest. Tanadgusix Corporation, St. Paul Island, Alaska.

Towell, R., R. R. Ream, J. Sterling, J. Bengtson, and M. Williams. 2018. Northern fur seal pup production and adult male counts on the Pribilof Islands, Alaska. Memorandum for the record, 8 November. AFSC, Marine Mammal Laboratory, NOAA, Seattle Washington.

Tran, J., L. M. Divine, and L. R. Heffner. 2020. "What are you going to do, protest the wind?" Community perceptions of emergent and worsening coastal erosion from the remote Bering Sea community of St. Paul, Alaska. Environmental Management 67(1):43–66. https://doi.org/10.1007/s00267-020-01382-6.

Trites, A. W., and P. A. Larkin. 1996. Changes in the abundance of Steller sea lions (*Eumetopias jubatus*) in Alaska from 1956 to 1992: how many were there? Aquatic Mammalogy 22:153–66.

US Global Change Research Program [USGCRP]. 2014. National climate assessment. https://nca2014.globalchange.gov/.

Veniaminov, I. 1984 [1840]. Notes on the islands of the Unalashka District. L. T. Black and R. H. Geoghegan, translators, Richard A. Pierce, editor. Limestone Press, Kingston, Ontario, Canada.

Worm, B., R. Hilborn, J. Baum, T. A. Branch, J. S. Collie, C. Costello, M. J. Fogarty, E. A. Fulton, J. A. Hutchings, S. Jennings, O. P. Jensen, H. K. Lotze, P. M. Mace, T. R. McClanahan, C. Minto, S. R. Palumbi, A. M. Parma, D. Ricard, A. A. Rosenberg, R. Watson, and D. Zeller. 2009. Rebuilding global fisheries. Science 31(325):578–85. doi:10.1126/science.1173146.

Young, R. C., A. S. Kitaysky, C. Carothers, and I. Dorresteijn. 2014. Seabirds as a subsistence and cultural resource in two remote Alaskan communities. Ecology and Society 19(4):40. http://dx.doi.org/10.5751/ES-07158-190440.

Zavadil, P. A., D. Jones, A. D. Lestenkof, P. G. Tetoff, and B. W. Robson. 2006. The subsistence harvest of Steller sea lions on St. Paul Island in 2005. Unpublished report available from the Aleut Community of St. Paul Island, Alaska.

Chapter 13

The Indigenous Guardians Network for Southeast Alaska

MICHAEL I. GOLDSTEIN, AARON J. POE, RAYMOND E. PADDOCK III, and BOB CHRISTENSEN

The Tlingit and Haida peoples of Southeast Alaska have histories of resilient economies that stretch back millennia. Marine and terrestrial ecosystem services continue to be critical cultural components, although current management policies and market economies do not always favor small-scale, stewardship-minded local harvesters and Indigenous people's needs. This contrasts with large land base designations in Southeast Alaska, which is principally managed by the USDA Forest Service (USFS).

The USFS and the Central Council of Tlingit & Haida Indian Tribes of Alaska (Tlingit & Haida), through a regional support network called the Sustainable Southeast Partnership (SSP; http://sustainablesoutheast.org/), have endorsed and supported the nascent Indigenous Guardians Network (IGN) in Southeast Alaska. The Guardians are dedicated to supporting Alaska Native tribal communities to express their inherent sovereignty and creating opportunities for these communities to conserve their homelands and traditional ways of life for future generations. Here we discuss the people, places, and partnerships behind this initiative, the history of how the program came to be, and some recent developments that appear to be long lasting. We also provide examples of ongoing work and our plans to build, nurture, and grow the Guardians to incorporate additional communities.

SOUTHEAST ALASKA

Southeast Alaska is a unique mosaic of islands, mainland, intercoastal land, and areas exposed to the open Pacific Ocean. This Alexander Archipelago is a 300-mile-long group of ~ 1,100 nearshore mountainous islands. Known locally as Tlingit Aani, the land is home to the Raven (Yéil) and Eagle/Wolf (Ch'aak'/Gooch) moieties. The Eagle/Wolf has two names to differentiate its primary crest between the north and south regions of Tlingit territory. Each moiety is further subdivided into clans and each clan into houses. The Haida peoples come from the islands to the south, the Haida Gwaii, also known as the Queen Charlotte Islands of British Columbia, Canada. Thus, the Tlingit and Haida peoples are spread across a unique region where deep channels and fjords separate the islands and mainland.

Mostly accessible by boat or plane, the rugged islands form a dense rainforest mixed with rock, snow, and ice. This coastal biome is categorized as northern temperate rainforest. The area is sparsely populated (~ 0.8 people/km²) and contains an abundance of fish and wildlife. Island and mainland communities generally are not connected by roads. The Alaska capital, Juneau, is the largest city in the region (~ 32,000 people). The largest cities on the islands south and west of Juneau are Ketchikan (~ 8,200 people) on Revillagigedo Island, and Sitka (~ 8,600) on Baranof Island. The region's main industries are tourism, fishing, mining, timber harvest, and government.

Much of Southeast Alaska is managed by the USFS Tongass National Forest. The IGN was developed to support Alaska Native communities in expressing their inherent sovereignty. The goal is to apply Traditional Ecological Knowledge and sustainable ways of living to monitor, protect, restore, and manage Tlingit Aani homelands and waters. The Guardians bring together new and existing partners from tribes and other Indigenous entities to work with the USFS to address community needs across Southeast Alaska (figure 13.1).

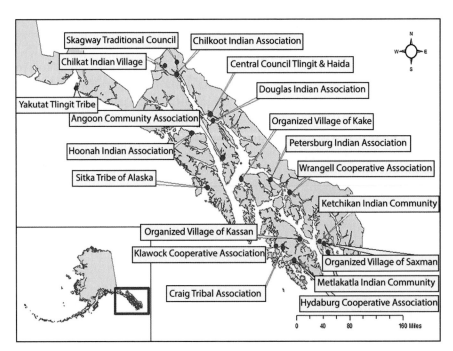

13.1. Southeast Alaska and the tribes invited to participate in the Indigenous Guardians Network formed by the Central Council of Tlingit and Haida Indian Tribes of Alaska, the USDA Forest Service, and the Sustainable Southeast Partnership in 2020. COURTESY OF THE TLINGIT & HAIDA TRIBES

We focused our partnership on four broad topic areas: adaptation planning, shared stewardship, heritage, and subsistence. The word "subsistence" is codified in regulations (see below), and our use in this chapter reflects that definition. Many Alaska Native people reject this term, however, because it can marginalize or trivialize the vital and central ways that the harvest, preparation, and sharing of wild foods fuel the life of Native communities. The term can be also seen as akin to mere "existence" or "survival" instead of conveying the true value of a life centered around traditional foods (figure 13.2) as an essential expression of connection to the natural world. Thus, in this chapter, subsistence also implies living a traditional way of life.

13.2. Traditional foods regularly harvested in Southeast Alaska. (*a*) Tree fungus, "chicken of the woods" (*Laetiporus conifericola*), here sautéed for vegetarian wontons; (*b*) Dungeness crab (*Metacarcinus magister*); (*c*) currants (*Ribes* spp.) and blueberries (*Vaccinium* spp.) jammed and jarred; (*d*) fiddlehead fern fronds(*Matteuccia* spp.) harvested in spring for stir-fry or salad; (*e*) roe of Pacific herring (*Clupea pallasii*) harvested in spring for personal consumption and festivals; (*f*) bull kelp (*Nereocystis luetkeana*); (*g*) salmon (*Oncorhynchus* spp.) harvested for subsistence, personal, sport, and commercial uses; and (*h*) Sitka black-tailed deer (*Odocoileus hemionus sitkensis*) harvested for subsistence and sport. PHOTOS COURTESY OF SUSTAINABLE SOUTHEAST PARTNERSHIP

ANCSA AND ANILCA

Resource exploration and extraction led to systematic persecution of the Tlingit and Haida. The Alaska Native Claims Settlement Act (ANCSA) became law in 1971 and was intended to resolve long-standing issues surrounding Aboriginal land claims in Alaska and to stimulate economic development throughout the state. In Alaska, there are no reservations like those found in the contiguous United States. ANCSA established Alaska Native "claims" to the land by transferring titles to 13 Alaska Native regional corporations and over 200 local village corporations. For example, Sealaska, the regional ANCSA corporation for Southeast Alaska, owns and manages 362,000 ac of land for > 23,000 shareholders.

The Alaska National Interest Lands Conservation Act (ANILCA) of 1980 added ~ 100 million ac to protected lands, including wilderness. Most important, ANILCA stipulated the designation of subsistence management. The Federal Subsistence Management Program is a multiagency effort to provide the opportunity for a subsistence way of life for rural Alaskans on federal public lands and waters while maintaining healthy populations of fish and wildlife.

RETURNING TO INDIGENOUS-LED STEWARDSHIP IN SOUTHEAST ALASKA

Though Tlingit people have resided in Southeast Alaska for over 10,000 years, current stewardship has principally been led by state and federal governments, including the regulation of commercial and sport hunting/fishing, recreation, small business development, and nature watching. This reality presents a few profound problems related to equity and environmental justice that we admittedly do not fully address in this chapter. Instead, we discuss the IGN as a functional example of how Southeast Alaska tribes and the US federal government can collaboratively manage lands and resources.

State and federal agencies have often worked at odds with local tribes, and the interpretation and integration of ANCSA/ANILCA have been difficult processes, such as in land exchange selections from federal to tribal or corporation management. In Southeast Alaska, the situation is further complicated by a profound lack of trust between the Tlingit and Haida people and the agencies that have perpetrated significant harm through colonization and resource extraction (Forbes 2019). When this history is compounded with rapid and accelerating climate impacts (Shanley et al. 2015), including larger storm and storm surge events, more overall precipitation although drier in summer, sea-level rise displacement, and loss of alpine habitats, it is easy to see that better cooperation is needed. Shared stewardship provides an opportunity to assist agencies in pivoting toward a diversified set of management approaches, which demonstrates a willingness to follow through on the government's commitment to support tribal self-determination. This can help rebuild previously damaged trust.

In 2011, leaders from tribes, Indigenous organizations, and conservation nonprofits came together to form the Salmon People Partnership, which exists today as the Sustainable Southeast Partnership. The SSP initially centered on helping people in the region create a new vision to move beyond the divisive "timber wars" of the previous decades and toward a vision of locally driven sustainability. As the SSP evolved, it focused on four pillars of sustainability: local business, energy independence, food security, and community fisheries and forestry (Forbes 2019). This helped set the stage for several initiatives focused on restoration, science, and monitoring in Southeast Alaska, which were led by the communities of Yakutat, Hoonah, Sitka, Kake, Klawock, Kasaan, and Hydaburg. Grant-funded "community catalysts" in each locale were supported by a "regional catalyst" for local fisheries and forestry to assist with large complex research projects that support traditional ways of life.

In 2014, a set of interviews with community leaders in the SSP identified common themes of interest on sustainable community development. Localizing benefits from natural resource management was a top priority, and there was interest in economic opportunities, protecting cultural practices, and localizing decision-making. In 2015, the SSP launched a community forest project known as the Hoonah Native Forest Partnership (HNFP). The HNFP is an alliance of landowners, organizations, and stakeholders, including the Hoonah Indian Association (a federally recognized tribe), the city of Hoonah, Sealaska, Huna Totem (Alaska Native corporation for Hoonah), the Nature Conservancy, Alaska Department of Fish and Game, and the USFS, all working together to study and protect the environment around Hoonah. The city of Hoonah, Sealaska, USFS, and Huna Totem are the primary landowners in the 205,000 ac project area. The HNFP works to assess resource conditions and identify projects intended to improve fish and wildlife habitat, ensure long-term timber production, and support sustainable watershed management. This collaboratively preserves valuable resources, creates economic opportunities for Hoonah residents, and improves community capacity through workforce development, community involvement, and land stewardship.

In Kake, the Keex' Kwaan Community Forest Partnership (KKCFP), launched in 2019, employs a similar approach to the HNFP by blending collaboration, local employment, and an emphasis on Indigenous values. Major partners and landowners include Sealaska, USFS, the Organized Village of Kake (a federally recognized tribe), and Kake Tribal Corporation. The strength of the KKCFP comes from involving tribal youth for resource inventories across a 150,000 ac project area, integrating traditional and cultural gathering practices as a primary emphasis in the analyses being conducted, and elevating tribal sovereignty as central to the overall purpose of the project.

Tribal Conservation Districts (TCDs) are Indigenous-led stewardship programs funded by the Natural Resources Conservation Service (NRCS) that provide resources to tribes (and ANCSA corporations) working together on shared stewardship priorities. The Prince of Wales TCD, launched in 2014, encompasses four tribes and four ANCSA corporations working on watershed and fisheries restoration and wildlife management. The Southeast Alaska TCD was launched in 2020.

Southeast Alaska Tribal Ocean Research (SEATOR, http://www.seator.org/) was formed in 2013, and its network now monitors harmful algal blooms, which are health risks to subsistence harvesters. Phytoplankton, shellfish, and water samples are collected and tested by each tribal partner. Data are used for early community detection and warning.

The successes of these local efforts, where tribes and Indigenous organizations have come together to lead stewardship initiatives across federal, state, and ANCSA corporation lands, inspire broader discussions about what might be possible.

A REGIONAL EFFORT FOR SHARED STEWARDSHIP MODELED AFTER BRITISH COLUMBIA GUARDIANS

Indigenous Guardians and Guardian Watchmen are terms used in Canada to identify people who monitor, manage, and steward their lands and waters. Indigenous Guardians serve as boots on the ground and act as the eyes and ears of the territory. Valérie Courtois, director of the Indigenous Leadership Initiative, said: "Indigenous Guardians programs strengthen our communities. They create jobs, lower crime rates and improve public health. But most importantly, they inspire our young people. They connect them to the land and their elders. They give them professional training tied to their language and culture. That offers hope that can combat the despair so many Indigenous youth feel today" (Indigenous Guardians 2021).

Indigenous Guardians programs in Canada developed an online toolkit, where Indigenous communities across Canada can learn, share, and connect to build and implement Indigenous Guardian programs. The toolkit includes practical information, tips, and resources; downloadable worksheets and templates to use and modify; stories of Guardians at work and quotes from people on the ground; and a map of Guardian programs across Canada.

In 2016, the SSP and the Nature Conservancy facilitated a cross-boundary community exchange with colleagues in British Columbia. The program had an emphasis on learning about their Coastal Stewardship Network and Guardian Watchmen programs. The SSP asked if localizing a version of the Coastal Stewardship Network in Southeast Alaska would provide better support to new and existing environmental programs in local communities and whether we could formalize community Guardian Watchmen. The SSP obtained funding to facilitate better collaboration between local stewardship efforts following the model of the Indigenous Guardians groups in British Columbia. Though the SSP worked closely with tribes in the region, it could not speak on behalf of tribal interests, so it organized a support network, which aimed to facilitate discussions toward the development of a Guardians Network in Southeast Alaska.

These efforts resulted in a series of meetings that helped strengthen the network. For example, a regional gathering in 2018 with participants representing 11 tribal communities and 20 organizations unanimously agreed that an Indigenous Guardians Support Network was an important effort worthy of their time and resources. Issues raised at a subsequent tribal environmental workshop facilitated by Tlingit & Haida included questions about whether non-Indigenous people and non-Indigenous conservation organizations were overly represented in program development; how an IGN can work to support and not overlap with other collaborative work, like that of SEATOR; and whether a future IGN could be centrally located for support and coordination without negatively affecting sovereign actions of individual tribes in the network.

More trust building between tribes and conservation organizations was needed, even among those that had worked closely with the SSP for years. Discussions also highlighted the need for IGN planning to be led by an Indigenous organization. Given its role as an Alaska Native regional nonprofit organization, the Tlingit & Haida assisted discussions at SEATOR, at SSP network meetings, and through other tribal outreach. Follow-up discussions resulted in a draft concept of a vision, mission, and guiding principles for a Guardians network in early 2020. This characterization was put forth by a working group from the SSP of approximately a dozen individuals from the Hoonah Indian Association, Organized Village of Kake, Tlingit & Haida, Sealaska, Spruce Root, and the Nature Conservancy. It largely reflected the multiyear learning around this topic and identified the "draft concept," which would be put forward for consideration by an eventual leadership body for the Guardians Network. Further, with the aim of centering Indigenous perspectives and leadership while being inclusive of the other entities interested in supporting this work, the SSP working group drafted a potential structure so that the IGN steering committee could improve upon it while developing the mission, vision, and guiding principles.

ESTABLISHING THE FIRST INDIGENOUS GUARDIANS NETWORK IN THE UNITED STATES

In 2020, supported by SSP partners Spruce Root, Sealaska, and the Alaska Conservation Foundation, the Tlingit & Haida and USFS formed the IGN in Southeast Alaska. Both parties contributed vision and funding to properly launch the network. This was aided by the long-term trust that had developed between the two organizational leaders (the Tlingit & Haida President and the USFS Alaska regional forester) who had worked together through several challenging issues over 20 years. The agreement is framed around four broad topic areas: adaptation planning, shared stewardship, heritage, and subsistence.

The multiyear, multiphase IGN agreement led to the hiring of two professionals at Tlingit & Haida to help coordinate the development of an IGN steering committee and charter, outreach to the complete set of tribes across Southeast Alaska (figure 13.1), and delineation of specific deliverables for new and existing stewardship and adaptation priorities. These efforts are supported by the SSP, Sealaska, and other agencies and partners to help grow accelerator events, provide facilitation services, and advance technology platforms. We envision a broad network of Guardians improving relationships and developing better, trusted collaborations among individuals, corporations, tribes, conservation groups, and agencies. We see people working together to protect wildlife habitat in watersheds relied upon for subsistence while also balancing other conservation outcomes and local economic benefits—all leading to collaborative co-management of lands and resources in Southeast Alaska. We believe in this so completely that the initiative in Southeast Alaska was renamed the Seacoast Indigenous Guardians Network in 2022.

On 15 July 2021, the USDA announced the Southeast Alaska Sustainability Strategy "to help support a diverse economy, enhance community resilience, and conserve natural resources" (USDA 2021). The overarching goals include limiting old-growth timber harvest to small and micro sales for community consumption and cultural uses, such as totem poles, canoes, tribal art, and musical instruments, and providing investments that are responsive to tribal and local priorities for sustainable economic development in Southeast Alaska. This specifically includes the complementary work of the IGN, SSP, HNFP, KKCFP, TCDs, the Southeast Conference, and the USFS and NRCS's Joint Chiefs' Restoration Initiative project.

When a trust deficit exists, then significant barriers to the development of shared stewardship models occur. This has happened between many organizations in Southeast Alaska, not just between federal agencies and tribes. Yet the agreement to establish the IGN, the support of the SSP and other tribal and nonprofit partners, and significant investment by the USDA and local corporations have provided the stimulus needed to jump-start an environmental justice program that is long overdue. It is important to note that conversations, visioning, and perspectives about any future IGNs must center around the Indigenous leaders and practitioners who are already doing this work. When establishing new and supporting programs, agencies and other organizations wanting to support Indigenous-led stewardship should be willing to share power and maneuver around their own preferences, prejudices, and traditions of doing business.

ACKNOWLEDGMENTS

We acknowledge that the efforts we describe in this chapter have taken place on Tlingit Aaní, which has sustained for millennia Tlingit, Haida, and Tsimshian peoples, the original stewards of the lands and waters of the region. Our hope in sharing this story is that we can complement their continuing efforts toward Indigenous ecological stewardship and improve control of and access to traditional lands and resources. We also acknowledge the numerous individuals who participated in the visioning of the Indigenous Guardians Network, including key inspiration offered by neighboring First Nations in Canada. For several years, tribes, Indigenous organizations, conservation groups, community advocates, agency managers, and scientists provided thoughtful insights about how such an effort might best come together. Along the way, we and they were inspired by elders and other culture bearers who emphasized the importance of doing this work "in a good way" that draws from Indigenous values and recognizes prior stewardship efforts. Finally, we acknowledge and appreciate the challenging nature of the conversations that led to the IGN. Its very formation is a result of people and organizations trusting one another enough to co-create something new despite deep and harmful past transgressions.

LITERATURE CITED

Forbes, P. 2019. Finding balance at the speed of trust: the story of the Sustainable Southeast partnership. http://sustainablesoutheast.net/wp-content/uploads/2021/02/Speed-of-Trust-2.pdf. Accessed 4 August 2022.

Indigenous Guardians. 2021. Toolkit. https://www.indigenousguardianstoolkit.ca/. Accessed 4 August 2022.

Shanley, C. S., S. Pyare, M. I. Goldstein, P. B. Alaback, D. M. Albert, C. M. Beier, T. J. Brinkman, R. T. Edwards, E. Hood, A. MacKinnon, M. V. McPhee, T. M. Patterson, L. H. Suring, D. A. Tallmon, and M. S. Wipfli. 2015. Climate change implications in the northern coastal temperate rainforest of North America. Climatic Change 130:155–70.

US Department of Agriculture [USDA]. 2021. USDA announces Southeast Alaska sustainability strategy, initiates action to work with tribes, partners and communities. Press Release No. 0157.21. https://www.usda.gov/media/press-releases/2021/07/15/usda-announces-southeast-alaska-sustainability-strategy-initiates. Accessed 4 August 2022.

Chapter 14

A Poem by **KIMBERLY BLAESER**

Glyph

i.
Onaabani-giizis/Snowcrust Moon (March 2021): A month in which Wisconsin allows the killing of 216 wolves and Deb Haaland is confirmed as Secretary of the Interior. *Ma'iingan*, our wolf relative. *Ogichidaakwe*, this woman warrior.

ii.
I remember the static of flannel pajamas halfway up my legs, my brother's dark flap of hair, and the scent of clover. I remember my Grandpa Blaeser's small clapboard house—but mostly, I remember the wolf pups.

In the wee hours of a late April morning, my uncles arrive in Ike's pickup—a rattle at the edge of the cracked sidewalk. My mom's voice coaxes us out of the heavy sleep of childhood. We wobble barefoot out the door, forget the danger of slivers from the two wooden steps. But the cold dew, the tickle of grass, finally wakes us. White clouds form from our breath, follow us as we giggle and shiver toward the glowing tips of cigarettes, toward the dark-skinned men with calloused hands, toward scratchy coughs and Uncle Emmett's echoing laughter.

After this day, a bundle passed in the dark will always stop time. The gunnysack of squirming. Musky puppy breath.

The morning comes with cinnamon pop-tarts, with coffee for the grownups. A real meal will follow. But first, the rescue story. Today, I can't listen carefully. I am nudged and gummed. Small-bodied into nestling. How long do we pile and unpile, collect the scent of yawns? I may have fallen asleep then, dreamed, misplaced pieces of the teachings.

iii.
In my nightmare, amorphous trophy hunters stalk, poison, and trap. Dogs chase down pregnant wolves. It is a half century later. Where the stories once lived, there is slaughter.

iv.
Some say "brother wolf." Their word balloons a comic-book creation.

Others place their hand there in the soft earth beside a four-inch print. Read this glyph like a text. The curved pads show us things about cycles and circles of kinship. We can trace the story words for cooperation there, for clan, loyalty, for balance. But a culture of capitalism makes words a commerce, sells the pelts of *Ma'iingan* to adorn Capitol rioters.

v.
Elders call *Ma'iingan* teacher. My dream mind watches. Can you see the hungry seasons when the songs and tracks of our wolf relatives helped us find game? See *mindimooyenh* trail the four-leggeds to learn medicines? Our fates paired—the Anishinaabeg and *Ma'iingan*—settler colonialism treated both the same. Feared us. Hated us. Placed on each a bounty.

Now by necessity, we learn the ideograms of legalese. 2012, we designate White Earth a *Ma'iingan* Sanctuary. 2019, Bad River adopts a formal plan for coexistence.

Memorize this premise: *Ma'iingan is a relative. You don't "manage" a relative; you build a relationship.*

vi.
You may not have settled fetal-deep on the rough brown cloth of belonging, slept poor but storied amid the mewing of orphaned wolf pups. But listen now to *aadizookaanag*, the sacred teachings.

This one tells of Seven Grandfathers, of gifts—and the animals who model them. Watch *Ma'iingan* bow. The ancient drama a ritual of humility—*dibaadendiziwin*. Throughout Anishinaabeg lands, live those who protect and work for the good of their community, their pack or their family. We call them Wolf Clan.

But some creatures, like mythic cannibal giants, live to destroy. They manage. They cull. They disturb natural balances. Despite their disguises, we have old names for them. *Wiindigoo. Maji-manidoo.*

vii.
For most of the year, I live far from the night howl of wolves. In the solitude of workdays, my small soft hands sometimes touch the bevel-edged photos. I do not need them to remember the cold noses of pups, the small kitchens of my youth, the voices of which I am woven.

But in the political caldron, slick-voiced spectres brew bad medicine—*maji-mashkiki*.

Again we wake. Beneath the ancient Snowcrust Moon—figures stilled in ochre.

Somewhere between wolf and warrior, I hold tight to the bundles of my past—to each warm-bodied talisman, to kinship.

Chapter 15

Case Studies of Species Recovery and Management of Trumpeter Swan and Leopard Frog on the Flathead Indian Reservation

KARI L. ENEAS, ARTHUR M. SOUKKALA, and DALE M. BECKER

INTRODUCTION

Since time immemorial, the ancestors of today's Confederated Salish and Kootenai Tribes (CSKT) have relied on the fish and wildlife that their homeland produces as resources for food, clothing, tools, and raw materials used to fill a variety of basic needs. The Tribes today are composed of descendants of the Salish (Flathead), Pend d'Oreille, and Kootenai Tribes that traditionally occupied 20 million acres stretching from central Montana to eastern Washington and north into Canada (McDonald et al. 2005). The CSKT are closely tied to our natural environment, and we believe everything in nature is embodied with a spirit. These spirits are woven tightly together to form a sacred whole (the Earth). We view these natural resources as essential components in sustaining tribal values and cultures.

Under the Hellgate Treaty of 1855, these Tribes ceded most of these ancestral lands to the US government, which designated the exterior boundary of the current Flathead Indian Reservation (FIR), which encompasses 1.34 million acres (542,279 hectares; figure 15.1). Under the treaty, tribal members retained the exclusive right to hunt and fish on the Reservation and to hunt, fish, and gather on traditional open and unclaimed lands outside the Reservation (Kappler 1904).

The FIR includes 91,000 acres (36,827 hectares) of protected habitat in the Mission Mountains Tribal Wilderness Area and other protected areas, such as the CSKT Grizzly Bear (*Ursus arctos horribilis*) Management Zone, Ferry Basin Elk (*Cervus elaphus*) Conservation Area, Hog Heaven Conservation Area, Little Money Bighorn Sheep (*Ovis canadensis*) Conservation Area, the Tribal Bison (*Bison bison*) Range, and the Kicking Horse Wildlife Mitigation Area.

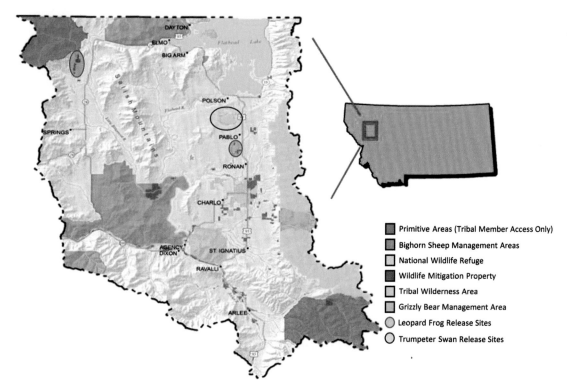

15.1. Map of the entire Flathead Indian Reservation in Montana with wildlife management areas, tribal wilderness areas, and release sites of trumpeter swans and northern leopard frogs. COURTESY OF CONFEDERATED SALISH AND KOOTENAI TRIBES

The Ninepipe and Pablo National Wildlife Refuges and seven federal waterfowl protection areas are managed by the US Fish and Wildlife Service (USFWS), and several wildlife management areas are managed by Montana Fish, Wildlife and Parks (figure 15.1).

The mission of the Tribal Wildlife Management Program (TWMP) is to protect, enhance and manage wildlife species and habitats to provide viable populations of all native wildlife species on the FIR for use by generations of today and tomorrow. The cultural importance of all species of native wildlife has been a cornerstone of the TWMP's wildlife management efforts. The program's long history in management and our initiatives to restore two locally extirpated species of wildlife have been strongly supported by tribal elders, the Tribal Council, and many local citizens.

TRUMPETER SWAN REINTRODUCTION AND MANAGEMENT

Trumpeter swans (*Cygnus buccinator*) were historically present in western Montana during the nesting season, with early observations by explorers and naturalists, including Lewis and Clark and David Thompson (1916) in the early

1800s (Banko 1960), who also noted an 1842 observation by Father Pierre-Jean De Smet of swan eggs being collected by an Indian hunting party near Flathead Lake (Thwaites 1906). Other references to trumpeter swans in western Montana include observations by E. S. Cameron in 1881, which described nesting trumpeter swans on the Thompson River in 1871, and along the South Fork of the Flathead River in 1889 (Coale 1915, Bent 1923).

After settlement by non-Indians in western Montana started in the late 1800s, documentation of resident summer trumpeter swans became rare, and the birds were extirpated as a local breeding species in the early 1900s, likely due to subsistence hunting and market hunting for pelts and feathers, as indicated by Hudson's Bay Company records (Linduska 1964). Between 1823 and 1880, approximately 108,000 swans were harvested, and skins and quills were exported. Similar large-scale harvests and shipments of swan parts (many of which were presumably trumpeter swans) were documented by Houston et al. (1997) in central Canada.

Habitat loss and degradation and the susceptibility of trumpeter swans to disturbance also played a role in local population declines, as many FIR wetlands were drained and converted to agricultural fields and pastures in the early 1900s. The history and decline of trumpeter swans in the Rocky Mountain population and the potential difficulties anticipated for restoration of the species in the northern Rockies were documented by Ball et al. (2000). The combined result was a decline in trumpeter swans and a constriction of their breeding range, leading to a population of fewer than 100 in the area around Yellowstone National Park. Early trumpeter swan conservation efforts resulted in the protection of the species from harvest and the establishment of Red Rock Lakes National Wildlife Refuge in southwestern Montana. However, trumpeter swans remained limited in number throughout much of their former range in the northern Rocky Mountains.

In the mid-1990s, the CSKT, Montana Fish, Wildlife and Parks, the USFWS, and local individuals began discussions on reintroducing trumpeter swans on the FIR. The Reservation's abundance and quality of wetland habitats were thought to provide a suitable location for successful reestablishment. After completion of an environmental assessment, work commenced on locating a source of swans.

The management plan for the Rocky Mountain population of trumpeter swans (Subcommittee on Rocky Mountain Trumpeter Swans 1992, 1998) recommended actions to reestablish the species throughout its original breeding range, conduct population surveys, begin population management (including augmentations and reintroductions), and begin public education and research. Interagency efforts to refine the focus of the plan resulted in the inclusion of the Flathead River drainage as a potential reintroduction site.

Initial reintroductions in 1996 were spearheaded by the CSKT and involved the release of 19 Canadian swans captured in southern Oregon. In 1998, 10 cygnets were transported from Northern Alberta to the FIR. However, these released trumpeter swans did not return to the release site in subsequent years.

The project was also limited by difficulties and uncertainties in obtaining wild swans for sustained releases (Becker and Lichtenberg 2007).

Evaluation of the project indicated a continuing strong interest by all partners and the public, but also a need to develop a means of ensuring a reliable source of swans from which to translocate birds. In 1999, the agency partners and the CSKT Tribal Council agreed to support the CSKT's development of a cooperative relationship with the Wyoming Wetlands Society (WWS). The WWS had a strong track record of captive reproduction of trumpeter swans and introduction of captive-reared swans in Wyoming. The CSKT provided fund-

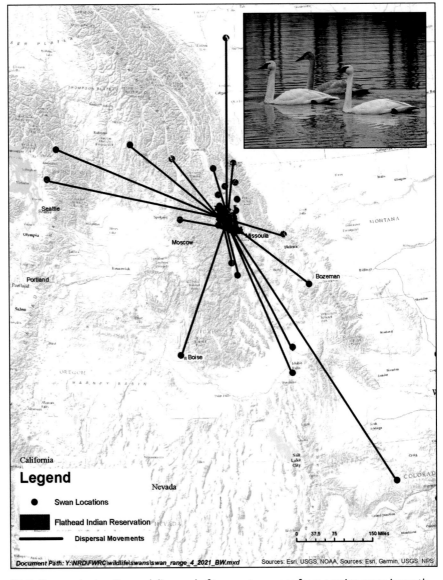

15.2. Seasonal migration and dispersal of trumpeter swans from nesting areas in northwestern Montana, 2002–2020. COURTESY OF CONFEDERATED SALISH AND KOOTENAI TRIBES

ing to obtain the swans, assist in upgrading its facilities, and expand captive breeding. The TWMP obtained 24 adult and subadult swans from a breeding facility in Montana (these swans' lineage was originally from the Rocky Mountain population in the northern Rockies). A similar contractual agreement for captive propagation with the Montana Waterfowl Foundation was also initiated. In 2008, two pairs of captive trumpeter swans were moved from the WWS facilities to those of the Montana Waterfowl Foundation to begin captive propagation there. All birds were examined and quarantined to ensure good health before potential contact with wild swans. Genetic relationships for each bird were evaluated, and some birds were traded for other captive trumpeter swans to reduce genetic duplication and to further diversify genetic stock.

The first 34 progeny were transported from WWS rearing areas in Wyoming to the FIR in 2002 and released at Pablo National Wildlife Refuge. Between 1996 and 2019, a total of 293 swans were released on the FIR; in 2020, 256 swans were documented, including 31 nesting pairs on the FIR and surrounding expansion areas in northwest Montana; and at least 199 swans wintered on the FIR in early 2021.

Since 2004, at least 221 nesting attempts have been recorded, and a minimum of 609 cygnets have been produced and survived to fledge (table 15.1). Some reintroduced swans have dispersed northward, and nesting pairs have established in other areas in northwestern Montana and southern British Columbia (Leighton 2015, Rachel Darvill, pers. comm. 2019).

TABLE 15.1. Trumpeter Swan Nesting Performance and Reproduction in Northwestern Montana, 2004–2020

YEAR	2004	2005	2006	2007	2008	2009	2010	2011*	2012*
Pairs	6	5	5	8	7	5	9	9	13
Nests	3	2	2	7	6	5	9	9	13
Productivity									
Nests	3	2	2	7	6	4	8	8	8
Cygnets	7	6	8	21	26	13	25	40	27
Fledged	7	5	8	17	19	13	20	37	20

YEAR	2013*	2014*	2015*	2016*	2017*	2018*	2019	2020
Pairs	22	18	33	25	26	25	30	31
Nests	16	13	24	19	26	23	25	19
Productivity								
Nests	13	12	21	17	22	21	22	17
Cygnets	41	45	78	71	75	80	82	57
Fledged	34	42	72	60	67	64	68	56

*2011–2018 data include areas outside of the Flathead Indian Reservation and elsewhere in northwestern Montana.

Trumpeter swans on the FIR are relatively isolated from the other nearest breeding populations in the Yellowstone National Park area, the Madison Valley in southwestern Montana, and the Blackfoot Valley in west-central Montana, but recent observations indicate some mixing with swans at those locations.

The autumn population in northwestern Montana comprises approximately 26% of the total number of trumpeter swans in the tristate area of Montana, Wyoming, and Idaho. Wildlife managers in the three states are now investigating ways to link these small and dispersed populations by seeking other potential reintroduction sites (Olson et al. 2021).

Extended migratory movements and dispersals have been documented by CSKT (CSKT Wildlife Management Program, unpub. data, 2021). Most birds have overwintered on the lower Flathead River on or adjacent to the southwestern part of the FIR. Swans from this project have also dispersed to other locations (figure 15.2). They have been observed as far as northwestern Colorado (960 km southeast); northwestern Washington (600 km west); Alberta (680 km north); Shelley, Idaho (460 km south); Creston, British Columbia (365 km north); and Fruitvale, British Columbia (440 km northwest).

Initial concerns about potential mortalities of released swans from ingestion of spent lead shotgun pellets in wetlands proved not to be a problem, likely due to tribal policies and bird hunting regulations, which have prohibited the use of lead shot for waterfowl hunting on the FIR since 1987. The primary cause of swan mortalities was collisions with overhead power lines, accounting for at least 121 deaths. A few collisions with fences, illegal shooting, and mortalities of unknown causes have also been documented. To address the power line collision mortality, we have worked with Mission Valley Power, a tribally operated electrical utility, to install line markers to add to the visibility of power lines in areas of documented swan mortality or areas heavily utilized by the swans. We also installed firefly flight diverters (P and R Technologies, Portland, Oregon).

Approximately 22.5 km of Mission Valley Power lines in and adjacent to wetland areas have been fitted with approximately 650 firefly flight diverters at seven sites, and only two additional collision-caused mortalities were noted at these sites. Power line collisions have declined, but sporadic collisions with unmarked lines still occur. Collision sites continue to be evaluated, and plans are developed for diverters at the sites.

Trumpeter swans are now a commonly observed species in many wetlands on the FIR and other areas in northwestern Montana during the summer months. In winter, a large proportion of the local swans migrates to the lower Flathead River on the FIR.

NORTHERN LEOPARD FROG REINTRODUCTION

Historically, northern leopard frogs (*Lithobates pipiens*) occurred throughout Montana at elevations below 6,000 feet (Maxell 2000), but they had been extirpated from most of western Montana (Reichel and Flath 1995, Werner 2003) and much of the western United States and Canada (Rorabaugh 2005). The Montana Natural Heritage Program has received no documented sighting of northern leopard frogs near Kalispell since 1998 or near Eureka since 2012. Three records from 2009 to 2016 near Missoula have been of single individuals. On the FIR, tribal elders remember there being many frogs in the valley, and they tell of two different types. We believe they are describing leopard frogs and Columbian spotted frogs (*Rana luteiventris*). On the FIR, the first recorded observation of leopard frogs was in 1892, which was followed by nine scattered records from 1942 to 1980. After 1980, there was only one record of a single individual in 2000, despite extensive surveys of amphibians conducted across the FIR lands since 1993.

The Tribal Wildlife Management Program Plan vision statement in part reads: "The Tribal Wildlife Management Program will strive to restore and enhance biological diversity of the ecosystems of the Flathead Indian Reservation" (Eneas 2022:1). Returning leopard frogs fills a missing ecological niche and provides a step toward restoring the diversity of native wildlife on the FIR.

Developing Methods

In 2001, we began to examine the feasibility of reintroducing leopard frogs into suitable habitat on the FIR. Localized populations occur in central and eastern Montana, and leopard frogs are abundant in southeastern Montana (Reichel and Flath 1995), but we were aware of only two populations of leopard frogs west of the Continental Divide in Montana: at Foy's Lake near Kalispell and in and around Timber Lake near Eureka. Because leopard frog populations in western Montana could have been introduced from elsewhere, we initiated DNA testing in 2001. Toe-clip samples were collected from the two western populations; from two populations in central Montana; and from museum specimens from Fort Steele, British Columbia, and Park City, Montana. Analyses revealed there was little genetic variation among Pacific Northwest populations (a divergence of only 0.6%), and the Montana populations west of the Continental Divide were genetically similar to populations east of the divide (a divergence of only 0.2%). These results supported our plan of reintroduction.

We chose to conduct our reintroduction using egg masses collected from the wild for several reasons:

- Eggs are available for longer periods of times and in much greater numbers than are adult frogs.
- Egg masses offer greater opportunities to increase survival during relocation whereas reintroducing adults in a new location would likely result in a decrease in survival.
- Relocating eggs increases the chances that adult frogs would return to the reintroduction sites to breed.
- Using eggs as opposed to adults decreases the chance of disease transmission from source sites to reintroduction sites.

Our initial methodology was tested in 2002 using Columbian spotted frogs as a surrogate species because they are abundant on the Reservation and occupy similar habitats to leopard frogs. Amphibian survey data and historical records were used to locate sources for egg masses. Egg mass searches were conducted from early April to mid-May. Early in the project, egg masses were obtained from the Blackfeet Reservation. Two Blackfeet Community College students were hired through a grant from Wildlife Conservation International to survey for leopard frogs and breeding activity. Later, the source area was expanded east to the Fort Belknap Reservation and to areas around the Bear Paw Mountains, Havre, and Malta. The Fort Belknap Tribal Council granted permission to collect egg masses from their Reservation, and their environmental staff assisted in locating breeding areas, monitoring breeding activity, and collecting egg masses.

We collected approximately 10% of egg masses present at source sites, hoping to minimize the effect on source populations (R. D. Semlitsch, pers. comm. 2001 or 2002) since only 1%–6% of eggs survive to metamorphosis (Hines et al. 1981, Merrell 1977). The procedure for transporting egg masses followed the Alberta Protocol (Kendell and Prescott 2007, Wendlandt and Takats 1999). We placed egg masses in a cooler with several gallons of source water. Additional clean source water was collected for use during transport. Water temperature was kept between 4°C and 16°C. Between egg collection and release, we rinsed egg masses at one- or two-hour intervals using a decreasing amount of source water and an increasing amount of untreated well water in 25% increments until egg masses were in 100% well water. We estimated the number of eggs in each mass using a volume displacement regression (Werner et al. 1999). The sequence of procedures varied from collection–transport–egg count estimate–rinse–placed into pens (site 1, SMC1) to collection–transport (rinse while in transport)–egg count estimate–placed in tubs for rearing (site 2, ULB1).

We monitored eggs daily during the two-week development period. We estimated hatching success based on nonhatched eggs remaining in the egg mass after hatching was complete (Corn and Livo 1989). Water temperature and pH were taken at both the source and introduction sites to provide baseline information.

We tested hatching and rearing pens to protect egg masses and hatchlings from predators. We constructed four enclosed pens with mesh hardware cloth on the outside and mesh nylon netting on the inside with an open bottom and a hardware cloth lid. Eggs were placed into nylon mesh baskets floating within the pens. After all viable eggs had hatched, 500 larvae were counted and released into each pen. The remaining larvae were released in shallow water along the shoreline near the enclosure. After the hatching, pens were checked once per week to monitor development. Occasionally, vegetation was picked and placed inside, but we made no concerted effort to increase the food supply on a regular basis. We modified our methods over time to incorporate our experience and thus improved methodologies. Ultimately, we hatched egg masses in plastic containers while maintaining water levels and water temperature.

Once hatching began, we removed hatchlings every two or three days and placed them into polyurethane tanks. Small numbers of bloodworms and algae were added to the water. We kept hatchlings in the tanks for approximately 10 days, while individuals shifted from yolk to algae feeding. After they converted to algae feeding, tadpoles were transferred to release sites. On occasion, some egg masses were placed directly into release wetlands, and some hatchlings were released earlier than 10 days from hatching.

Three methods were used to monitor juvenile and adult frog persistence, dispersal, and breeding. We used time-constrained methods when surveying dispersed areas and during exploratory surveys. We walked a set linear distance along the banks of a wetland or stream after repeated activity was noted so year-to-year surveys could be compared. Breeding was surveyed by stopping at likely spots and listening for calling. If calling was noted, we searched areas to pinpoint breeding pods (small areas with several frogs calling) and to count egg masses.

Learning and Adapting

From 2003 to 2015 a total of 153 egg masses were translocated from source areas. We transported a low of only 3 egg masses in 2004 compared with a high of 19 in 2007. Estimated embryos ranged from 3,328 to 63,748 (mean = 34,325) per year.

Timing egg mass collecting trips from source populations that were three to eight hours' driving time away proved difficult. Breeding would start, stop, and restart depending on weather. It was generally preferable to have someone local to the source populations monitor breeding. Working with Blackfeet students on the Blackfeet Reservation and with the tribal Environmental Protection Agency staff of the Fort Belknap Indian Community was invaluable. They were able to monitor local breeding and determine when egg laying was beginning, resulting in much more efficient collecting.

Hatching success ranged from 23% to 92% (mean = 72%). The estimated number of tadpoles released per year ranged from 750 to 58,631 (mean = 25,893). Hatchling survival in pens to metamorph ranged from 21% to 88% (mean = 68%).

Hatching and rearing eggs, hatchlings, and tadpoles in pens proved to be a great learning tool but was impractical as a method for reintroduction. It allowed us to monitor hatching success and survival and to learn about the metamorphosis process. However, even at the low numbers put into pens (500), the resources of space and food seemed limiting. Outside the pens, although hatchling survival could not be determined, young metamorphs were larger and were observed two to three weeks prior to seeing metamorphs inside the pens.

Hatching eggs in captivity and rearing them for up to 10 days was much more successful. Using this method, hatching success varied from 52% to 92% (mean = 82%), which was similar to estimates made in Colorado and Wyoming (Corn and Livo 1989). Also, protecting the hatchlings from predators and harsh weather events for even the first few days while they fed primarily on the yolk sacs proved successful. It also allowed us to distribute young hatchlings in a dispersed area within a wetland or among several wetlands.

Overwintering appeared to be a major bottleneck in the successful reintroduction in the SMC1 area. In the four years of egg and hatchling releases at SMC1, only two potential overwintering frogs (juveniles) were observed. We surveyed for several years after releases ended but never observed any leopard frogs. We were never able to determine the problem. Spotted frogs overwintered and bred in the area successfully throughout the project.

Our attempts to study overwintering by trying to prevent young metamorphs from dispersing using fenced-in wetland areas failed. In two instances, cattle destroyed the fences. In the third instance, there was an opportunity for metamorphs to escape through a drain pipe that was not known about.

Successful Translocations

Despite minimal success during the first four years, we decided not to give up. In 2006 we had restored a large backwater wetland off the Little Bitterroot River that had been ditched and partly drained near Niarada, Montana. The restoration produced a 20-hectare shallow wetland that seemed ideal as a reintroduction site. As with the first area (SMC1), spotted frogs bred and were abundant in the area as were Sierran treefrog (*Pseudacris sierra*).

We began releases into the ULB1 wetland in 2006, releasing 7,355 hatchlings that year, followed by 58,631 hatchlings in 2007. In July 2007, we observed some leopard frogs that, based on size, appeared to be juveniles from the previous year. It was a promising sign. In 2008, we released 16,802 hatchlings into the ULB1 wetland. On 3 August 2008 we definitely observed adult frogs from previous releases, confirming overwinter survival. From 2009 to 2015 we released 205,038 hatchlings (mean = 29,291/year) into the wetland and surrounding area.

In 2010 we heard our first calling frogs, indicating breeding activity. Over the next 2 years we heard an increasing number of frogs calling (2–3 in 2011 and 6–10 in 2012) but were unable to locate egg masses. In 2013 we heard up to 10 calling individuals and located three egg masses. This represented a significant milestone in the project. It was the first documented breeding of northern leopard frogs on the FIR in over 30 years.

From 2013 through 2015 we continued to document breeding activity and egg laying. In 2009 we had begun releasing some hatchlings in wetlands outside our main ULB1 wetland, some up to five river miles away. This allowed us to test whether other wetlands in this riverine wetland complex would support tadpoles, metamorphs, and potential overwintering. If successful, it would also promote expansion of the breeding population. Ultimately both objectives were met. In 2014 the largest leopard frog we had observed (10.2 cm snout to vent length) was recorded, an individual that had survived for several years.

In 2016 we stopped releasing additional hatchlings to enable the monitoring of the natural reproduction and survival of the newly established population. Subsequent surveys showed continued persistence with calling and egg masses documented within the release areas every year from 2016 to 2021. In 2016 we also documented egg masses 3.5 river miles downstream from the main release wetland. In 2017, a tribal land lessee reported frogs in his hayfield 4.0 river miles from the nearest release site. Follow-up surveys found young metamorphs in substantial numbers, which indicated that breeding likely occurred nearby. During the spring of 2018, we documented breeding activity over a 9-mile stretch along the Little Bitterroot River. These annual surveys have confirmed breeding along the same part of the river through 2022, and in the spring of that year we documented additional breeding activity 6 miles up Sullivan Creek, a tributary to the Little Bitterroot River.

Annual summer surveys have also documented frogs from the same 9-mile stretch of the river as the breeding surveys showed. In addition, we have documented leopard frogs up to a mile off the river bottom wetlands in hayfields and at an irrigation reservoir.

Fall surveys centered around frogs staging on the banks of the Little Bitterroot River. Frogs were common on the banks of the river, indicating that at least part of the population was using the river to hibernate. Results varied widely because of seasonal timing and weather conditions, but during our best surveys we observed 4 frogs (2.5 leopard frogs and 1.5 spotted frogs) per 100 meters of linear distance surveyed.

Next Steps

Our monitoring of leopard frogs will continue to document persistence and, we hope, expansion of the population spatially.

We will monitor the interaction of bullfrogs and leopard frogs. As there is little hope of controlling the expansion of nonnative bullfrogs, we hope that leopard frogs will be able to coexist with this nonnative invader. Our ultimate goal is to establish more than one leopard frog population on the FIR. We plan to begin transplanting egg masses to an area on the eastern side of the Reservation in the near future.

CONCLUSION

These two projects illustrate the CSKT's commitment to restoring formerly extirpated species to the FIR ecosystem. Along with other tribal wildlife management efforts, including proactive management and restoration of native wildlife and habitats and the recovery of listed wildlife species such as the bald eagle (*Haliaeetus leucocephalus*), northern gray wolf (*Canus lupus*) peregrine falcon (*Falco peregrinus*) and increased populations of grizzly bears (*Ursus horribilis*), the CSKT has enhanced the bio-fauna of northwestern Montana for generations to come.

LITERATURE CITED

Ball, I. J., E. O. Garton, and R. E. Shea. 2000. History, ecology and management of the Rocky Mountain population of trumpeter swans: implications for restoration. Pages 46–49 in R. E. Shea, M. D. Linck, and H. K. Nelson, editors. Special edition: the proceedings and papers of the Seventeenth Trumpeter Swan Society Conference. Trumpeter Swan Society, Maple Plain, Minnesota.

Banko, W. E. 1960. The trumpeter swan. North American Fauna 63. US Fish and Wildlife Service, Washington, DC.

Becker, D. M., and J. S. Lichtenberg. 2007. Trumpeter swan reintroduction on the Flathead Indian Reservation. Pages 100–105 in M. H. Linck and R. E. Shea, editors. Selected papers of the Twentieth Trumpeter Swan Society Conference. Trumpeter Swan Society, Maple Plains, Minnesota.

Bent, A. C. 1923. Life histories of North American wild fowl: ducks, geese, swans, scoters and others. United States Museum Bulletin 126. Smithsonian Institution, Washington, DC.

Coale, H. K. 1915. The present status of the trumpeter swan (*Olor buccinator*). Auk 32:82–90.

Corn, P. S., and L. J. Livo. 1989. Leopard frog and wood frog reproduction in Colorado and Wyoming. Northwest Naturalist 70:1–9.

Eneas, K. 2022. Tribal Wildlife Management Program Plan, fiscal year 2022. Confederated Salish and Kootenai Tribes, Pablo, Montana.

Hines, R. L., B. L. Les, and B. F. Hellmich. 1981. Leopard frog populations and mortality in Wisconsin, 1974–1976. Technical Bulletin No. 122. Department of Natural Resources, Madison, Wisconsin.

Houston, S. C., M. I. Houston, and H. M. Reeves. 1997. The 19th century trade in swan skins and quills. Blue Jay 55(1):24–34.

Kappler, C. 1904. Treaty with the Flathead: Hellgate Treaty 1855. US Government Printing Office, Washington, DC. https://www.fws.gov/pacific/ea/tribal/treaties/flatheads_1855.pdf.

Kendell, K., and D. Prescott. 2007. Northern leopard frog reintroduction strategy for Alberta. Technical Report T-2007-002. Alberta Conservation Association, Edmonton, Alberta, Canada.

Leighton, D. 2015. Trumpeter swan recolonization of the Columbia River wetlands in southeastern British Columbia. Biodiversity Centre for Wildlife Studies: Wildlife Afield 12(2):142–83.

Linduska, J. P. 1964. Waterfowl tomorrow. US Government Printing Office, Washington, DC.

Maxell, B. A. 2000. Management of Montana's amphibians: a review of factors that may present a risk to population viability and accounts on identification, distribution, taxonomy, habitat use, natural history and the status and conservation of individual species. Report to Northern Regional Office (Region 1), US Forest Service, Missoula, Montana.

McDonald, T., T. Tanner, L. Bigcrane, and D. Rockwell. 2005. Mission Mountains tribal wilderness: a case study. Confederated Salish and Kootenai Tribes, Pablo, Montana.

Merrell, D. J. 1977. Life history of the leopard frog (*Rana pipiens*) in Minnesota. Occasional Paper No. 15. University of Minnesota, Bell Museum of Natural History, St. Paul.

Olson, D., B. Smith, S. G. Brooke, and S. Armitage. 2021. Trumpeter swan survey of the Rocky Mountain population, US breeding segment. US Fish and Wildlife Service, Denver, Colorado.

Reichel, J., and D. Flath. 1995. Identification of Montana's amphibians and reptiles. Montana Outdoors, May–June.

Rorabaugh, J. C. 2005. *Rana pipiens* (Schreber 1782): northern leopard frog. Pages 570–77 in M. Lannoo, editor. Amphibian declines: The conservation status of United States species. University of California Press, Berkeley.

Subcommittee on Rocky Mountain Trumpeter Swans. 1992. Pacific Flyway management plan for the Rocky Mountain population of trumpeter swans. Pacific Flyway Study Committee, US Fish and Wildlife Service, Office of Migratory Birds, Portland, Oregon.

Subcommittee on Rocky Mountain Trumpeter Swans. 1998. Pacific Flyway management plan for the Rocky Mountain population of trumpeter swans. Pacific Flyway Study Committee, US Fish and Wildlife Service, Office of Migratory Birds, Portland, Oregon.

Thompson, D. 1916. David Thompson's narrative of his explorations in western America, 1784–1812. Pages 420–21 *in* J. B. Tyrell, editor. Publications of the Champlain Society, vol. 12 Champlain Society, Toronto, Ontario, Canada.

Thwaites, R. G. 1906. Early western travels. Arthur Clark, Cleveland, Ohio.

Wendlandt, M., and L. Takats. 1999. Northern leopard frog reintroduction: Raven River pilot year (1999). Alberta Environment, Fisheries and Wildlife Management Division, Edmonton, Alberta, Canada.

Werner, J. K. 2003. Status of the northern leopard frog (*Rana pipiens*) in western Montana. Northwestern Naturalist 84(1):24–30.

Werner, J. K., J. Weaselhead, and T. Plummer. 1999. The accuracy of estimating eggs in anuran egg masses using weight or volume measurements. Herpetological Review 30(1):30–31.

Chapter 16

Co-Management in Alaska
A Partnership among Indigenous, State, and Federal Entities for the Subsistence Harvest of Migratory Birds

PATRICIA K. SCHWALENBERG, LILIANA C. NAVES, LARA F. MENGAK, JAMES A. FALL, THOMAS C. ROTHE, TODD L. SFORMO, JULIAN B. FISCHER, *and* DAVID E. SAFINE

INTRODUCTION

The Alaska Migratory Bird Co-Management Council (AMBCC) brings together Indigenous, federal, and state partners to manage the subsistence harvest of migratory birds in Alaska. The AMBCC was created in 2000 to provide a meaningful role for Alaska Native peoples in harvest management and conservation of migratory birds. This was prompted by unintended shortcomings of the 1916 migratory bird treaty with Canada, its implementing law (1918 Migratory Bird Treaty Act, MBTA), and a treaty with Mexico, which set a hunting closure in the spring and summer to protect nesting birds and restore bird populations depleted by commercial hunting. However, for millennia, spring bird harvests had been key to supporting food security for northern Indigenous peoples. The spring-summer hunting closure brought hardships for Alaska Native peoples. Following requests by Native leaders, the treaties with Canada and Mexico were amended in 1997 to recognize the spring-summer subsistence harvest of migratory birds in Alaska.

    The history of bird harvest management has greatly influenced relations between Alaska Native peoples and resource management agencies. A 2018 joint federal-state apology to Alaska Native peoples for unintended harm caused by the MBTA illustrates efforts to mend relations (USFWS 2018). However, there are still challenges regarding trust since Alaska Native peoples endured hardships and intergenerational trauma from the spring-summer harvest closure, which lasted for nearly a century.

    Alaska is the only US state where the spring-summer harvest of migratory birds and their eggs is legally authorized. From a Western harvest management perspective, this harvest is controversial. Compared to fall-winter harvests, spring-summer harvests include a higher proportion of breeding-age birds, which have lower mortality and higher reproductive rates than immature birds

(Martin 1995). Harvest regulations for the spring-summer season do not include some tools commonly used in harvest management, such as daily and possession limits and special permits with required harvest reporting (table 16.1). The set of species opened for the spring-summer harvest includes dozens of species considered nongame by most state and federal management organizations (50 CFR 20.11), including species of conservation concern such as the yellow-billed loon and bar-tailed godwit (Naves and Zeller 2017, Naves 2018, Naves et al. 2019). In addition, achieving compliance with harvest regulations and law enforcement in rural Alaska is often challenging due to the complex and contentious harvest management history, diverse cultural and socioeconomic contexts, and the logistics of covering a vast and remote land.

From the perspective of Alaska Native users, some Western bird harvest management tools and regulations impose constraints to traditional harvest practices (table 16.1). In some regions, Alaska Native users oppose requirements, such as the need to procure a state hunting license and waterfowl stamp, which entail paperwork and economic burdens. Complex harvest regulations and requirements can cause hardships and misunderstandings in sociocultural contexts, such as in rural Alaska, where bureaucracy and paperwork have become widespread only relatively recently.

TABLE 16.1. Overview of Seasons and Regulations for Harvest of Migratory Birds and Their Eggs in Alaska

FALL-WINTER GENERAL SEASON (1 SEPTEMBER-22 JANUARY)	SPRING-SUMMER SUBSISTENCE SEASON (2 APRIL-31 AUGUST)
• Eligibility: no restrictions	• Eligibility: residents of qualified regions and communities in rural Alaska and other Alaska residents formally invited to hunt by immediate family members residing in qualified regions and communities
• Bird hunting only	
• About 35 species opened for harvest in Alaska (ducks, geese, tundra swans, sandhill cranes, Wilson's snipes, northwestern crows, cormorants, snowy owsl)[a]	
	• Bird and egg harvest
• Defined daily and possession limits	• 90 species opened for harvest (ducks, geese, tundra swans, sandhill cranes, seabirds, shorebirds, loons, grebes, snowy owls)[b]
• Defined shooting hours based on daylight	
• Requirements: state hunting license, federal and state waterfowl stamps	
• Special permits with required harvest reporting: tundra swan and emperor goose	• 30-day closure during the egg incubation and chick-rearing period
	• No daily and possession limits
	• No shooting hours
	• Requirements: state hunting license, state waterfowl stamp
	• No special permits with required harvest reporting

Sources: ADF&G 2021, USFWS 2021.
[a]Regulations for the fall-winter season specify bird families open for harvest (50 CFR 20.11), with exceptions in annual regulations in subpart K (e.g., species of conservation concern).
[b]Regulations for the spring-summer season define species open for harvest (50 CFR 92.22).

This complex and difficult history has resulted in varying degrees of acceptance of and compliance with harvest regulations in rural Alaska. Alaska Native peoples still express resentment regarding the traumatic events in their relationships with Western governments and organizations, such as being used as slaves during the Russian fur trade, the deadly evacuation of their Aleutian Islands homelands during World War II, mistreatment of youth in involuntary boarding schools, prohibition of the use of Native languages, and the 1918–2002 closure of the spring-summer bird harvest (Keeler et al. 1962, Yupiktak Bista 1974, Burwell 2005, Langdon 2014).

Since 1970, bird populations in North America have decreased by 30% (Rosenberg et al. 2019). Decreased continental bird abundance negatively affects harvest sustainability at the flyway scale, including in Alaska. In sharing the AMBCC history and context with a broader audience, we hope to encourage new partnerships and support collaboration among Indigenous and non-Indigenous individuals to ensure sustainable harvest opportunities, healthy fish and wildlife populations, and community well-being. By joining forces, we will be able to overcome the societal and environmental challenges we now face to ensure healthy bird populations that can sustain harvest by multiple users.

ALASKA NATIVE PEOPLES IN A FAST-CHANGING WORLD

The only time I really feel I am myself is when I am hunting. Every year I must return to the tundra if for only a few days.
—Anonymous quote, undated (in Yupiktak Bista 1974)

Hunting, fishing, and gathering wild resources have allowed Indigenous peoples to flourish and develop complex cultures in Alaska's various and, at times, extreme climates. Alaska Native peoples include six main groups—Inupiaq; St. Lawrence Island Yupik; Central Yup'ik; Athabascan; Alutiiq-Sugpiaq and Unangax̂-Aleut; and Eyak, Tlingit, Haida, and Tsimshian—that speak 20 languages and have distinct cultures and histories (Langdon 2014). Within this cultural diversity, Alaska Native peoples share core values, such as a strong connection to ancestral lands, reliance on traditional knowledge, physical and cultural dependence on wild foods and materials, and sharing of resources. Today, 22% of Alaska's population is American Indian or Alaska Native, the highest rate for this ethnic group among all US states (US Census Bureau 2019, AFN 2021). There are 228 federally recognized tribes in Alaska (BIA 2021). While 83% of Alaskans live in nonsubsistence (urban) areas, residents of about 260 rural communities (villages) are an intrinsic part of Alaska's unique socioeconomic and cultural diversity (Fall 2019). However, these rural communities face substantial challenges, including high costs of living, few economic and educational opportunities, and competition for fishing and hunting resources.

In 1971, following the discovery of oil on the North Slope of Alaska, the Alaska Native Claims Settlement Act (ANCSA) was passed. Alaska tribes negotiated a settlement of $962.5 million and surface and subsurface rights to 44 million acres of land in exchange for renouncing all unresolved land claims, including Aboriginal hunting and fishing rights along the proposed Trans-Alaska Pipeline System corridor. The ANCSA also created 13 regional for-profit corporations and over 200 village corporations to provide economic opportunities, employment, and health, education, and housing services to their shareholders (Arnold 1978, Berger 1985). In 1980, the Alaska National Interest Lands Conservation Act (ANILCA, Title VIII) attempted to address issues related to providing adequate access for Indigenous peoples to fishing and hunting resources by establishing a priority for subsistence uses by rural Alaska residents on federal lands. Ambiguity and disagreement between state and federal laws in Alaska regarding subsistence reflect the competition for access to fish and wildlife (Wheeler and Thornton 2005).

Besides socioeconomic and cultural changes, large-scale climate and environmental changes are also affecting life in Arctic and subarctic communities (Norton-Smith et al. 2016). Harvesting and sharing wild harvests provide both food and necessary cultural stability in a time of rapid change in all aspects of life (USFWS 1980, Moerlein and Carothers 2012).

A SUBSISTENCE-CASH MIXED ECONOMY IN RURAL ALASKA

In Alaska, state and federal laws recognize subsistence as the noncommercial customary and traditional uses of wild resources for food, clothing, fuel, transportation, construction, art, crafts, sharing, and customary trade (Fall 2018). Subsistence hunting in Alaska is associated with customary and traditional patterns of resource use typical of Indigenous cultures (5 AAC 99.010, 99.025-12), which tend to occur in rural areas where the population includes a high proportion of Alaska Native residents (in the late 2010s, about 123,000 people—17% of Alaska's population—resided in 260 rural communities; Fall 2019).

In the late 2010s, the estimated harvest of wild resources by rural residents for subsistence uses was about 34 million pounds of food per year, or 176% of the daily protein needs and 25% of the caloric needs of the rural Alaska population (Fall 2018). About 83% of rural households harvest fish and 60% harvest game. As harvests are shared with nonharvesting households, about 95% of rural households use fish and 86% use game. Harvests follow the seasonal cycle of availability of animals and plants. Statewide,

16.1. Alaska Native families harvesting wild foods.
PHOTO BY CLYDE TREFON, USED WITH PERMISSION.

the rural harvest is composed of fish (54%), land and marine mammals (22% and 14%), wild plants (4%), shellfish (3%), and birds and eggs (3%). Birds are a small proportion of the total harvest but have cultural and practical value as foods that traditionally have alleviated hunger and starvation (Fall 2018).

In the 21st century, cash economies and modern technologies have been integrated with harvesting activities. Extended families collectively invest cash incomes in harvesting gear and supplies, such as motorized skiffs, snowmobiles, fishing nets, ammunition, and fuel, and directly support people in need just as harvests have been shared for generations (USFWS 1980, Berger 1985). Combined with kinship-based labor, cash investments generate a greater output in wild harvests than the equivalent amount of money spent on store-bought foods (Wolfe and Ellanna 1983).

Many successful households generate both cash and harvest incomes. A family's subsistence production is usually supported by and complemented by cash, so these two economic sectors are mutually supportive. High reliance on wild resources and vibrant sharing networks persist in rural communities (BurnSilver et al. 2016). Harvesting and sharing wild foods support communities' sustainability and offer opportunities for youth to learn key skills and values (figure 16.1).

BIRDS IN ALASKA NATIVE CULTURES

What us Natives know is that everything is connected to each other. Even if there is one species we see that is thriving, it is depending on another species, and that species is depending on another. In turn, we are depending on that species and many other species. We are all connected, we try to respect that animal so it will return more.
—Gerald Kameroff, Kalskag, 2017 (in Naves and Keating 2019)

Bird Harvesting and Uses as Food and Materials

Traditionally, hunted birds have alleviated hunger and starvation during times of scarce resources, especially in spring (Wolfe et al. 1990, Fienup-Riordan 1994). Modern socioeconomic conditions prevent famines in rural Alaska communities, but food security still is a concern. In spring, arriving migratory birds are the first fresh food available after a long winter of depending on preserved foods. Birds and eggs diversify the subsistence diet, where many meals consist of fish. The cultural, emotional, and dietary value of resources that traditionally alleviated hunger is an enduring part of the practice of subsistence bird hunting (USFWS 1980).

In the past, Alaska Native peoples used diverse traditional tools and ingenious methods to harvest birds (Pratt 1990, Russell and West 2003, Fienup-Riordan 2007, Corbett 2016). Until the mid-1900s, birds provided skins for clothing, bones for tools, and food for people and sled dogs (Hughes 1984,

Fienup-Riordan 2007). In the 21st century, Native peoples use birds and eggs mostly as human food and cultural resources. Although some bird hunts are specialized, bird hunting is often done in conjunction with other activities, such as moose or marine mammal hunting and berry picking. Besides the meat, bird parts, such as the skin, organs, and fat, are eaten for calories, vitamins, and other nutrients (Unger 2014, Naves and Fall 2017).

The total harvest of migratory birds in rural Alaska is about 370,000 birds/year and includes ducks (54%), geese (33%), seabirds (7%), swans (3%), cranes (2%), and shorebirds (< 1%) (Naves and Alaska Department of Fish and Game, unpub. data; Naves, Knight, and Mengak 2021). Seabird eggs represent a large proportion of the total egg harvest (Paige and Wolfe 1997, 1998, Naves 2018). Sixty percent of the annual bird harvest occurs in spring, 10% in summer, and 30% in fall-winter. Traditionally, bird harvesting stops or is much reduced in summer to allow birds to breed, and then harvesters switch their focus to fisheries, yielding high harvest productivity. Fall and winter harvests of birds occur in the southernmost coastal regions, which are migratory pathways and wintering grounds for many species of seabirds and waterfowl (Wolfe et al. 1990).

Birds Are Messengers

It brings you back to health when it comes alive on the tundra. When [birds] leave [in the fall] I always say to them goodbye and come again. My grandmother said the same thing. I did not know her, but my sister told me. Some other people I know do the same thing because we miss them when they are gone.
—Grace Anaver, Quinhagak, 2017 (in Naves and Keating 2019)

Birds communicate and alert people to weather conditions, ecological processes, and environmental features and are viewed as omens of good or bad events, all of which involves understanding bird ecology and behavior (Russell and West 2003, Fienup-Riordan 2020). In the spring, the arrival of migratory birds is a special event for northern Indigenous peoples and a reliable sign that the long winter is losing its grip on the land (Nelson 1983). Birds' singing and breeding displays are joyful and intrinsic parts of the soundscape.

Birds' abilities to provide information relate to their knowing and responsive character. Harlequin ducks swimming downstream in a fast river show travelers a safe path. Ravens flying in circles or crying out indicates the presence of moose or another animal that ravens will eat if a hunter is successful. People give thanks to ravens and gulls for indicating the remains of missing people that could otherwise not be found (Russell and West 2003).

Unusual bird behaviors are also of significance and can foretell good or bad events. A small bird landing on a person takes away illness. If a bird repeatedly flies over a village, someone will die. A loon that takes flight or cries out when being chased is a bad sign because loons usually dive when being pursued. Birds that give a bad sign are not inherently bad; they just show what is in the future (Fienup-Riordan 2020).

Learning to Hunt and Share

When I was growing up, I used bow and arrow and slingshot, those were my weapons, my hunting tools. It teaches a child how to stalk a bird without spooking it, try to get as close to it as possible and that is a challenge. It teaches patience to get to the bird, how to crawl, or stoop down, or hide behind a knoll, it teaches those skills.
—John O. Mark, Quinhagak, 2017 (in Naves and Keating 2019)

From an early age, an Alaska Native child's education focuses on the knowledge and skills needed to become a successful hunter and fisher. Despite increased Western influence on Indigenous communities in Alaska, harvesting, sharing, and consuming wild foods continue to foster tradition, identity, social structure, recreation, and self-worth (Fienup-Riordan 1990:167–91, 2000). Traditionally, birds have played a key role in this education. Small birds are the first quarry for young children learning hunting skills (Brandt 1943, Irving 1958, Nelson 1980, Fienup-Riordan 2020). Today, children usually start hunting with a BB gun and are still taught to give their catches to grandparents and other elders, even if their catches are small (Fienup-Riordan 2007).

Oral Tradition and Material Culture

Birds feature in traditional ceremonies, stories, place names, songs, and material iconography, such as masks, hunting tools, and domestic objects. Traditional harvesting gear was finely decorated to please and honor the animals taken. In Yup'ik culture, bird iconography has been associated with marine mammal hunting because of a belief that these animals perceive hunters as birds (Fienup-Riordan 1996, 2007, Russell and West 2003).

Small birds carved from ivory or walrus teeth were fashioned into pieces for a traditional dice game played across the Arctic. In Yup'ik and Inupiat dances, men wear fans with five feathers that amplify the movements of their hands and arms. Birds often are featured in Yup'ik dance masks as shamans' helping spirits, hunted animals, story characters, or aspects of the traditional worldview (Fienup-Riordan 1996, Mooney and McIntyre 2019). Today, mask making is important in Native art and economies, and masks are made to be displayed and worn.

Knowledge, culture, and language are intertwined. The representation of birds in tangible (materials, iconography in tools, art objects) and intangible (stories, songs, place names) cultural aspects of Alaska Native peoples exhibits knowledge, values, and worldviews that are important to guide and support bird research, management, and conservation.

MIGRATORY BIRD TREATIES AND WESTERN HARVEST MANAGEMENT

In the 1800s and 1900s, the commercial overuse of migratory birds was widespread. Worldwide, bird populations plummeted, and several species went extinct. The protection of migratory birds required international cooperation, leading to the 1916 Convention for the Protection of Migratory Birds in the United States and Canada, which was enacted in the United States in 1918 as the Migratory Bird Treaty Act (MBTA; 16 USC 703–12). The treaty with Canada, which was aimed at stopping commercial hunting and protecting nesting birds, established a hunting closure between 10 March and 1 September and limited hunting to 3.5 months annually. Subsequent migratory bird treaties between the United States and other countries (Mexico in 1936, Japan in 1972, and Russia in 1976) included similar durations for the hunting season authorized during the nonbreeding period. Together with the Lacey Act of 1900 (16 USC 3371–78), which addressed the possession and transport of wildlife, the MBTA established the primacy of the US federal government to manage migratory birds and work with the states and territories to establish consistent management programs and regulations across the United States (Anderson et al. 2018).

The MBTA was crucial for bird conservation (Williams et al. 2021). However, the treaty largely failed to consider harvests by northern Indigenous peoples (Mitchell 1986). The fall-winter hunting season provides limited harvest opportunities for Alaska Natives because birds' southern migration begins in the late summer. After the MBTA, subsistence bird harvests in Alaska continued to follow the seasonal cycles of availability of the resources, but such harvests were then technically illegal.

When Alaska became a state in 1959, its new Constitution (Article I, § 1) established that all people of the state were entitled to equal rights, regardless of ethnicity. Title 16 of the Alaska statutes followed federal laws and regulations pertaining to the harvest of migratory birds, including the spring-summer closed season as defined in the MBTA (AS 16.05.930:302.01, 308.01, 308.06, Klein and Seim 1965, Argy 1967, Mitchell 1986).

In 1960, the Bureau of Sport Fisheries and Wildlife (BSFW) of the US Fish and Wildlife Service adopted waterfowl law enforcement guidelines. In the Bethel area of Southwest Alaska, hunters were apprehended and fined for taking waterfowl in the closed season. The following year, the BSFW held public meetings across the state, expecting that information about harvest regulations would increase compliance. In the spring of 1961, law enforcement patrols encountered strong resistance from Native hunters, which included shooting at BSFW personnel and aircraft near Bethel. At Utqiaġvik (formerly Barrow) in northern Alaska, two hunters were apprehended for shooting birds during the closed season. In response, about 150 Alaska Native people, each with a duck in hand, protested to the local game warden (Burwell 2005). In 1962–1963, guidelines were modified to require BSFW staff to notify local entities in advance of law enforcement patrols. To avoid further escalation of the conflict, the

BSFW discontinued the law enforcement patrols and commissioned a study to appraise socioeconomic and cultural aspects related to bird hunting in the Yukon-Kuskokwim delta region (Klein and Seim 1965). Resistance by harvesters and a lack of clarity on if and how law enforcement would affect harvesters had aggravated the conflict.

Many Indigenous people continued to hunt birds and gather eggs to feed their families, while many others felt compelled to halt, modify, or disguise their traditional harvest practices. Women plucked and processed birds in hiding, fearful of being cited for illegal harvests (Siikauraq Martha Whiting, Northwest Arctic Borough, pers. comm., 2017). This state of deep unease continued for nearly four decades, causing hardship for the subsistence communities and building long-lasting barriers to conservation collaborations including Indigenous Alaskans (USFWS 1980). A complex socioeconomic and political context continued to brew diverse perspectives from Alaska Native hunters, sport and non-Native hunters, conservationists, scientists, and government agencies in Alaska, other US states, and abroad (Argy 1967, Day 1969).

In January 1979, the United States and Canada signed a first protocol to amend the Convention between the United States and Great Britain for the Protection of Migratory Birds (Canada Treaty) to allow for a spring-summer subsistence harvest in Alaska and northern Canada. The decades of perseverance of Alaska Native leaders had played a key role. However, the ratification of the protocol amendment was a complex process due to opposition in the United States and abroad and the need for diplomatic negotiations with the four countries with which the United States had signed migratory bird treaties (Canada, Mexico, Japan, and the Soviet Union; Mitchell 1986).

In 1997, the US Congress ratified amendments to the Canada and Mexico treaties that recognized the traditional spring-summer subsistence bird harvest by northern peoples. The treaty amendment states: "(b) in the case of the United States (i) Migratory birds and their eggs may be harvested by the Indigenous inhabitants of the state of Alaska. Seasons and other regulations implementing the non-wasteful taking of migratory birds and the collection of their eggs by Indigenous inhabitants of the state of Alaska shall be consistent with the customary and traditional uses by such Indigenous inhabitants for their own nutritional and other essential needs" (US Congress 1996, protocol article II.4.b).

However, the Alaska Constitution (Article I, § 1) guarantees that all people of the state are entitled to equal rights regardless of ethnicity, so the treaty amendments and the AMBCC founding documents were not based on ethnicity. The use and definition of "Indigenous inhabitants" in the amendment language was deliberately crafted in federal documents to avoid racial references that could have conflicted with federal and state laws. The term "Indigenous" was redefined as "Alaska Natives and permanent resident non-Natives with legitimate subsistence hunting needs" (*Congressional Record*, Senate, 23 October 1997, S11167). In the regulations for the Alaska subsistence harvest of migratory birds (67 FR 53518, 16 August 2002) "Indigenous inhabitant" means "a permanent resident of a village within a subsistence harvest area, regardless of race."

The goal of the amended treaties was to promote the conservation of migratory birds by including subsistence hunting in the regulatory process. The amendment authorized the USFWS to open regulated spring-summer subsistence hunts of migratory birds in Alaska. It mandated that Alaska's Indigenous peoples have a meaningful role in harvest management. The amendment also specified that (1) subsistence harvests should remain at traditional levels relative to bird population sizes; (2) subsistence harvest data should be integrated with flyway and national harvest management programs; and (3) regulatory processes for all migratory bird hunting should be inclusive of users and responsive to conservation needs.

THE ALASKA MIGRATORY BIRD CO-MANAGEMENT COUNCIL

Creation of a Co-Management System

In 1998, the USFWS, the Alaska Department of Fish and Game (ADF&G), and members of the Native Migratory Bird Working Group initiated consultation and public meetings across Alaska to define the structure for the meaningful involvement of federal, state, and Indigenous Alaskan partners as equals (63 FR 49707, 17 September 1998). Native leaders, who were instrumental in the pursuit of the amendment, had encouraged tribal participation in this process, resulting in the formation of the Native Migratory Bird Working Group. A Yup'ik mask design was chosen as the logo (figure 16.2). The mask depicts a goose surrounded by eight feathers representing the steps to implement a legal spring-summer subsistence hunt: (1) notify people of the intent to form management bodies; (2) meet to share ideas; (3) send out ideas and listen; (4) choose the form of management bodies; (5) start rule-making; (6) recommend rules for Alaska; (7) link with management in other US flyways; and (8) link with the Nation.

16.2. Logo of the Alaska Migratory Bird Co-Management Council

After the consultation process was completed in 2000 (65 FR 16405, 28 March 2000), the Native Migratory Bird Working Group (which later became the AMBCC Alaska Native Caucus) and representatives from the ADF&G and USFWS met in Girdwood, Alaska, to draft operating procedures and bylaws. The founding members selected the name Alaska Migratory Bird Co-Management Council to highlight equity among the three groups in the decision-making process. Co-management can be defined as "two or more entities, each having legally established management responsibilities, working collaboratively to achieve mutually agreed upon or compatible objectives to protect, conserve, use, enhance, or restore natural and cultural resources" (USFWS 2016).

The AMBCC held its first official meeting in late 2000 and began drafting regulations for the legal spring-summer subsistence harvest season based on the traditional and cultural practices of Alaska Native peoples. In 2002, the USFWS published final procedural rules for establishing spring-summer bird hunting regulations with the AMBCC (67 FR 53511, 16 August 2002). After nearly 80 years of efforts to establish legal spring-summer hunting, the first regulated harvest season was opened in 2003 (68 FR 43010, 21 July 2003).

Roles of the AMBCC: Providing Meaningful Participation in Harvest Management

As established in the amendments to the Canada and Mexico treaties, the AMBCC's overarching role is to ensure that Alaska's Indigenous inhabitants have an effective and meaningful role in the conservation of migratory birds. The AMBCC is committed to

- ensuring the recognition of subsistence hunting and the conservation of migratory birds;
- promoting co-management;
- developing recommendations for regulations that are consistent with the customary and traditional uses of migratory birds and their eggs;
- recommending law enforcement policies;
- conducting migratory bird research that incorporates Indigenous knowledge;
- providing education and information to the public; and
- developing cooperative programs to monitor migratory bird populations and gather harvest data. (67 FR 53511, 16 August 2002)

Governance Structure and Processes

Ten AMBCC regional management bodies gather input from subsistence users across Alaska on topics related to bird harvest and conservation, and they coordinate management functions at the regional and local levels (AMBCC 2021a). The regional bodies are administered by Alaska Native nonprofit organizations and have their own bylaws, local representation structures, and meeting procedures: Ahtna Intertribal Resource Commission (upper Copper River), Aleutian Pribilof Islands Association, Association of Village Council Presidents (Yukon-Kuskokwim Delta), Bristol Bay Native Association, Chugach Regional Resources Commission (Gulf of Alaska and lower Cook Inlet), Kawerak Inc. (Bering Strait–Norton Sound), Maniilaq Association (northwestern Arctic), North Slope Borough, Sun'aq Tribe of Kodiak, and Tanana Chiefs Conference (interior Alaska).

Several standing and temporary committees and working groups convene experts from partner organizations. Each committee focuses on a specific topic to explore and seek solutions and to advise the full council. The Technical Committee, for example, provides scientific and Indigenous knowledge, advises on ecological and socioeconomic factors related to proposed harvest regulations and other management and conservation topics, and develops management plans and agreements. Other committees focus on topics such as law enforcement; handicrafts, including those using bird parts; outreach and communication; harvest assessment; and the council's budget. The AMBCC has formal representation on the Pacific Flyway Council.

The council holds two regular meetings per year. In the spring, the council considers proposed regulatory changes. In the fall, the council considers technical reports and emerging management and conservation topics. The meetings are open to the public, and meeting transcripts are publicly available (AMBCC 2021a, b).

REGULATORY CONTEXT

Since 1947, bird harvest management and conservation in North America has been coordinated by four administrative flyways (Pacific, Central, Mississippi, and Atlantic). Alaska belongs to the Pacific Flyway along with states in the lower 48 west of the Rocky Mountains. However, birds that breed in Alaska migrate across all four flyways (and to other parts of the world), warranting the coordination of migratory bird management in Alaska with all four flyway councils.

Waterfowl harvest in the Pacific Flyway is about 2.5 million birds/year (Olson 2020). The yearlong harvest of migratory birds in rural Alaska (about 370,00 birds/year; Naves and ADF&G Division of Subsistence, unpub. data; Naves, Knight, and Mengak 2021) represents about 15% of the waterfowl harvest in the Pacific Flyway and about 81% of the migratory bird harvest in Alaska. Harvest in rural Alaska includes large proportions of the total Pacific Flyway harvest for individual waterfowl species, such as emperor goose, black brant, and sea ducks.

Regulations for the spring-summer season were first authorized in 2003 and are still being fine-tuned to achieve a balance between addressing bird conservation needs and minimizing impacts of regulations on subsistence users and Alaska Native sovereignty (Title 50, CFR 92, US National Archives and Records Administration 2021). Harvest regulations for the spring-summer season have been developed with the intent of preserving customary and traditional patterns of harvesting, sharing, and using birds and their eggs (Wolfe et al. 1990, Kelso 1982). Spring-summer harvest regulations do not include daily and possession limits because traditional patterns of harvest and use involve taking advantage of harvest opportunities and minimizing the costs of harvesting, which are essential for successful food gathering and extensive sharing of resources between harvesters and nonharvesters. The ability to formally invite immediate family members residing in ineligible regions and communities supports kinship-based harvesting and processing traditions, providing food for eligible residents and the transmission of knowledge and culture across generations. In northern latitudes, long daylight hours in the spring and summer preclude the need for the definition of shooting hours.

Throughout the years, the legal definition of Indigenous inhabitants as "Alaska Natives and permanent resident non-Natives with legitimate subsistence hunting needs" (see the section "Migratory Bird Treaties and Western Harvest Management" above) has resulted in challenges in developing harvest regulations based on Alaska Native traditions and cultural practices. For instance, regulations concerning the sale of handicrafts to non-Natives and the ability to invite urban family members back to their home villages to assist in the harvest had to be carefully crafted and shepherded through the regulatory process in order to maintain the integrity of customary and traditional practices while considering the inclusion of non-Natives.

HARVEST ASSESSMENT

Harvest Management Requires Harvest Data

Harvest data are needed to document the socioeconomic contributions of harvests, assess harvest effects on populations, solve allocation issues, and track changes in harvests. The quantification of bird subsistence harvests in Alaska started in the 1950s largely to address issues related to the MBTA closure of the spring-summer harvest (Hansen 1957, USFWS 1957, Thompson and Person 1963, Klein and Seim 1965) and to inform land use planning (Patterson 1974, USFWS 1980). As considerations evolved for the MBTA amendment to legalize the spring-summer subsistence harvest, it became imperative to address concerns about potential impacts of this harvest on bird populations and about resource allocations between users in Alaska and elsewhere (USFWS 1980, 48 FR 10101, 10 March 1983). Addressing these concerns required additional data on the amount, timing, and geography of harvest patterns, prompting further studies on the ecological and socioeconomic aspects of the bird subsistence harvest in Alaska (Loranger 1985, Wolfe et al. 1990, Paige and Wolfe 1997, 1998).

Amending the treaty with Canada to legally authorize the Alaska spring-summer subsistence harvest of migratory birds required harvest assessment to inform harvest management. The amendment protocol stipulated: "It is the intention of DOI/FWS and the Alaska Department of Fish and Game that management information, including traditional knowledge, the number of subsistence hunters and estimates of harvest, will be collected cooperatively for the benefit of management bodies" (US Congress 1996, letter of submittal). The 2001 bylaws define AMBCC's roles to include "develop[ing] cooperative programs to monitor migratory bird populations and harvests" (Article I.e) and "provid[ing] technical and harvest information to the flyways' councils and the Service Regulations Committee" (Article IV.c).

AMBCC Harvest Assessment Program

In 2000, the AMBCC created a committee to design an Alaska-wide harvest survey (AMBCC 2003, Reynolds 2007). The AMBCC Harvest Survey Committee convenes federal, state, and Alaska Native partners to provide guidance on harvest assessment, accounting for ecological and socioeconomic factors as well as Western and Indigenous knowledge. The AMBCC harvest survey underwent two rounds of review to fine-tune methods and attempt to balance objectives and priorities with the available funding (Naves et al. 2008, George et al. 2015, Otis et al. 2016).

The AMBCC survey documents the yearlong harvest of dozens of bird species (Naves, Knight, and Mengak 2021). Data collection relies on the participation of regional and local Native partners. Developing trust and collaboration in harvest assessment involves community and household informed consent; confidentiality of household data; participation in development of survey methods, data collection, and data review; and sharing of results in audience-specific formats (AFN 2013, NSF-IARPC 2018). The AMBCC Harvest Assessment Program also conducts research to address information needs, including the characterization of geographic and seasonal harvest patterns, ethnography, traditional knowledge, and culture- and place-based outreach and communication related to harvest management and bird conservation (Rothe et al. 2015, Naves and Fall 2017, Naves and Zeller 2017, Naves 2018, Naves et al. 2019).

CO-MANAGEMENT ON THE GROUND

Examples of actions implemented by the AMBCC include:

- Collaboration across the Pacific Flyway, which started with the Yukon-Kuskokwim Delta Goose Management Plan. This achieved recovery of populations of cackling goose, black brant, greater white-fronted goose, and emperor goose (Raveling 1984, Mitchell 1986, Pamplin 1986).

- Development of regulations to allow residents of excluded areas to participate in the spring-summer migratory bird harvest based on the invitation of family members residing in eligible areas (50 CFR 92.5). Many Alaska Natives have migrated to urban areas for work, education, health services, and other reasons, yet they maintain strong ties to their culture and traditions, including the need to participate in the seasonal round of subsistence activities.

- Development of regulations establishing strict salvage requirements (including definitions of "edible meat" and "nonwasteful taking") for the spring-summer subsistence season based on the Alaska Native value of not wasting any part of animals that allow themselves to be taken (81 FR 18781, 1 April 2016).

- Development of regulations allowing the sale to non-Natives of handicrafts made by Alaska Native artisans that include inedible parts of migratory birds (82 FR 34263, 24 July 2017, 50 CFR 92.6). Prior to this regulation, the sale to non-Natives of such handicrafts was illegal.

- Collaboration on and support of reopening the emperor goose harvest in 2017 after a 30-year hunting closure, including the development of the first management plan by the AMBCC (AMBCC 2016, PFC 2016, 83 FR 13684, 30 March 2018, 50 CFR 92.22).

- Establishment of the flexible timing of the summer harvest closure based on input from local subsistence users (50 CFR 92.31).

- Presentation of a federal-state joint apology to Alaska Native peoples for unintended harm caused by the 1918–2002 MBTA closure of the spring-summer harvest (USFWS 2018).

- Support for migratory bird culture camps and festivals in the Copper River and North Slope regions to foster the transmission of traditional knowledge and outreach and communication related to harvest management.

- Research on the harvest of and traditional knowledge related to shorebirds to inform place- and culture-based outreach and communication in western Alaska communities (Naves et al. 2019).

- Development of an ethnographic bird identification guide by Chugach tribes (south-central Alaska) complemented by traditional stories, Indigenous knowledge, and recipes.

The AMBCC Native Caucus has provided a venue for Alaska Native peoples to pursue collaboration and resolution on topics that are important for them. For example, the AMBCC Native Caucus has accomplished the following:

- Worked with the Alaska congressional delegation to obtain an exemption for subsistence hunters to the requirement to purchase a federal duck stamp.

- Written a letter to the California Coastal Commission urging protection of wintering habitat important for black brant, shorebirds, and other migratory birds. The letter highlighted the Native Caucus's concerns with a permit application for a broad expansion of shellfish aquaculture operations in Humboldt Bay, which threatened eelgrass beds.

CO-MANAGEMENT IN THE CURRENT BIRD CONSERVATION CONTEXT

Collaboration among Indigenous, state, and federal partners for bird subsistence harvest management in Alaska has achieved substantial progress, though work remains in meaningfully engaging Alaska Native partners in bird conservation. Co-management needs to anticipate emerging conservation needs.

Continent-wide declines in bird populations negatively affect harvest sustainability, including in Alaska (Rosenberg et al. 2019). Since 2014, seabird die-offs in the Bering and Chukchi Seas have been unusually impactful (Romano et al. 2020, Van Hemert et al. 2021). Emerging bird conservation issues and associated reduced harvest sustainability can potentially raise the need for restrictive harvest regulations for the spring-summer subsistence harvest of migratory birds in Alaska. Strong collaboration among all AMBCC partners will be crucial to tackling such potential regulatory issues.

Indigenous peoples have begun to play a more prominent role in the stewardship of natural resources (Wells et al. 2020). In many places, Indigenous peoples now have a powerful voice and the endurance to advocate and work for the protection of their communities, cultures, and the fish, wildlife, and land they depend on. Supporting the further engagement of Alaska Native partners in bird conservation will require leadership and proactive work to collaboratively devise and implement processes, increase the funding needed for this work, and expand partnerships. State and federal support will be needed to ensure a fair distribution of the bird conservation burden so that subsistence hunting is not overly taxed. Work is needed to better evaluate the impacts of diverse human activities on bird populations, minimize wasteful bird mortality (e.g., bycatch in fisheries, collision with human-made structures), and protect and restore habitats—including extensive Alaska Native landholdings—that meet birds' critical ecological needs.

CONSERVATION THROUGH CO-MANAGEMENT: WORKING TOGETHER IN CULTURAL DIVERSITY

Knowledge and understanding about the AMBCC's history, roles, processes, and goals are critical for meaningful involvement by all partners. Persistent internal outreach and education among partners are needed to foster a common understanding of and commitment to a true co-management regime. Collective agreement on and support for the goals and foundations of the program foster trust and cooperation. By engaging in planning and co-learning, AMBCC partners find common ground in the basic information and build up from there to take action and improve management and conservation outcomes.

Developing trust requires deliberate efforts. Despite substantial progress at the AMBCC, relations are still fragile between subsistence users and bird management and conservation entities. Engaging in meaningful dialogue early and often on issues of concern to all partners, especially seeking input from Alaska Native partners on issues that directly affect them, goes a long way toward promoting trust, cooperation, and goodwill. Meaningful consultation is part of the co-management routine and should involve identifying and following up on both trust boosters and busters for all partners. Working together on specific topics is an effective approach to build trust because it allows individuals to know each other and ensures diverse perspectives in planning and interpreting research, the equitable distribution of decision-making power, and that the needs of diverse partners are met.

It takes time to build co-management. Reiteration of the commitment to and ownership by all partners to the co-management process are key for the AMBCC to continue evolving into an effective, united force in the bird management and conservation arena. In a similar context, the national flyway system took decades to coalesce into a cooperative effort. This long-term endeavor involves periodically reaffirming the commitment by all partners, adjusting structures and process, acknowledging achievements and challenges, and clarifying actions needed to achieve goals (table 16.2). For example, following breakdowns in communication in 2012, the AMBCC convened a summit gathering for all partners to honestly discuss concerns and to collaboratively devise solutions. Under the theme "how to get to yes," partners were asked to craft solutions that would work for most people, such as adjustments to proposed harvest regulations so they could move forward. Such periodic gatherings are important to keep the AMBCC relationships strong and effectively working for all partners. Only by genuinely joining forces will we be able overcome the societal and environmental challenges we now face to ensure a future where we are all graced by healthy bird populations that can sustain harvest by diverse user groups along the flyways.

ACKNOWLEDGMENTS

First and foremost, we acknowledge the AMBCC founders, whose foresight and perseverance led to a legal spring-summer subsistence harvest of migratory birds. These include all members of the Alaska Native Migratory Bird Working Group (the precursor to the AMBCC Native Caucus) as well as USFWS and ADF&G representatives. Additionally, we express our gratitude to all past and current AMBCC representatives, staff, and direct partners for their dedication and willingness to acknowledge and embrace the cultural diversity of the AMBCC, overcome challenges, and work together, sharing the common goal of migratory bird conservation. We also thank members of the AMBCC regional management bodies for playing the crucial role of bringing the traditional knowledge, perspectives, and concerns of subsistence users and communities into the co-management process. Finally, and most important, we pay honor to our elders and ancestors, Native and non-Native, for their strength, wisdom, and knowledge, which have guided us along the way.

TABLE 16.2. Challenges and Actions to Support Co-Management

CHALLENGES TO CO-MANAGEMENT	ACTIONS THAT SUPPORT CO-MANAGEMENT
• Intergenerational historic trauma, which makes it difficult for Alaska Native partners to respect and trust non-Native partners and Western knowledge • Lack of trust in the co-management processes and motivations of partners • Agencies' lack of trust in Indigenous knowledge as a reliable source of information • Unfamiliarity with cross-cultural collaboration • Unfamiliarity with incorporating Indigenous cultures and values into research and management projects • Actions that position agencies as the expert on many topics, causing Indigenous partners to feel disempowered • Legal mandates or policies that limit sharing of decision-making authority or prevent an agency's ability to support a proposed regulation • Fear of losing control • Insufficient funding for all partners to meaningfully participate in the co-management process, invest in working together, and build relations • Lack of motivation by partners	• Seeking ways to reconcile historic trauma and conflicts and to develop trust • Genuine and attentive interactions • Asking for and respecting the opinions and ideas of others • Increasing awareness and skills for all partners on cross-cultural conflict resolution • Consistent and regular information sharing • Developing effective processes for shared decision-making and supporting diverse ideas and solutions • Combining diverse ways of knowing to support ownership by all partners and encourage wider understanding of ecological and sociocultural systems • Reliance on local partners for on-the-ground knowledge and capacity • Developing long-lasting relations that continue beyond the life of a single project • Supporting technical capacity building across all partners • Expanding partnerships to include nongovernmental organizations • Pursuing diverse funding sources besides federal agencies, including sources accessible to tribes only • Supporting Native partners as stewards of bird resources

LITERATURE CITED

Alaska Department of Fish and Game [ADF&G]. 2021. Alaska 2020–2021 migratory game bird hunting regulations. https://www.adfg.alaska.gov/static/regulations/wildliferegulations/pdfs/waterfowl.pdf. Accessed 14 July 2021.

Alaska Federation of Natives [AFN]. 2013. Alaska Federation of Natives guidelines for research. http://www.ankn.uaf.edu/IKS/afnguide.html. Accessed 25 June 2021.

Alaska Federation of Natives [AFN]. 2021. Alaska Native peoples. https://www.nativefederation.org/alaska-native-peoples/. Accessed 23 June 2021.

Alaska Migratory Bird Co-Management Council [AMBCC]. 2003. Recommendations for a statewide Alaska migratory bird subsistence harvest survey. Division of Migratory Bird Management, US Fish and Wildlife Service, Anchorage, Alaska.

Alaska Migratory Bird Co-Management Council [AMBCC]. 2016. Management plan for the emperor goose. https://www.alaskamigratorybirds.com/images/regulations/management/AMBCC_Emperor_Goose_Management_Plan_Sept2016.pdf. Accessed 18 July 2022.

Alaska Migratory Bird Co-Management Council [AMBCC]. 2021a. Alaska Migratory Bird Co-Management Council: conservation through co-management. http://www.alaskamigratorybirds.com/. Accessed 23 June 2021.

Alaska Migratory Bird Co-Management Council [AMBCC]. 2021b. Alaska Migratory Bird Co-Management Council. https://www.fws.gov/alaska/pages/alaska-migratory-bird-co-management-council-ambcc. Accessed 23 June 2021.

Anderson, M. G., R. T. Alisauskas, B. D. Batt, R. J. Blohm, K. F. Higgins, M. C. Perry, J. K. Ringelman, J. S. Sedinger, J. R. Serie, D. E. Sharp, and D. L. Trauger. 2018. The Migratory Bird Treaty and a century of waterfowl conservation. Journal of Wildlife Management 82:247–59.

Argy, N. T. 1967. A summary of the waterfowl–Alaska Eskimo conflict. Bureau of Sport Fisheries and Wildlife, Alaska. In possession of author.

Arnold, R. D. 1978. Alaska Native land claims. Alaska Native Foundation, Anchorage.

Berger, T. R. 1985. Village journey: the report of the Alaska Native Review Commission. Hill and Wang, New York.

Brandt, H. 1943. Alaska bird trails: adventures of an expedition by dog sled to the delta of the Yukon River at Hooper Bay. Bird Research Foundation, Cleveland, Ohio.

Bureau of Indian Affairs [BIA]. 2021. Alaska region. https://www.bia.gov/regional-offices/alaska. Accessed 25 June 2021.

BurnSilver, S., J. Magdanz, R. Stotts, M. Berman, and G. Kofinas. 2016. Are mixed economies persistent or transitional? evidence using social networks from Arctic Alaska. American Anthropologist 118:121–29.

Burwell, M. 2005. "Hunger knows no law": seminal Native protest and the Barrow duck-in of 1961. Paper presented at the 32nd Annual Meeting of the Alaska Anthropological Association, 12 March 2005, Anchorage.

Corbett, D. 2016. Saĝdaĝ: to catch birds. Arctic Anthropology 53:93–113.

Day, A. 1969. Northern Natives, migratory birds, and international treaties. Bureau of Sport Fisheries and Wildlife, US Department of the Interior, Washington, DC.

Fall, J. A. 2018. Subsistence in Alaska: a year 2017 update. Division of Subsistence, Alaska Department of Fish and Game. https://www.adfg.alaska.gov/static/home/subsistence/pdfs/subsistence_update_2017.pdf. Accessed 14 July 2021.

Fall, J. A. 2019. Alaska population trends and patterns, 1960–2018. Division of Subsistence, Alaska Department of Fish and Game. https://www.adfg.alaska.gov/static/home/library/pdfs/subsistence/Trends_in_Population_Summary_2019.pdf. Accessed 14 July 2021.

Fienup-Riordan, A. 1990. Eskimo essays: Yup'ik lives and how we see them. Rutgers University Press, New Brunswick, New Jersey.

Fienup-Riordan, A. 1994. Boundaries and passages: rule and ritual in Yup'ik Eskimo oral tradition. Civilization of the American Indian Series. University of Oklahoma Press, Norman.

Fienup-Riordan, A. 1996. The living tradition of Yup'ik masks: Agayuliyararput, our way of making prayer. University of Washington Press, Seattle.

Fienup-Riordan, A. 2000. Hunting tradition in a changing world: Yup'ik lives in Alaska today. Rutgers University Press, New Brunswick, New Jersey.

Fienup-Riordan, A. 2007. Yuungnaqpiallerput: the way we genuinely live. Masterworks of Yup'ik Science and Survival. University of Washington Press, Seattle.

Fienup-Riordan, A. 2020. Nunakun-gguq ciutengqertut: they say they have ears through the ground: animal essays from Southwest Alaska. University of Alaska Press, Fairbanks.

George, T. L., D. Otis, and P. Doherty. 2015. Review of Alaska migratory bird subsistence harvest survey. Department of Fish, Wildlife, and Conservation Biology, Colorado State University, Fort Collins.

Hansen, H. A. 1957. Utilization of wildlife by Alaska Natives. US Fish and Wildlife Service, Washington, DC.

Hughes, C. C. 1984. Saint Lawrence Island Eskimo. Pages 262–77 in Handbook of North American Indians. Volume 5. Smithsonian Institution, Washington, DC.

Irving, L. 1958. On the naming of birds by Eskimos. Anthropological Papers of the University of Alaska 6:61–77.

Keeler, W. M., H. J. Wade, and J. E. Officer. 1962. Report to the secretary of the interior by the Task Force on Alaska Native Affairs. Bureau of Indian Affairs, Washington, DC.

Kelso, D. D. 1982. Subsistence use of fish and game resources in Alaska: considerations in formulating effective management policies. Transactions of the 47th North American Wildlife and Natural Resources Conference, 26–31 March 1982, Portland, Oregon. Wildlife Management Institute, Washington, DC.

Klein, D. R., and D. E. Seim. 1965. Availability and utilization of migratory waterfowl in western Alaska. Alaska Cooperative Wildlife Research Unit, University of Alaska, Fairbanks. Report to the US Fish and Wildlife Service, Bureau of Sport Fisheries and Wildlife, Portland, Oregon.

Langdon, S. 2014. The Native people of Alaska: traditional living in a northern land. 5th edition. Greatland Graphics, Fairbanks, Alaska.

Loranger, A. 1985. Historical and contemporary subsistence harvest of migratory birds in Alaska. US Fish and Wildlife Service, Anchorage, Alaska.

Martin, K. 1995. Patterns and mechanisms for age-dependent reproduction and survival in birds. American Zoologist 35:340–48.

Mitchell, D. C. 1986. Native subsistence hunting of migratory waterfowl in Alaska: a case study demonstrating why politics and wildlife management don't mix. Transactions of the 51st North American Wildlife and Natural Resource Conference 51:527–34.

Moerlein, K. J., and C. Carothers. 2012. Total environment of change: impacts of climate change and social transitions on subsistence fisheries in Northwest Alaska. Ecology and Society 17:10.

Mooney, S., and C. McIntyre. 2019. Yua Henri Matisse and the inner Arctic spirit. Heard Museum, Phoenix, Arizona.

National Science Foundation Interagency Arctic Research Policy Committee [NSF-IARPC]. 2018. Principles for conducting research in the Arctic. https://www.nsf.gov/geo/opp/arctic/conduct.jsp. Accessed 14 July 2021.

Naves, L. C. 2018. Geographic and seasonal patterns of seabird subsistence harvest in Alaska. Polar Biology 41:1217–36.

Naves, L. C., and J. A. Fall. 2017. Calculating food production in the subsistence harvest of birds and eggs. Arctic 70:86–100.

Naves, L. C., and J. M. Keating. 2019. Shorebird subsistence harvest and Indigenous knowledge in Alaska. Draft report. Division of Subsistence, Alaska Department of Fish and Game, Anchorage.

Naves, L. C., J. M. Keating, T. L. Tibbitts, and D. R. Ruthrauff. 2019. Shorebird subsistence harvest and Indigenous knowledge in Alaska: informing harvest management and engaging users in shorebird conservation. Condor 121:1–19.

Naves, L., A. J. Knight, and L. F. Mengak. 2021. Alaska subsistence harvest of birds and eggs, 2004–2020 data book, Alaska Migratory Bird Co-Management Council. Division of Subsistence, Alaska Department of Fish and Game, Special Publication No. 2021-05.

Naves, L. C., D. Koster, M. G. See, B. Easley, and L. Olson. 2008. Alaska Migratory Bird Co-Management Council, migratory bird subsistence harvest survey: Assessment of the survey methods and implementation. Division of Subsistence, Alaska Department of Fish and Game, Special Publication No. 2008-05.

Naves, L. C., and T. K. Zeller. 2017. Yellow-billed loon subsistence harvest in Alaska: challenges in harvest assessment of a conservation concern species. Journal of Fish and Wildlife Management 8:114–24.

Nelson, R. 1980. Shadow of the hunter: stories of Eskimo life. University of Chicago Press, Chicago, Illinois.

Nelson, R. 1983. Make prayers to the raven: A Koyukon view of the northern forest. University of Chicago Press, Chicago, Illinois.

Norton-Smith, K., K. Lynn, K. Chief, K. Cozzetto, J. Donatuto, M. H. Redsteer, L. E. Kruger, J. Maldonado, C. Viles, and K. P. Whyte. 2016. Climate change and Indigenous peoples: a synthesis of current impacts and experiences. General Technical Report PNW-GTR-944. US Department of Agriculture Forest Service, Pacific Northwest Research Station, Portland, Oregon.

Olson, S. M. 2020. Pacific Flyway data book, 2020: migratory bird population indices, harvest, and hunter participation and success. Division of Migratory Bird Management, US Fish and Wildlife Service, Vancouver, Washington.

Otis, D., T. L. George, and P. Doherty. 2016. Comparison of alternative designs for the Alaska migratory bird subsistence harvest survey. Department of Fish, Wildlife, and Conservation Biology, Colorado State University, Fort Collins.

Pacific Flyway Council [PFC]. 2016. Management plan for the emperor goose. Pacific Flyway Council, care of US Fish and Wildlife Service, Division of Migratory Bird Management, Vancouver, Washington. https://pacificflyway.gov/Documents/Eg_plan.pdf. Accessed 18 July 2022.

Paige, A. W., and R. J. Wolfe. 1997. The subsistence harvest of migratory birds in Alaska: compendium and 1995 update. Division of Subsistence, Alaska Department of Fish and Game, Technical Paper No. 228.

Paige, A. W., and R. J. Wolfe. 1998. The subsistence harvest of migratory birds in Alaska, 1996 update. Draft report. Division of Subsistence, Alaska Department of Fish and Game.

Pamplin, W. L. 1986. Cooperative efforts to halt population declines of geese nesting on Alaska's Yukon-Kuskokwim delta. Transactions of the 51st North American Wildlife and Natural Resources Conference, 21–26 March 1986, Reno, Nevada.

Patterson, A. 1974. Subsistence harvests in five Native regions. Joint Federal-State Land Use Planning Commission for Alaska, Resource Planning Team. In possession of author.

Pratt, K. L. 1990. Economic and social aspects of Nunivak Eskimo cliff-hanging. Arctic Anthropology 27:75–86.

Raveling, D. G. 1984. Geese and hunters of Alaska's Yukon delta: management problems and political dilemmas. Transactions of the 49th North American Wildlife and Natural Resources Conference, 23–28 March 1984, Boston, Massachusetts.

Reynolds, J. 2007. Investigating the impact of sampling effort on annual migratory bird subsistence harvest survey estimates. Final report for USFWS MBM Order No. 701812M816. Solutions Statistical Consulting, Anchorage, Alaska.

Romano, M. D., H. M. Renner, K. J. Kuletz, J. K. Parrish, T. Jones, H. K. Burgess, D. A. Cushing, and D. Causey. 2020. Die-offs, reproductive failure, and changing at-sea abundance of murres in the Bering and Chukchi Seas in 2018. Deep Sea Research Part II: Topical Studies in Oceanography 181–82:104877.

Rosenberg, K. V., A. M. Dokter, P. J. Blancher, J. R. Sauer, A. C. Smith, P. A. Smith, J. C. Stanton, A. Panjabi, L. Helft, M. Parr, and P. P. Marra. 2019. Decline of the North American avifauna. Science 366:120–24.

Rothe, T. C., P. I. Padding, L. C. Naves, and G. J. Robertson. 2015. Harvest of sea ducks in North America: a contemporary summary. Pages 440–91 *in* J.-P. L. Savard, D. V. Derksen, D. Esler, and J. M. Eadie, editors. Ecology and conservation of North American sea ducks. Studies in Avian Biology. CRC Press, Boca Raton, Florida.

Russell, P., and G. C. West. 2003. Bird traditions of the Lime Village area Dena'ina: upper Stony River ethno-ornithology. Alaska Native Knowledge Network, Fairbanks.

Thompson, D. Q., and R. A. Person. 1963. The eider pass at Point Barrow, Alaska. Journal of Wildlife Management 27:348–56.

Unger, S. 2014. Qaqamiiĝux: traditional foods and recipes from the Aleutian and Pribilof Islands. Aleutian Pribilof Islands Association, Inc., Anchorage, Alaska.

US Census Bureau. 2019. Quick facts: Alaska. https://www.census.gov/quickfacts/AK. Accessed 25 June 2021.

US Congress. 1996. Protocol amending the 1916 Convention for the Protection of Migratory Birds. Senate Treaty Document 104–28. https://www.congress.gov/104/cdoc/tdoc28/CDOC-104tdoc28.pdf. Accessed 18 July 2022.

US Fish and Wildlife Service [USFWS]. 1957. Progress report no. IV: fish and wildlife resources of the Lower Yukon River (Marshall to mouth), based on 1956 field investigations. Branch of River Basin Studies, US Fish and Wildlife Service, Juneau, Alaska.

US Fish and Wildlife Service [USFWS]. 1980. Subsistence hunting of migratory birds in Alaska and Canada: final environmental assessment for the 1979 protocol amendment for the Convention for the Protection of Migratory Birds in Canada and the United States of America. Alaska Region, Migratory Bird Management, Anchorage.

US Fish and Wildlife Service [USFWS]. 2016. Native American policy: working with Native American tribes. https://www.fws.gov/nativeamerican/pdf/Policy-revised-2016.pdf. Accessed 14 July 2021.

US Fish and Wildlife Service [USFWS]. 2018. State of Alaska joins service in formal apology for harmful impacts of past bird harvest prohibitions. https://www.fws.gov/alaska/stories/state-alaska-joins-service-formal-apology-harmful-impacts-past-bird-harvest-prohibitions. Accessed 23 June 2021.

US Fish and Wildlife Service [USFWS]. 2021. Regulations for the 2021 Alaska subsistence spring/summer migratory bird harvest. Alaska Migratory Bird Co-Management Council. https://www.fws.gov/alaska/pages/alaska-migratory-bird-co-management-council-ambcc. Accessed 14 July 2021.

US National Archives and Records Administration. 2021. Code of federal regulations: Title 50: wildlife and fisheries; Part 92: migratory bird subsistence harvest in Alaska. https://ecfr.io/Title-50/Part-92. Accessed 14 July 2021.

Van Hemert, C., R. J. Dusek, M. M. Smith, R. Kaler, G. Sheffield, L. M. Divine, K. Kuletz, S. Knowles, J. S. Lankton, D. R. Hardison, R. W. Litaker, T. Jones, H. K. Burgess, and J. K. Parrish. 2021. Investigation of algal toxins in a multispecies seabird die-off in the Bering and Chukchi Seas. Journal of Wildlife Disease 57:399–407.

Wells, J. V., N. Dawson, N. Culver, F. A. Reid, and S. M. Siegers. 2020. The state of conservation in North America's boreal forest: issues and opportunities. Frontiers in Forests and Global Change 3:90.

Wheeler, P., and T. Thornton. 2005. Subsistence research in Alaska: a thirty-year retrospective. Alaska Journal of Anthropology 3:69–103.

Williams, C. K., R. D. Applegate, and P. M. Coppola. 2021. Why are there so many waterfowl and so few northern bobwhites? rethinking federal coordination. Journal of Wildlife Management 85:665–73.

Wolfe, R. J., and L. J. Ellanna. 1983. Resource use and socioeconomic systems: case studies of fishing and hunting in Alaskan communities. Division of Subsistence, Alaska Department of Fish and Game, Technical Paper No. 61.

Wolfe, R. J., A. W. Paige, and C. L. Scott. 1990. The subsistence harvest of migratory birds in Alaska. Division of Subsistence, Alaska Department of Fish and Game, Technical Paper No. 197.

Yupiktak Bista. 1974. Does one way of life have to die so another can live? a report on subsistence and the conservation of the Yupik life-style. Alaska Native Knowledge Network, University of Alaska, Fairbanks. https://www.uaf.edu/ankn/indigenous-knowledge-syst/does-one-way-of-life-have-2/. Accessed 29 July 2022.

Chapter 17

Research with Tribes
A Suggested Framework for the Co-Production of Knowledge

CALEB R. HICKMAN, JULIE THORSTENSON, ASHLEY CARLISLE,
SERRA J. HOAGLAND, *and* STEVEN ALBERT

INTRODUCTION

Tribes own, co-manage, and have authority over a vast land area within the United States. Native American land may collectively include more acreage than all of the national parks and conservation areas in North America combined (Nabhan 2000), and they support healthy and well-protected populations of fish and wildlife while enabling people to live and thrive on the landscape. For spiritual, practical, and other reasons, tribes have always placed a high value on supporting ecological integrity. Where tribal lands are still intact, the ecosystems they harbor often have equal or higher species diversity and densities of protected species than surrounding nontribal lands (Corrigan et al. 2018, Schuster et al. 2019). The first peoples of this land possess intimate relationships to place, acquired and passed on through hundreds of generations. For these reasons, tribal lands offer excellent opportunities to study fish, wildlife, and their habitats.

Some of the first studies of the flora and fauna of tribal lands were done under the auspices of documenting Traditional Ecological Knowledge (TEK) by Western scientists (see part IV of this volume for much more about TEK). TEK first gained notoriety in the field of ethnobotany, as early researchers queried Native people about the edible, medicinal, and other properties of local plants (Huntington 2000, Berkes et al. 2000; for an excellent history of ethnobotany in North America, see Schlag 2015).

In our experience, most tribal natural resource departments support the ethical acquisition of knowledge and the active study of fish, wildlife, and their habitats. Tribal survival depends on successfully integrating and passing along a deep understanding of animals' ecology, movements, and behavior (Parlee et al. 2005). As one Cherokee elder remarked, "We have always thought like scientists" (C. Hickman, pers. comm.). Unfortunately, there is a long history of studies carried out by outside researchers under less than collaborative conditions (Brunet et al. 2014, 2016, Wilson et al. 2020). To tribes, where memories are long, the modern mining of intellectual property is little different from the exploitation of the preceding five centuries of property theft and human rights violations. In the 21st century, however, a more inclusive approach to the acquisition, dissemination, and use of knowledge and the sharing of intellectual property has gained ground. "Co-production" aims to put principles of empowerment into practice, allowing tribal communities greater control over the research process (definition adapted from DuRose et al. 2012 and Beier et al. 2016).

In this chapter, we explore the co-production of knowledge in the context of fish and wildlife research on tribal land, offer examples of where this approach is succeeding (and a few cautionary tales), and provide suggestions for meaningful and successful co-produced research. While tribal consultation—the process of the federal government reaching out to tribes when considering actions or policies that affect tribal communities (e.g., endangered species consultation)—is generally a distinct process from research, we believe there is much in this chapter that can inform consultation as well.

This chapter is based on a review of current literature and the experiences of the authors, who are or have been employees or are citizens of seven tribal nations and have worked with scores of others. In addition to working with outside researchers, we have conducted our own investigations on tribal and federal lands across the United States and in other parts of the world. Several of us have served on institutional review boards (IRBs) and other types of committees for evaluating outside research with tribal entities, and most of us have collaborated with federal and state governments to implement conservation research, planning, and management. Some of us have had to balance our professional training and our traditional cultural ways of knowing in order to work with outside researchers. We met several times over the course of a year to discuss our positive experiences and the troublesome issues we have encountered, and we spoke to colleagues to broaden our perspectives. Our intent here is not to provide a comprehensive overview of how to conduct research with tribal communities, which is a topic worthy of an entire book. Rather, we hope to start a larger, ongoing conversation to help researchers and tribal citizens work together in a productive and respectful way. This information is intended for investigators pursuing research on tribal lands; for tribal natural resource staff with whom they partner; for tribal leaders who want to work with outside researchers while safeguarding tribal citizens, natural resources, and intellectual property; and for students and future conservation professionals who plan to work in the field of natural resources and may work for or interact with tribes.

As is noted many times throughout this volume (though it is impossible to overstate), the sovereignty of Indian tribes has been recognized by the United States and, before that, the British, Spanish, Dutch, French, and other colonial powers. Tribal sovereignty is enshrined in the US Constitution and has been upheld by numerous court cases (see part II of this volume). As sovereign nations, tribes and the land they steward are not subject to the same access and

> **TRANSITIONAL TERMINOLOGY**
>
> We define **Traditional Ecological Knowledge** as a cumulative, evolving, and living body of information, beliefs, and practices related to the environment that has been handed down through generations via songs, stories, and other instruction. TEK is concerned with the relationship of living beings, including humans, with their environment (Kimmerer 2002). Although some have argued that use of the term "traditional" tends to relegate this information to the past (Powys Whyte 2013), in our context it embraces both ancient and modern knowledge that was developed and continues to be developed by tribes and tribal individuals. TEK differs from individual experiential knowledge primarily in the much longer time frame under which TEK has been tested and confirmed or modified. TEK also comes from an entire society rather than from individual experiences alone.
>
> **Indigenous knowledge** is a body of preserved information built up by a group of people through generations of living in close contact with nature (Johnson 1992). Tribal citizens often see the term "Indigenous" as referring to historical contexts that the current social and cultural society has moved away from. This perspective may not include some of the more adaptive processes during colonial periods.
>
> **Co-production of knowledge** refers to the contributions of multiple sources and the capacities of different stakeholders, who work together on an equal footing and with mutual respect with the goal of co-creating information to improve environmental understanding and decision-making (adapted from Lemos and Morehouse 2005). Co-production stands in distinction to collaboration in that co-production offers a higher level of transparency, and the power, knowledge, decision-making, benefits, and finances are shared equally among participants from the outset of a project.

public laws that govern other public or private lands in the United States. Where research is concerned, tribes have the clear right (indeed, the responsibility) to set their own rules and guidelines for access to land and knowledge and to determine the protocols about how that knowledge is disseminated and used. Most of all, tribes deserve the right to tell their own stories without corruption or appropriation by outside entities.

WHY WORK WITH TRIBES?

Tribal and nontribal wildlife researchers share the same goal: an understanding of the ecology of fish and wildlife so that healthy populations of species and their habitats can be protected, managed, and used for future generations. This obvious objective should serve as a bond that unites tribal and nontribal researchers in the acquisition of information. In addition, many federal laws, executive actions, treaties, and agreements mandate meaningful collaboration with tribes for any action that directly *or indirectly* affects or could potentially affect tribal land and resources (see chapter 9). Other reasons for productive collaboration or co-production include:

- Tribally held knowledge can provide information about seasonal and annual movements of species, trends in short- and long-term population change, landscape-level habitat shifts, and the impacts of climate change (Polfus et al. 2014, Armstrong and Veteto 2015).

- Indigenous people have a deep connection with natural resources and their homelands (including places where they are no longer active residents) and a deep understanding of local ecology (Blicharska et al. 1981). Tribal professionals, elders, hunters, fishers, and other community members offer novel perspectives, ideas, conservation techniques, and ecological knowledge that can inform conservation management in holistic ways.

- Environmental and general research ethics require that all stakeholders have a meaningful voice in projects that may affect their resources (Aggestam 2017).

- Co-production often leads to more successful and sustainable outcomes.

- Natural resources are often considered by tribes (and legally under the National Historic Preservation Act) as traditional cultural properties, a status conferring federal protection and protocols that may be unfamiliar to those working in the realm of natural resources.

- Many tribal lands are relatively pristine and can serve as reference areas and natural laboratories (Hoagland et al. 2017, Ramírez and Santana 2019).

- Tribal lands share extensive borders and wildlife populations with federal, state, and private lands. Effective conservation of the fish, birds, large mammals, and other wildlife species that move across these boundaries requires collaboration.

- Co-production with tribes can open new avenues for grants and other financial support.

From a tribe's perspective, work with outside researchers has the potential to

- make use of the expertise of highly trained professionals to protect habitats and species of cultural and economic importance;

- strengthen tribal sovereignty by supplying information tribes can use for more effective resource management;

- make tribal voices more prominent in land use planning and in species management and recovery;
- preserve cultural and ecological information; and
- provide jobs, income, training, professional development, and technology transfer to tribal staff.

THE ORIGINS OF DISTRUST

With all the voluntary and mandated reasons for collaboration, why is there so much distrust among tribes regarding non-Indian researchers? To answer this, it is important to look at our collective history. From the earliest years of the "Columbian exchange," death and disease seemed to follow Europeans in their interactions with Native peoples, resulting in numerous pandemics that within little more than a century killed an estimated 90% of the people living in the Americas before contact, and wiped out entire cultures (Koch et al. 2019). This loss was exacerbated as tribes were forced to suppress their cultures and adopt European systems of philosophy, language, and capitalism. Tens of thousands of living tribal members, including several contributors to this volume, retain the emotional scars of the period that ended only a few decades ago when children were forcibly removed from their parents, sent to school hundreds or thousands of miles away, forced to cut their hair and adopt non-Indian forms of dress and manners, and beaten for speaking their own language (Adams 1995, Davis 2001, Reyhner 2018). "Kill the Indian to save the child" is a quote attributed to Richard Henry Pratt, who developed the infamous Carlisle Indian School in Pennsylvania. Current revelations surrounding the deaths of nearly 6,000 Indigenous children, including 751 in one unmarked mass grave at a boarding school in British Columbia (BBC 2021), attest to the brutality of these places. Tribes were encouraged to see their land as a bargaining tool and their resources as currency, a process that deteriorated the landscape, put dozens of tribes on the brink or over the edge of extinction, and left tens of millions of acres permanently degraded (Koch et al. 2019).

Tribal distrust with Western science can be traced back to attitudes toward tribes as novelties or objects of research, which much of the society at large now realizes were unethical. Some early scientists espoused racist beliefs or viewed tribal peoples as a different species. In the early 20th century, approximately 25% of Native American women were sterilized under the Family Planning Services and Population Research Act of 1970 (Carpio 2004), an effort initiated through researchers and the federal Indian Health Service (Torpy 2000). At the Pueblo of Zuni, the names of Frank Hamilton Cushing and Matilda Coxe Stevenson—to many, obscure 19th-century anthropologists—are known by nearly every Zuni tribal member above school age and universally spoken of with contempt more than 130 years after they last set foot on tribal lands (see, for example, LaCapra 2018). Most tribes in the United States have comparable stories of unwelcome scientists poking and prodding them while stealing their physical or intellectual patrimony and sacred objects. For many tribes, the disinterment and theft of human remains and ceremonial items and their removal to study centers thousands of miles away or overseas are part of living memory, and tribal leaders are still actively fighting to have the bodies of their ancestors returned to their homelands from the museums, universities, and warehouses where they are stored (Guilford 1996, Colwell-Chanthaphonh et al. 2010, Kretzler 2015). A former director of the Hopi Cultural Preservation Office, Leigh Kuwanwisiwma, estimated that over the course of his career he took part in the repatriation and reburial of 8,000 Hopi ancestors and 16,000 funerary objects (L. Kuwanwisiwma, pers. comm. 2021). The eponym of the Smithsonian Institution's National Museum of the American Indian, George Gustav Heye, was a convicted robber of Native American graves, who once remarked, "It is strange how scientific work sanctions everything."

If one is tempted to think of these actions as all in the past, one need only consider the controversy generated around the Ancient One (generally referred to by non-Natives as Kennewick Man), an ancient Native American male skeleton (Rasmussen et al. 2015) found in Washington state in 1996. The disposition of his remains created considerable acrimony between tribes, who wanted to rebury their ancestor, and Western researchers, who were keen to glean information (Thomas 2001, Owsley and Jantz 2014).

Subtle forms of intellectual property theft can be found on thousands of internet sites or by strolling around many cities in the western United States, where Native American–themed jewelry and crafts—often made by non-Native people—turn a handsome profit for someone other than the person who created the original design.

Tribal distrust in medical research arose again in the 1990s, when Arizona State University was asked to conduct a study on the Havasupai Tribe's diabetes epidemic. It was later revealed that one researcher kept blood from tribal members in order to work on new lines of investigation, something that only came to light when a Native American undergraduate student asked if the researcher had obtained Havasupai tribal permission. It was eventually revealed that researchers had been using unapproved blood samples on many projects completely unrelated to the original research. Nearly everyone at Havasupai, and tribal communities throughout the region, are well aware of the story, though they are looking for avenues to move forward with researchers in a productive way (Pacheco et al. 2013).

THE COLUMBUS EFFECT AS A FORM OF PLAGIARISM

It is by now a well-worn joke—among Native people, an especially sardonic one—that Columbus "discovered" America and its people. What is less amusing is that the practice of non-Indian researchers "discovering" and publishing tribal information as novel ideas persists today. Researchers are trained to strive to uncover new information, and many careers depend on an increasing tally of publications. The unwanted sharing of knowledge can be especially harmful in the realm of TEK when researchers describe a previously unpublished fact or system of belief in order to advance their professional standing. Academic libraries are full of works written about tribes to which tribal members, often lacking academic credentials, do not have access. For many tribal traditionalists, there is considerable power in keeping information within a trusted circle, ensuring it is not misused.

Another, perhaps more insidious form of "discovery" is when the researcher claims the academic currency of an Indigenous identity. Grim humor aside, despite prolific family myths, there is no such thing as "part Cherokee" to a Cherokee citizen. And the apocryphal "Cherokee princess" never existed. For Cherokees, either you are or you are not one of them. Yet these family stories of misidentity persist, occasionally coming from research professionals themselves. Over a million people self-identified as Cherokee in the last US Census, which cannot mean the Tribe misplaced over 60% of its people. Whether from ignorance, white guilt, or perhaps nefarious reasons, false identities are rampant, and they are dangerous to sovereignty. Under a false identity, researchers, many with academic influence, can gain access to funding, cultural knowledge, and other sources of data not meant for them. Currently, there is no protection for tribes against false identity claims, so they persist in the ranks of academia and in federal agencies.

Those who fraudulently claim an Indigenous identity or inappropriately gather knowledge do so in order to leverage authority they have no right to assert (see Viren 2021). As academics, we all agree that plagiarism is the cardinal sin of research. Why then do we tolerate fellow professionals who plagiarize identities and Indigenous knowledge, resulting in delegitimizing research, careers, and entire bodies of work?

AVOIDING PAST MISTAKES

Even with good intentions, co-production may falter for a variety of reasons. Failures often occur during the initial consultation, when nontribal researchers do not consider tribal sovereignty, lumping tribes with other stakeholders, such as states or private landowners. Here are some other pitfalls to avoid:

Proper tribal authorities are not consulted. Who speaks for a tribe? Each tribe is unique, and many have multiple layers of religious and secular authorities or ad hoc groups, such as livestock committees, hunting or fishing committees, research review boards, and cultural resource advisory teams. Ignoring one of these groups may lead to project failure because the tribe has been inadequately consulted.

Choosing the wrong liaison. Use of tribal liaisons is almost always a good idea, but researchers should avoid working through a single person unless they are certain of their authority. The system of finding any willing tribal member to give "permission" for research or searching for the "token" collaborator has been widely abused by outside investigators. A new consensus is developing around the concept of informed community consent (see below).

The initial consultation process is poorly planned or rushed. It cannot be stressed enough: this process takes time. Trust needs to be earned before information is shared. Researchers should plan on numerous in-person meetings, which should be followed up by continual relationship building.

Communication is flawed. Tribes often operate with different methods or expectations of communication than nontribal researchers. When researchers explain a misunderstood communication or a lack of tribal representation or review, we have often heard, "We sent an email to the Tribal Chairman's office." Speaking *at* a tribe does not constitute communication. Most tribes have several layers of administration, and it takes time for communication to filter through the proper channels. Tribal staff are often small in number and overworked, and they field numerous requests for consultation and participation.

The goals of a project have little or no concrete benefit for a tribe. Researchers too often approach tribes with ideas for studies that fit their own interests, expertise, or funding requirements without much regard to whether the tribe or its citizens will receive any benefit. While tribes value all species, for example, some are more central to tribal economies, have higher cultural significance, or face greater threats on tribal land than others.

Academic or career advancement takes precedence over co-production. Tribal natural resource professionals are more likely to be concerned with gaining information that can help manage fish, wildlife, and habitat than with publishing academic papers. In addition, tribes closely guard information related to the locations of big game populations, federally listed or culturally sensitive species, and sacred sites, and thus may oppose open publication.

There is a failure of equity and inclusion or an unequal distribution of resources and credit. In research settings, there is often a division of tasks, such as research design, consultation, fieldwork and data collection, mapping, data analysis, writing, review, and editing. Academic institutions and publishers may restrict authorship credit to the lead scientists that conducted and wrote up the work. Over 500 years of scholarship, this has led to thousands—possibly tens or hundreds of thousands—of publications that were based on tribal knowledge without proper attribution.

Guidelines at some notable institutions, however, have begun to shift. Harvard University's Academic Research Guidelines now indicate that authorship should be considered for those who have "made substantial contributions to the conception or design of the work; or the acquisition, analysis, or interpretation of data" (Harvard University 2021). A researcher should ask: Would this acquisition of information have been possible without person X's participation? If the answer is no, then the tribal citizens who aided the acquisition or interpretation of data, facilitated logistics, conducted fieldwork, or were otherwise essential to the project should be considered for authorship. (All contributors should of course review and approve a manuscript.) The creative opportunities to develop means of eliciting inputs are manifold. For example, manuscript review could include a discussion about the paper, which could be transcribed or summarized.

Equity also includes project finances. If a researcher is supported by a grant or home institution or if the work generates funding (e.g., book deals) to conduct the research, tribal staff or the tribe itself should be compensated for their time, effort, and intellectual property. This is another reason to engage tribal co-producers early, to accommodate their time constraints and financial needs. Be aware that authors are not just credibility builders but honest co-producers.

TRIBAL INSTITUTIONAL REVIEW BOARDS

The US Code of Federal Regulations outlines the formation of IRBs composed of professionals and citizens responsible for safeguarding human research subjects. An IRB reviews and approves or denies research based on whether it may cause physical, psychological, or social harm. Institutions with IRBs (or with similar research review boards, RRBs) include hospitals, universities, federal research agencies, and increasingly, tribes. An IRB is now seen as essential in many quarters if research involves human subjects, even just for interviews. Gachupin et al. (2019) list 14 tribes with IRBs/RRBs. The National Park Service (n.d.) maintains a roster of tribal research policies, processes, and protocols, many of which outline IRBs. One benefit to tribes of an IRB/RRB is that the board can include ad hoc specialists, such as elders, cultural advisors, or language experts. Tribes without IRBs may wish to develop an agreement with a trusted external partner, such as a university, that does have one.

SOME POSITIVE EXAMPLES

Several chapters in this book (see chapters 12, 13, 15, 16, 21, 25, and 26) provide excellent examples of conscientiously co-produced research. Below are some additional examples.

Small Mammal Monitoring on the Cheyenne River Sioux Reservation

During coauthor Julie Thorstenson's time as wildlife biologist for the Cheyenne River Sioux, we received a federal grant to conduct small mammal inventories. We determined the need for the project after reviewing a local field guide with significant data gaps from tribal lands. We also wanted to gather traditional, cultural, and language information about specific species. We partnered with local high schools to recruit Native students, who learned research methodology and general concepts of natural resource management. Each student was also asked to choose one study species and work with a tribal elder to learn the Native word and cultural significance of the animal. At the end of the project, we published a field guide.

Co-Production of Hopi Ethnographic Studies

The Hopi Tribe has been one of the most intensively studied in North America. Anthropologists and archaeologists have written about the Tribe since shortly after contact with the Spanish in the early 16th century. Since the 1990s, the Tribe has made impressive strides in taking control of ethnographic research conducted on its trust and nontrust ancestral lands on the Colorado Plateau.[1] The Hopi make little distinction between cultural resources and natural resources, especially in the Tribe's collective and individual knowledge about animals and plants, accumulated over millennia.

Much of this work was started by the first director of the Hopi Cultural Preservation Office (HCPO), Leigh Kuwanwisiwma, and is currently being carried on by Stewart Koyiyumptewa and his staff. The HCPO sees benefits from collaborating with scientists and academics from across the region, as long as the objectives and methods of the research are in alignment with the Tribe's overall vision for collaborative research and co-production. Much of this vision is outlined in a groundbreaking book, *Footprints of Hopi History: Hopihiniwtiput Kukveni'at* (Kuwanwisiwma et al. 2018). In an interview on 15 July 2021, Leigh noted:

> When I got into office, it was open season on Hopi from the standpoint of research and federal and state agencies asking the tribe to participate in projects. I ran into a backlog of federal agency notifications. We still had IBM typewriters, so I was busy writing form letters to [applicants] letting them know that we were in receipt of their notice, and we had some initial comments. I called the researchers in, one by one, and learned about their research and, in a couple of cases, I had to tell them that they needed to stop their research . . . and turn over all their field notes and photographs with their negatives. Another project was on Hopi language. I was interested in that, but I told them the Tribe needed to be involved. Over time, one by one, year after year, we got in contact with the local universities, putting them on notice that any research needed to

17.1. Hopi cultural advisors Gilbert Naseyowma and Terrance Outah at Chaco Culture National Historical Park, an area of great historic and cultural importance and deep ties, located within Hopi ancestral lands. PHOTO BY STEVE ALBERT

go through the Hopi Cultural Preservation Office. We would [review] proposals [and sometimes approve] research, perhaps with modifications. That became the standard, and we were able to come up with some very good research. One of the things we told the researchers was, "Hey, the Hopi Tribe has its own research goals, too."

Co-Production of Sicklefin Redhorse Restoration in the Southern Appalachians

The sicklefin redhorse (*Moxostoma* sp.) is a large sucker, relatively little studied until the 1990s. It was thought to be extirpated from the upper Tennessee River system when large portions of its habitat were destroyed by dams' production. The Cherokee have a name that fits the fish perfectly—*ugidatli* / ᎤᏲᏞ, which translates as "wearing a feather," a reference to its dorsal fin shape. Multiple issues continue to threaten the species, including pollutants, stream channelization, in-stream mining, and predation. In 2014, several public and private entities in North Carolina and Georgia, including the Eastern Band of Cherokee Indians (EBCI), signed a memorandum of understanding to study and co-manage the species' recovery. This project is particularly important to the EBCI because redhorse were once an important food fish for the tribe, particularly during the fall when other food sources were scarce. The committee has prioritized reintroduction areas, produced research on population ecology and habitat requirements, and collected gametes for the propagation and annual release of tens of thousands of fish (figure 17.2). A noteworthy part of the co-production is that the EBCI has led much of the restoration and research. Finances and knowledge were shared equally across agencies, and it has been common to see tribal employees working on state and federal lands.

17.2. Sicklefin redhorse restoration in the southern Appalachians was a collaborative initiative between the Eastern Band of Cherokee Indians and state and federal partners.
THE SICKLEFIN REDHORSE CANDIDATE CONSERVATION AGREEMENT PROJECT

WORKING TOWARD CO-PRODUCTION

Co-production involves much more than consulting with tribes or individual tribal citizens. The essence of co-production is working *with* tribes, not using them as a means to gather information (Podestá et al. 2013). Co-production means a tribe is helping drive the research objectives and study design and has been meaningfully engaged in all aspects of the project (Houde 2007). Most of the successful and enduring collaborations inhabit a space where the cooperating partners are allies—equals supporting each other (see, for example, Ludwig and El-Hani 2020)—which, for tribes, means enhancing tribal sovereignty and the capacity to self-manage their natural resources.

One of the best recent examples of co-production comes from Canada, where Canadian government scientists and conservation professionals are proactively and thoughtfully working with First Nations on the management of many species, especially seabirds. Pillars of their approach include a respect for and an incorporation of TEK, co-produced research, and the transparent sharing of project funds. Their collective approach to management also includes a recognition of Indigenous sovereignty over research sites, community direction of ongoing research, and the promotion of Inuit goals of self-determination while charting a course on how to achieve local, national, and international conservation and research needs in the Canadian Arctic (Mallory et al., in press).

RECOMMENDATIONS FOR CO-PRODUCED RESEARCH

We have developed several suggestions for the successful co-production of research. The following suggestions on "how to be an ally" are adapted from Chambers and Nicholson (2022):

- Educate yourself about the historical and current contexts of the local community. Learn about the cultural practices, values, and beliefs of the tribe to help you better understand their needs and what may drive their priorities and land management practices. Listen actively, without an inner dialogue or composing your next question. Often, no response is necessary.

- Introduce yourself as a whole person, including where you grew up, who your ancestors are, and the names of your parents, spouses, and children (note that your title comes last). Visit often, follow through, do not overpromise, and attend community events (both happy and sad celebrations). But take care and understand that ethical participation, even if welcomed, means that some cultural activities might be outside the scope of your work (see section about IRB above). Come back, even years later, because if you were welcomed, it can be expected you are now part of their life.

- Encourage your home organization to think at a policy level. NGOs and government agencies can be allies by fostering welcoming environments for Native people to feel included and thrive, and by supporting Native youth initiatives, such as the American Indian Science and Engineering Society or the National Congress of American Indians youth programs.

- Organizational co-production includes sharing research ideas and outcomes as well as management responsibility. Ask yourself where the center of power and finances for a project is, and if resources are being shared equitably. Review the US Fish and Wildlife Service's Native American Policy (USFWS 2016).

Additional recommendations include:

- Be sure you understand who owns or manages the land on which you are working. Apply early for all necessary permits (see the section about IRBs above). Even with permits in hand, ask before you go anywhere on tribal lands, collect data, or initiate questioning.

- Meet with tribal officials, administrators, and staff *before* you have your research thesis and methods developed. Make sure your partners have the capacity, the willingness, and the administrative and financial support they need to carry the project through, and make sure that local staff have key (not token) roles.

- Be cognizant of different styles of communication. Some tribes and some Native people may not feel comfortable immediately and directly saying no to a proposal. Sometimes "maybe" is a polite way of saying no. If you feel like you are pushing too hard to make a project happen, you probably are. Other times, the lack of communication can be an answer too.

- Be sure you are working with authorized representatives of the tribe. A single tribal member may not necessarily speak for the tribe, nor do they necessarily understand the sensitivities of cultural information. Engage tribal staff with expertise in the field of focus, even though they may not have formal academic credentials (but you can attempt to find those that do). Many federal agencies have tribal liaisons. Researchers should determine if their work needs to be brought before the tribal leadership or a cultural review board. A consensus is developing around the concept of informed *community* consent, that is, obtaining the consent of an entire community that has had time to be fully informed of any potential research and to carefully consider all the implications (see, for example, Dickert et al. 2021).

- Define clear objectives and responsibilities with the tribe and, if necessary, create a co-production and communication plan. Larger institutions should consider having these reviewed by an ethics specialist.

- Be patient, and expect that tribally co-produced projects will take much longer. Many tribes are inherently conservative (in the nonpolitical sense of that word), making decisions deliberatively, by consensus, and approving them only when the benefits are clear. This has helped us survive for millennia. Decision makers and tribal councils may meet infrequently. Schedule time in the area that is not directly related to your research, and get to know your co-producers to build trust. When the project is over, return to share your results with the community and the individuals with whom you worked.

- Many tribal elders are more comfortable with their Native language. Expect to use and pay for translation at every step of a tribal elder's involvement.

- Be sensitive to concerns about intellectual property and academic imperialism (see, e.g., Rakowski 1993). Be explicit about who will own the data, where it will be stored, and who has the rights to disseminate and publish the research. If appropriate, develop an intellectual property rights agreement. If some data are sensitive, consider two versions: a confidential report, which local staff can use as a resource, and a public one, which can be shared more widely or published.

- At all times, be transparent about everything, including project finances. Tribal natural resource departments are often overtaxed and underfunded. Don't make your research a financial or time burden. Find out what the tribe's financial and training needs are and include tribal staff in your budget. Make provisions for capacity building, travel stipends, and compensation or honoraria for local staff, experts, guides, and interview subjects. Consider providing scholarships, internships, or fellowships for local students and young researchers. Leave something positive behind.

- Be sensitive to the types of deliverables desired by the tribe. Most academic or professional researchers prefer peer-reviewed publications, but these may not be as important or accessible to tribal staff, who may prefer results they can use immediately and can show to tribal officials and the tribal public. Consider creating reader-friendly reports or summaries in addition to peer-reviewed papers. Several journals, such as the *Canadian Journal of Zoology*, have ethical standards for working with Indigenous peoples. Their guidelines state that "submitted manuscripts examining TEK will have to demonstrate that the research was ethically conducted with respect of Indigenous Peoples. The submissions will have to follow these principles: (i) research has received appropriate Indigenous Research Ethics approval, (ii) true and informed consent was a prior sought before the start of the research, (iii) data were returned to the People where they belong for the protection and preservation of their Traditional Ecological Knowledge, and (iv) Indigenous Peoples have approved the submission of the manuscript" (Canadian Journal of Zoology 2021).

- Consider ways in which access might be provided to your tribal partners and what open access means for tribes not able to read research that was conducted with them. Many tribes lack access to the large quantity of peer-reviewed literature behind membership or financial paywalls. The same issue applies to tribal researchers wishing to publish their work.

- Expand authorship definitions. Academic institutions have expanded definitions of authorship-worthy contributions beyond the old roles of data analysis and manuscript writing. Harvard University's revised guidelines describe authors as collaborators with "substantial contributions to the conception or design of the work; or the acquisition, analysis, or interpretation of data; . . . or have drafted the work or substantively revised it" (Harvard University 2021). A physics paper in *Physical Review Letters* (Aad et al. 2015) listed more than 5,000 authors—more authors than there were words in the text. For projects with tribes, consider *all* the members who made the research possible, including translators, facilitators, and technicians.

- For sensitive research or collaborative studies involving many entities, consider the use of a professional facilitator or a specialist in the field of human dimensions. Many potential pitfalls can be avoided through sensitive and thoughtful communication. (See, for example, Bennett et al. 2017, Redpath et al. 2013.)

In closing, we'd like to provide a somewhat surprising example of how thoughtful, groundbreaking research can be ethically co-produced. In 1993, a few years before the controversy around the Ancient One was in the news, a much quieter bit of similar research was taking place. On Prince of Wales Island, Alaska, another ancient human skeleton was unearthed by archaeologist Timothy Heaton and colleagues. However, in contrast to the approach taken for the Ancient One, the scientists and managers of this find immediately contacted and began consulting with local tribal governments, seeking their ideas about what should be done with/for the man, who became known as Shuká Káa (The One Before Us).

The parties came to a quick consensus that the rightful owners were the Tlingit people in Southeast Alaska. Although the Tlingit had some reservations about allowing scientists to test the remains, after considerable deliberation within the Tribe—a process that took years—the Tlingit agreed to allow analyses of the bones and some limited further excavation, as long as they were a part of the process and were kept abreast of additional discoveries. This led to a 12-year partnership between the Tribe and the scientific community. After important research was conducted, all remains were returned to the Tlingit people for reburial and a celebration festival.

NOTE

1. Contrary to what was until very recently taught in schools, described in scientific and popular literature, and listed on interpretive signs and pamphlets in national and state parks across the Southwest, ancestral Puebloan peoples (formerly called Anasazi, though that term is, appropriately, no longer used) did not mysteriously "vanish" from Mesa Verde, Chaco Canyon, and thousands of other sites across the region. During a prolonged drought in the late 13th century, they relocated to more secure, wetter environments, joining their relatives and friends in places like the Hopi mesas; the Pueblos of Zuni, Acoma, and Laguna; and dozens of sites in the Rio Grande valley. "We're still here" is a refrain commonly heard among Pueblo people when describing their ancestors who built these places.

LITERATURE CITED

Aad, G., et al. 2015. Combined measurement of the Higgs Boson mass in pp collisions at √s=7 and 8 TeV with the ATLAS and CMS experiments. Physical Review Letters 114:191803.

Adams, D. W. 1995. Education for extinction: American Indians and the boarding school experience, 1875–1928. University Press of Kansas, Lawrence.

Aggestam, F. 2017. Wetland restoration and the involvement of stakeholders: an analysis based on value-perspectives. Landscape Research 39:680–97.

Armstrong, C. G., and J. R. Veteto. 2015. Historical ecology and ethnobiology: applied research for environmental conservation and social justice. Ethnobiology Letters 6:5–7.

BBC. 2021. Canada: 751 unmarked graves found at residential school. https://www.bbc.com/news/world-us-canada-57592243. Accessed 31 July 2021.

Beier, P., L. J. Hansen, L. Helbrecht, and D. Behar. 2016. A how-to guide for co-production of actionable science. Conservation Letters 10:288–96.

Bennett, N. J., R. Roth, S. C. Klain, K. Chan, P. Christee, D. A. Clark, G. Cullman, D. Curran, T. J. Durbin, G. Epstein, A. Greenberg, M. P. Nelson, J. Sandlos, R. Stedman, T. L. Teel, R. Thomas, D. Veríssimo, and C. Wyborn. 2017. Conservation social science: understanding and integrating human dimensions to improve conservation. Biological Conservation 205:93–108.

Berkes, F., J. Colding, and C. Folke. 2000. Rediscovery of Traditional Ecological Knowledge as adaptive management. Ecological Applications 10:1251–62.

Blicharska, M., R. J. Smithers, M. Kuchler, G. K. Agrawal, J. M. Gutiérre, A. Hassanali, S. Huq, S. H. Koller, S. Marjit, H. M. Mshinda, H. H. Masjuki, N. W. Solomons, J. Van Staden, and G. Brownrigg. 1981. Conserving the natural heritage of Latin America and the Caribbean: the planning and management of protected areas in the Neotropical realm. Proceedings of the 18th working session of the International Union for Conservation of Nature's Commission on National Parks and Protected Areas, Lima, Peru, 21–28 June. International Union for Conservation of Nature, Gland, Switzerland.

Brunet, N. D., G. M. Hickey, and M. M. Humphries. 2014. The evolution of local participation and the mode of knowledge production in Arctic research. Ecology and Society 19:69.

Brunet, N. D., G. M. Hickey, and M. M. Humphries. 2016. Local participation and partnership development in Canada's Arctic research: challenges and opportunities in an age of empowerment and self-determination. Polar Record 52:345–59.

Canadian Journal of Zoology. 2021. Statement on Indigenous traditional knowledge. https://cdnsciencepub.com/journal/cjz/about. Accessed 31 March 2021.

Carpio, M. V. 2004. The lost generation: American Indian women and sterilization abuse. Social Justice 3:40–53.

Chambers, C. L., and K. L. Nicholson, editors. 2022. Women in wildlife science: building equity, diversity, and inclusion. Johns Hopkins University Press, Baltimore, Maryland.

Colwell-Chanthaphonh, C., R. Maxson, and J. Powell. 2010. The repatriation of culturally unidentifiable human remains. Museum Management and Curatorship 26:27–43.

Corrigan, C., H. Bingham, Y. Shi, E. Lewis, A. Chauvenet., and N. Kingston. 2018. Quantifying the contribution to biodiversity conservation of protected areas governed by Indigenous peoples and local communities. Biological Conservation 227:403–12.

Davis, J. 2001. American Indian boarding school experiences: recent studies from Native perspectives. Organization of American Historians Magazine of History 15(2):20–22.

Dickert, N. W., K. Metz, M. D. Fetters, A. N. Haggins, D. K. Harney, C. D. Speight, and R. Silbergleit. 2021. Meeting unique requirements: community consultation and public disclosure for research in emergency settings using exception from informed consent. Academic Emergency Medicine 10:1183–94.

DuRose, C., Y. Beebeejaun, J. Rees, J. Richardson, and L. Richardson. 2012. Connected communities: towards co-production in research with communities. Arts and Humanities Research Council, Manchester, England.

Gachupin, F. C., B. Lameman, and F. Molina. 2019. Guidelines for researchers. Department of Family and Community Medicine, College of Medicine, University of Arizona, Tucson.

Guilford, A. 1996. Bones of contention: the repatriation of Native American human remains. Public Historian 18:119–43.

Harvard University. 2021. Faculty of Arts and Sciences guidelines on authorship and acknowledgement. https://research.fas.harvard.edu/links/guidelines-authorship-and-acknowledgement. Accessed 5 July 2021.

Hoagland, S. J., R. Miller, K. Waring, and O. Carroll. 2017. Tribal lands provide forest management laboratories for mainstream university students. Journal of Forestry 115:484–90.

Houde, N. 2007. The six faces of Traditional Ecological Knowledge: challenges and opportunities for Canadian co-management arrangements. Ecology and Society 12:34.

Huntington, H. 2000. Using Traditional Ecological Knowledge in science: methods and applications. Ecological Applications 10:1270–74.

Johnson, M. 1992. Lore: capturing Traditional Environmental Knowledge. International Development Research Centre, Ottawa, Canada.

Kimmerer, R. W. 2002. Weaving Traditional Ecological Knowledge into biological education: a call to action. BioScience 52:432–38.

Koch, A., C. Brierley, M. M. Maslin, and S. L. Lewis. 2019. Earth systems impacts of the European arrival and Great Dying in the Americas after 1492. Quaternary Science Reviews 207:13–36.

Kretzler, I. 2015. Repatriation of "culturally-unidentifiable" human remains: the view from Fort Vancouver. SAA Archaeological Record 2015:21–24.

Kuwanwisiwma, L. J., T. J. Ferguson, and C. Colwell, editors. 2018. Footprints of Hopi history: Hopihiniwtiput Kukveni'at. University of Arizona Press, Tucson.

LaCapra, D. 2018. Understanding others: peoples, animals, pasts. Cornell University Press, Ithaca, New York.

Lemos, M. C., and B. J. Morehouse. 2005. The co-production of science and policy in integrated climate assessments. Global Environmental Change 15:57–68.

Ludwig, D., and C. N. El-Hani. 2020. Philosophy of ethnobiology: understanding knowledge integration and its limitations. Journal of Ethnobiology 40:3–20.

Mallory, M. L., J. Toomasie, S. Emond, G. Lamarche, L. Roberts, L. Pirie-Dominix, and J. F. Provencher. In press. Community-scientist collaboration in the creation, management and research for two national wildlife areas in Arctic Canada. Advances in Ecological Research.

Nabhan, G. P. 2000. Interspecific relationships affecting endangered species recognized by O'odham and Comcáac cultures. Ecological Applications 10:1288–95.

National Park Service. n.d. Tribal research policies, processes and protocols. https://www.nps.gov/subjects/tek/tribal-research-policies-processes-and-protocols.htm. Accessed 20 July 2021.

Owsley, D. W., and R. L. Jantz. 2014. Kennewick Man: the scientific investigation of an ancient American skeleton. Smithsonian Institution Press, Washington, DC.

Pacheco, C. M., S. M. Daley, T. Brown, M. Filippi, K. A. Greiner, and C. M. Daley. 2013. Moving forward: breaking the cycle of mistrust between American Indians and researchers. American Journal of Public Health 103:2152–59.

Parlee, B., M. Manseau, Å. K. É. Dene, and F. Nation. 2005. Using traditional knowledge to adapt to ecological change: Denésôåiné monitoring of caribou movements. Arctic 58:26–37.

Podestá, G. P., C. E. Natenzon, C. Hidalgo, and F. R. Toranzo. 2013. Interdisciplinary production of knowledge with participation of stakeholders: a case study of a collaborative project on climate variability, human decisions, and agricultural ecosystems in the Argentine Pampas. Environmental Science and Policy 26:40–48.

Polfus, J. L., K. Heinemeyer, and M. Hebblewhite. 2014. Comparing Traditional Ecological Knowledge and Western science woodland caribou habitat models. Journal of Wildlife Management 78:112–21.

Powys Whyte, K. 2013. On the role of Traditional Ecological Knowledge as a collaborative concept: a philosophical study. Ecological Processes 2:7.

Rakowski, C. A. 1993. The ugly scholar: neocolonialism and ethical issues in international research. American Sociologist 24:69–86.

Ramírez, F., and J. Santana. 2019. National parks and biodiversity conservation. Pages 33–38 in F. Ramírez and J. Santana, editors. Environmental education and ecotourism. Springer Nature, Cham, Switzerland.

Rasmussen, M., M. Sikora, A. Albrechtsen, T. S. Korneliussen, J. V. Moreno-Mayar, G. D. Poznik, C. P. E. Zollikofer, M. S. Ponce de León, M. E. Allentoft, I. Moltke, H. Jónsson, C. Valdiosera, R. S. Malhi, L. Orlando, C. D. Bustamante, T. W. Stafford Jr., D. J. Meltzer, R. Nielsen, and E. Willerslev. 2015. The ancestry and affiliations of Kennewick Man. Nature 523:455–58.

Redpath, S. M., J. Young, A. Evely, W. M. Adams, W. J. Sutherland, W. Whitehouse, A. Amar, R. A. Lambert, J. D. C. Linnell, A. Watt, and R. J. Gutiérrez. 2013. Understanding and managing conservation conflicts. Trends in Ecology and Evolution 28:100–109.

Reyhner, J. 2018. American Indian boarding schools: what went wrong? what is going right? Journal of American Indian Education 57:58–78.

Schlag, J. 2015. Historiography of North American ethnobotany. US Studies Online. https://usso.uk/historiography-of-north-american-ethnobotany. Accessed 23 July 2021.

Schuster, R., R. R. Germain, J. R. Bennett, N. J. Reo, and P. Arcese. 2019. Vertebrate biodiversity on Indigenous-managed lands in Australia, Brazil, and Canada equals that in protected areas. Environmental Science and Policy 101:1–6.

Thomas, D. H. 2001. Skull wars: Kennewick Man, archaeology, and the battle for Native American identity. Basic Books, New York.Torpy, S. J. 2000. Native American women and coerced sterilization: on the Trail of Tears in the 1970s. American Indian Culture and Research Journal 24:1–22.

US Fish and Wildlife Service [USFWS]. 2016. Native American policy for the US Fish and Wildlife Service. Department of the Interior, National Archives and Records Administration, Washington, DC.Viren, S. 2021. The Native scholar who wasn't. New York Times, 25 May.

Wilson, K. J., T. Bell, A. Arreak, B. Koonoo, D. Angnatsiak, and G. J. Ljubicic. 2020. Changing the role of non-Indigenous research partners in practice to support Inuit self-determination in research. Arctic Science 6:127–53.

Chapter 18

Thoughts of an Anishinaabe Poet on Wildlife Biology

| MARCIE RENDON

There is no word for "wildlife" in the Ojibwe language. Nanabazho, the trickster spirit, was sent to Earth by the Creator to walk the Earth and name all the animals and living creatures. There were no *wild* animals. There were, and still are, *owayseug*. Animals. *Ahmik, gawg, maengun, nigig, migizi, ahjijawk, ziizhiib, gigoon*—to name a few. Each animal had a name and purpose. Food or helper or protector or healer. Nanabazho named each in recognition that we are all related. One cannot exist without the other.

Nanabazho named the plants also. *Mahnomen, weegwas, ininaig, asema*. The rivers (*michi ziibii*) and rocks (*asin*) and weather (*animikii, geezheebasun*). All of which coexist. All of which enable the existence of all others. And all are life. All enable all life to continue.

Merriam-Webster Dictionary: Definition of *wildlife*: living things and especially mammals, birds, and fishes that are neither human nor domesticated.

Christianity creates a separation between all living beings by a decree from the Christian God in the book of Genesis:

> [26]Then God said, "Let us make mankind in our image, in our likeness, so that they may rule over the fish in the sea and the birds in the sky, over the livestock and all the wild animals, and over all the creatures that move along the ground."
>
> [27]So God created mankind in his own image, in the image of God he created them; male and female he created them.
>
> [28]God blessed them and said to them, "Be fruitful and increase in number; fill the earth and subdue it. Rule over the fish in the sea and the birds in the sky and over every living creature that moves on the ground."

Mahnomen Waboose Moose Wawashkeshi

Shunned in favor of Swift and Company
or Armour Meats
Where sheep, hogs and cattle are herded in and
processed out as sides of beef, leg of lamb, cubes of lard, shanks of ham
Clear-cut forests and urban sprawl diminish Wild life
A people's food is disappeared, the people too
Commercial industries dependent on domestication
Create palates tamed to blandness
Until capitalist hunger creates haute cuisine
The gentrification of Indigenous foods
On menus reading
"Perfect wild rice soup begins with rice harvested in the wild"
"tout le lapin" and rabbit terrine
Moose cheese from Sweden
Venison—the "luxury" meat

you may dam the river
but can you still the tide?

The Red River of the North

A river
Gulps mud
Persistent rush
Roots bared
Asphalt and concrete
No match for nature's cleansing
Springtime water eats the banks
Spits it out to feed the Hudson Bay

Willow arms caress
Summer muddy water
Catfish lurk
Bullheads scour
Mosquito buzz
A covert cover
For cottonwoods to gossip
Up and down the river

Leaves drop
Until the trees are bare enough
For red-tinged water to catch
The harvest moon
The river feeds on perfect light
While fish and other water creatures
Dance through silted clay
Down, down on the riverbed

Solid hoofs on snowy hills
Grasp for purchase
Finally, at river's edge
At water holes drilled
Into two-feet-thick ice
Cattle drink
The moisture of their labored breath
Mingles with the water flowing slowly north

Desert dreams . . .

i am the soul of all the ancient warriors
Born of the desert

Scorpion

Time travels the horizon
Each day a century,
Each minute a year
Yesterday is tomorrow stretching back to just moments ago.
When i was a child i saw two-leggeds retreat from war singing songs of grief.
Nature gives and takes, a natural order to the universe.
Man-made barriers propagate a sea of sorrow.
Yesterday, or was it seven years ago?
Rain fell, water rushed through creek bed.
It has been so dry that even we who live on drops of water thirst.

Prickly Pear Cacti

Ecologically efficient
i move across the desert floor.
Seeking one drop of water
      My throat rasps
      Parched leather
Wind evaporates moisture
This dust that i become is your future

Cholla Cacti

i am the blinding mirror image of near-sighted dreams.
i was born of the desert when no concrete wall stopped the sun or outshone
    the moon.
i am no boundaries.
Try stop me, i go where i want
i am the nightmare of short-sighted vision
Foretelling, no water, strip-mined mountains, scorching heat.
i am the sharp pain of overheated global reality

Desert

Earthmaker, Gather up these souls, as Puma's speed and spirit are needed
    once again for the two-leggeds' journey home.
In this graveyard of dreams dying is not a gentle passage and living is the
    greater challenge.
Rattlesnake poison bleaches bones, which speak of lost family
and forgotten prayers drop like rain
An altar forms from east to west, south to north
A constant day of dead

Puma, i offer tobacco to the spirits of this desert,
that you, Puma, Protector of the Universe,
That you be welcomed to the world that exists parallel to this.
Your muscle becomes the supple, subtle shifting of the desert floor.
No more suit of camouflage to hide on desert cliffs. You are the desert now.
Waiting. Waiting for redemption, your killing bite becomes the cacti thorns, your claws cholla's revenge. Your grace becomes the sun sliding down the horizon, darkness bringing respite from the heat. All the powers of your nocturnal wanderings combine with centuries-long energy in the making. Puma, Protector of the Universe, may your spirit be released to guide two-leggeds on their journey home.
Ah ho

◆

The Anishinaabe worldview is in direct opposition to the Christian worldview. These diametrically opposed worldviews and patterned ways of behavior—where the Anishinaabe worldview is to live in balance with all other living beings of the world while the Christian has a policy of consumption, domestication, subjugation, and destruction—has led to the need for wildlife biologists "to conserve, protect and enhance fish, wildlife and plants and their habitats for the continuing benefit of the American people."

Chapter 19

Protecting What We've Been Blessed With

Big Game and Other Wildlife Programs of the Navajo Nation

| *An Interview with* **GLORIA TOM**

The Navajo Nation is the largest Tribe in the country, in terms of both population and land base. The Navajo Nation Fish and Wildlife Department manages a vast area of forests, woodlands, canyons, lakes, and rivers larger than the state of West Virginia. Gloria Tom has been director of the department since 1998 and is a leader in conservation issues that affect tribes throughout the country. She oversees a staff of more than 50 and has fought for decades for equitable funding and resources for tribes. This interview was conducted by the editors, Serra Hoagland and Steven Albert, on 2 December 2020.

Editors: Tell us a little bit about the history of your department.

Gloria: The department was established in the 1950s, when there was also an interest in bringing back mule deer. Because it wasn't regulated, they had been nearly extirpated. The department started a transplant program where they brought back the deer. A hunting program was developed in the '70s and well established by the '80s. We had tribal wildlife codes and cooperative agreements with the state of Arizona and the Fish and Wildlife Service for law enforcement coordination at that time. Ours was one of the first tribes with hunting codes. The program really began with a predator control program because farmers and ranchers had lost livestock due to predators, and sometimes the game would get into the fields. In the '60s and '70s the focus shifted from predator control to a more comprehensive program. By the late '70s they had full-time biologists on staff.

Editors: What sorts of programs do you run today?

Gloria: Obviously, the need is greater than what we have. We currently have about 50 employees in the department, big game management being one of five programs under the department. Big game has biologists and several technicians, with their primary focus on managing big game and habitat, including mule deer, elk, antelope, black bear, bighorn sheep, and turkey. The emphasis is to monitor populations and protect habitat. Unfortunately, we're losing a lot of habitat to encroachment where individual tribal members want to move to remote locations on the Reservation, and with that comes utilities and roads.

Our law enforcement program is only five officers covering 18 million acres. We've been successful at securing funding to get two more soon. We can regulate tribal laws, but sadly the federal government often lacks the ability to address cases through prosecution and investigation due to [lack of] funding and lack of resources. Our department presents the cases anyway. Our endangered species program has about eight people. We recently acquired some ranches in Colorado with about three tribal staff and about five on contract to manage those properties. The ranches are intended for livestock, big game (bison), and they have a good partnership with Colorado Parks and Wildlife for big game management.

Editors: One of your more successful initiatives is the trophy bighorn sheep program.

Gloria: The bighorn sheep program is one of the most highly sought-after hunting licenses in the West, through auction sales. One of our tags recently went for $56,000. We are monitoring the sheep population and were able to identify what was limiting their population, and we implemented management strategies to address those limitations. Some of the limiting factors were diseases from domestic sheep and predation. We strive to keep them separated from domestic sheep by fencing and coordinating with land users in the area. From one small herd along the San Juan River, we were able to establish two other herds along Lake Powell. The total number of animals is now about 300. We were able to fund a telemetry project and other management initiatives using the proceeds from auction sales. A few years ago, we also started offering a once-in-a-lifetime hunt for tribal members. We are really trying to protect that herd in the river corridor. There are poaching issues as well. It's hard to get law enforcement patrols down the river.

Editors: What were some of the challenges you had to overcome and what were some of your successes?

Gloria: It's been a continual challenge to bring wildlife and fisheries to the forefront of the attention of the Tribe. Most of the time we're generating revenue to keep people employed and wildlife and fish thriving, but to put fish and wildlife as a priority with the Navajo Nation really has not happened because there are pressing social issues. The other big challenge we face is habitat loss due to people wanting to move to remote areas.

We also contend with a rapidly rotating leadership. Leadership at the local level changes every four years. There is also a large range of values of community leaders, and we are continually sharing information and educating tribal leaders. We have considered zoning sensitive wildlife areas, such as mule deer wintering ranges and raptor nesting areas. Our natural heritage program that oversees threatened and endangered species is in the process of updating a policy for land use. This is a good educational tool to share with communities and local chapter councils. Some of the successes involved [in] that zoning policy have allowed them to bridge the gap between the department and the community since it is so user-friendly. This was one of our first attempts to show why wildlife habitat is important for tribal members' health and the health of the ecosystem.

Another successful program is our youth hunting program that started in 2005 with a mentoring program for youth who did not have the skills for hunting. We recruit about 15–20 kids every year who go through an application process, explain why they are interested and why they don't have that opportunity at home to learn basic hunting and camping skills. Sometimes we have up to 60 kids applying, so it's competitive. The kids go through hunter education and then two weekends of firearm range practice, shooting rifles. They're provided an adult mentor. Once they complete the course, they can participate in a three-day mule deer hunt. Over the past 10 years, every kid has been successful. This program is sponsored by nontribal organizations. The kids get fully outfitted by the sponsors with all the stuff they need for camping.

Finally, I'll mention our climate change program. Five years ago we received a grant from the BIA to develop a climate change vulnerability assessment. We completed an adaptation plan for the Navajo Nation with help from people on the ground who work with our local communities. We hosted workshops to come up with natural resource management concerns and climate change impacts. The adaptation plan is now in place, and we have a new climate change program established by our drought insurance program through USDA. We have two years of committed funding for the climate change program, and their goal is to start working with communities on implementing strategies and addressing climate change.

Editors: What is the biggest misconception about your program or project?

Gloria: People always ask us, "Are you under the states?" *Absolutely not!* Tribes are not monitored by states. I would tell people, tribes are doing wonderful things with their wildlife, we don't have to ask for permission from the states. We have our own laws and regulations; you don't need a state permit to hunt at Navajo.

Chapter 20

Shash

| *A Story by* RAMONA EMERSON

In 1916, a brown bear kidnapped my grandmother Minnie from the shaded crests of the Chuska Mountains.

She was just a baby then, cooing and moving inside the wraps of her cradleboard. As the spring winds blew, her mother and father, my great-grandmother and great-grandfather, prepped their summer grazing grounds above the heat of the flats. They had a small herd that they walked up every year, my grandmother and her older sister in tow. It was not an easy life, but it was what they did every year, every season. That year was no different, even with an infant.

Grandma came from the Two Grey Hills and Toadlena area, surrounded by the peaks of the Chuska Mountain range, where they kept their hogan and summer fields. A walk up to the mountain was an all-day excursion. On foot and with their horses, they left their modest home in Toadlena when they could still see the stars. A lantern lit their way early in the journey, laying out the familiar trail they always followed. Grandma was wide awake, looking up into the night sky, hearing the clang of water bottles, shovels, and tools that hung from the horse saddle. As the sun began to crest, they stopped and prayed for a safe journey.

My grandmother watched the whole journey tied into her cradleboard and fastened to her mother's back with a thick strap of leather. She watched her father and his sheepdog wrangle a few stray sheep between the juniper bushes and piñon trees, the universe rocking left to right with her mother's stride. They walked for hours, stopping only to tend to my grandma, who cried for changing and nursing, and to alert her mother to the sun in her eyes. A little over halfway into the journey, they stopped at a familiar rise, a grassy flat surrounded by tall cottonwood trees and a small brook where the sheep could drink.

They laid out a thick blanket under the shade of the largest tree and propped my grandma's cradleboard up against it. Grandma sat and watched her family pull food from their bags and eat. They had canned peaches and apricots in mason jars, roasted pieces of mutton, and still-warm tortillas. They talked to Grandma and sang to her in Navajo. Her sister, Rose, toddled around the grass with a couple of the spring lambs, a piece of tortilla gripped in her hand. My grandma felt her cradleboard gently pressing her to sleep as a breeze pulled through, and sleep took hold. My great-grandmother draped the rainbow of her cradleboard with a thin white cloth.

While she slept, my great-grandmother and great-grandfather could hear the cries of one of the sheep, caught in something down by the creek. My great-grandfather's voice boomed from the water, as my great-grandmother held Rose on her hip and chased a few of the lambs back to the grass. She walked with Rose over to the edge of the water and watched her husband carry another lamb in his arms, back up the steep ridge to the meadow to join his crying mother. They walked back to their horses to finish their lunch and check on my sleeping grandmother.

When they returned to the shade, they saw everything as they left it—the horses swatting the occasional fly with their tails, the leaves of the cottonwood trees rubbing against each other in the branches above their heads, the sheep moving slowly over the fresh grass. But there was no cradleboard. Her mother saw the emptiness beneath the shade where she had propped my grandma, the earrings of the board clenching the cottonwood bark. Her baby was gone within the five minutes that it took for her to walk to the water's edge. She never heard her baby cry. She was just gone.

My great-grandmother clenched Rose in her arms as she walked up through the woods behind the cottonwood tree, calling out for her daughter as if she could answer. But no sound returned to her. There under the tree, she could see two distinct drag marks where the cradleboard had been, that trailed off into the forest floor. Next to the marks were the distinct and deep prints of a bear, the claws pushed into the soft clay. Once the clay met the forest, the prints and the drag marks disappeared into the floor of old pine needles and last fall's leaves and cones.

My great-grandfather pulled his rifle from his horse and ran after the tracks as my great-grandmother followed with Rose, both of their cries filling the valley. They called Minnie's name for hours until the weight of Rose labored their crying mother to the ground. They returned to their horses, my great-grandfather riding to the hogan of the man who lived along the ridgeline to ask for help. He and his horse left a quick rising trail of dust as Rose and my great-grandmother tended to their small herd and cried. The herd felt the fear, many of them laying in the cool grass waiting for the baby in the cradleboard to come back.

Great-grandfather returned with the man from the hogan and his three sons, all on horseback, their rifles at their shoulders. One of the sons took my great-grandmother and Rose back to the hogan to wait along with their small herd. The young man waited with Rose as my great-grandmother was stricken with grief, unable to stop looking toward the ridgeline where her happy baby was taken. They all eventually stood looking in the same direction until it was too dark to see anything but constellations. They never heard gunshots or the cries of a baby. But they waited all night in the silence, praying for a miracle.

The men had made it all the way to the top of the mountain and had only picked up an occasional sign of bears. It had been a long winter that year, and many of the bears were just now rising from their winter rest. They were hungry. It was something that haunted my young great-grandfather as they searched through the night for his baby. All four of them with lanterns had split into all directions, deciding to meet back at the center before dawn. By the time the sky was turning blue, my great-grandfather had finally started to cry, making his way over the sharp rocks on the ridge with his aching feet. The four men stood in silence as the light of the rising sun hit them, lighting up the ground beneath their feet with the golden light of frost.

In the distance, the light of morning revealed a perfect trail of tracks that ran along the carpet of pine needles on the forest floor. The tracks glistened in the sunlight as the men followed them in silence. As they continued, they made their way to the springs near the top of the mountain. The sun became hotter and brighter, the tracks harder and harder to see, until they disappeared altogether. They stopped and stood in silence, their rifles in hand, looking through the tree lines. Then they heard it.

My grandmother screamed from the bottom of her lungs. She was cold and hungry and missed the constant sway of my great-grandmother's gait. All she could see was the sky and the band of cedar above her head. All she could do was cry.

The men followed the crying right to the edge of the ridge, where a pool of water had formed in the rocks. There on the edge of the rocks, my grandmother remained tied into her cradleboard, crying a deep and hoarse cry. My great-grandfather dropped to his knees and held her, pulling the cradleboard to his chest. As the old Navajo man and his sons blocked the morning sun, her father untied the leather straps of her cradleboard and pulled my grandmother from her blankets. He hugged her so hard and for so long that she stopped crying. He raised her above his head and looked at her little body, running his fingers through her thick, black hair. She was untouched. None of them could believe it.

My great-grandmother had never slept. She waited at the road, watching the ridgeline, waiting for the clap of horses. As the sun rose, she had prayed for the safety of her baby, but she had almost resigned herself to the fact that there was little chance her baby would survive an encounter with a bear. She had heard as a child that bears were both magical and taboo—a mountain human who could see the horror in taking a baby and, at the same time, the thing that could give you a sickness deep in your soul. She prayed that the bear would spare her daughter, or that her sturdy cradleboard and their prayers would save her.

The men appeared at the end of the road, keeping a quick pace. Rose ran to the corral and watched their trail of dust grow from the horizon. My great-grandmother did not wait. She ran toward the men as fast as she could, her moccasins wet with the morning dew from the sagebrush. My great-grandfather jumped from his horse to meet her, my grandmother in blankets in his arms, sleeping. They both dropped to the ground with my grandma and stared at her.

Some Navajos would believe that she could be struck with sickness from the bears—even my great-grandfather, who was studying traditional medicine. But they chose to believe that their baby had something much stronger. She was special. She was a light that could not be extinguished. Even the bears could see the medicine inside my grandmother. They had no idea what their baby had seen, or what the bears had done to their little girl. But their baby was healthy and strong, and even after her ordeal, my grandmother astonished my great-grandmother and great-grandfather with her resilience and with her smarts. She was returned for something much bigger.

Grandma didn't have the easiest of lives. She was taken and sent to boarding school when she was six years old and never saw her mother again. But my great-grandfather and great-grandmother never worried about her. She was beautiful and strong, and she raised a whole family of children by herself after she lost her husband. She built her own house, stone by stone, wire by wire. She went back to college when she was in her 70s, graduating with a bachelor's degree in education. She filled all her grandchildren with the gift and importance of education.

Grandma was my grandmother and my mother, the one who guided her fingers across pages to teach me to read and steadied my arms when I showed weakness. She taught all of us to live as if we could depend on absolutely no one but her and ourselves. I still believe it. I think the bears saved her because she was the one and only truly pure soul on this planet and the best human in the universe. When Grandma could still walk, we adventured together everywhere. I held her hand through every forest and empty trail around our house on the Reservation. She taught me everything I needed to know to survive.

More than anything, we had a connection that was strong and unbreakable. When I eventually moved away to live with my mom in the city, I counted the days before I would see Grandma again. She wrote me letters about the rain and about the rows of corn in her garden. I called her and told her about the crazy things I was seeing in the world. I waited for her in my dreams. I still do.

Up on the mountain all those years ago, the bears had spared her because she brought the world a gentleness and a sense of duty that only furthered the journeys of others. For years, she worked as a dorm matron at one of the local Navajo boarding schools. The boys she cared for became great men, running businesses, leading in their communities, and taking care of their families. Even until her last days, they came to visit her, to hear and feel her kindness again.

Grandma had told us the bear story before, but she had never shared any details until she was 90 years old, and no longer able to speak with her mouth. Strokes had lingered like flies over her the last 10 years of her life. They had taken her from the strongest woman I ever knew, to the strongest woman who was trapped by her body's rebellion. She talked to me with a pen and a pad of paper.

"Did you know what the bears said to me when I was a baby?" Grandma handed me the pad. I held her hand, and she smiled.

"No, Grandma. What did they say?" She scribbled on the pad and handed it back to me.

"That my granddaughter was waiting for me."

20.1. The author and her grandmother. PHOTO COURTESY OF THE AUTHOR

Chapter 21

A Model for Stewardship
The Lower Brule Sioux Tribal Wildlife Department

| *An Interview with* SHAUN GRASSEL

Shaun Grassel is an enrolled member of the Lower Brule Sioux Tribe [LBST] in South Dakota and the first member to earn a doctoral degree. He now works for his Tribe's wildlife department, where he focuses on habitat restoration, species reintroduction, and species preservation. This interview was conducted on 19 November 2021 by the editors, Serra Hoagland and Steven Albert, and has been lightly edited.

Editors: Tell us a little bit about the history of your tribal wildlife program.

Shaun: The Tribe has been in front of tribal wildlife management for decades. We restored bison populations in the '70s and elk in the '80s through agreements with different agencies, and we're a part of the Intertribal Bison Cooperative [now known as the InterTribal Buffalo Council]. Initially, the Bureau of Indian Affairs had a wildlife management component at the local agency office, and in the mid-'80s the Tribe contracted that program. Initially, it morphed into an outfitting and guiding service. At that time, the Tribe built a rustic lodge and cabin to house hunters, and we have camping pads for RVs. The Tribe capitalized on the bison and elk that had been brought in.

There used to be a lot of [non-Indigenous] commercial goose hunting operations, and the Tribe had already bought land that had a commercial hunting operation on it. We continued that for several years. This was before South Dakota got very commercial with hunting. Around the same time, the Tribe started its own wildlife program called Lower Brule Sioux Tribe Wildlife Enterprises. The Tribe was going to build an economy around wildlife where hunting and lodging were a big part of it. The biologist at the time started to work with other tribes in the Great Plains so eventually the department itself became a wildlife contractor. They would go to other reservations, conduct wildlife surveys, and complete a report for the tribe. That lasted until the early '90s. Around 1992, a new biologist came on board and started to standardize survey methods. He computerized a lot of things. Now they have great records that go back to at least 1992, and some go further back.

Editors: What else was going on around that time?

Shaun: We were getting more into federal and state collaborations regarding jurisdiction and authority, and we attended technical meetings, like flyway meetings. We got creative and set waterfowl seasons to benefit tribal members. There were lots of changes in those early years. We started doing mitigation work for the Army Corps of Engineers for land and habitat that were flooded when hydropower dams were built in the 1950s. The state and the army corps were working together, and then the Tribe started to do it themselves—things like planting trees—and that program grew when I came on board in 1995.

Editors: What were some of the first things you worked on in your position?

Shaun: At the time, I was being mentored by Joel Bich, our habitat biologist. I started doing things like estimating harvest rates, season lengths, etc., while Joel went more toward the habitat mitigation work. Then around 1997, federal legislation went through that included the state of South Dakota, the LBST, and Cheyenne River Sioux Tribe. This became part of a Water Resources Development Act. First, the WRDA returned corps land along the Missouri River back to the Tribe. The remaining land outside of the reservations went to South Dakota Game and Parks. But other tribes still have army corps land within their reservations.

Second, there was a habitat trust fund created for habitat mitigation. We got congressional appropriations to implement mitigation plans. This included recreational areas and facilities like boat ramps. We built a new office complex, purchased machinery—this all came together around the mid-2000s. We planned to replace lost habitat, but soil types weren't always great, and it's often hard to grow trees on the prairie. We have trimmed that program back, but still focus on wildlife food plots.

Editors: How else is your work partitioned?

Shaun: Joel is good at acquiring and restoring land for the Tribe to leave open for wildlife. Often the lands are owned by nontribal members or are put out to bid. With the trust fund, nonmembers can't just purchase land, but they can lease it for wildlife management. The LBST will purchase land from a nontribal member, and then they lease the land back from the Tribe at the same amount the Tribe needs to pay. Then they restore the land, using something like NRCS [Natural Resources Conservation Service] program money or BOR [Bureau of Reclamation] funding. We have lots of land in Conservation Reserve Program status. Some of our effort is dedicated to weed control, and our program is unique in that half of the program is mitigation work.

Editors: We have heard a lot about your black-footed ferret and swift fox restoration programs. Can you tell us a little bit about them?

Shaun: In the late '90s, the black-tailed prairie dog was up for ESA listing. There was a lot of uncertainty regarding how the listing could impact the Tribe and our land use. The Tribe drafted a management plan—they may have been the first tribe to do so. At that time, the potential to reintroduce black-footed ferrets was evaluated. They looked at other tribes and federal partners for examples of black-footed ferret recovery and worked extensively with USFWS. At the same time, we were considering reintroducing swift fox. We implemented feasibility studies to reintroduce both species and worked to get both local support and political support. In 2006 we reintroduced ferrets and the swift fox. We worked with an outside organization to help with habitat suitability modeling.

At the beginning, it did not look like the ferrets were going to thrive because there were not enough prairie dogs to support them. We didn't think we could get funding for swift fox, but we could get ferret money, so we decided to marry the two projects together. We reintroduced both species within 20 days of each other. The monitoring involved working nights for several years. The first year of the swift fox project did well, the second and third years didn't go so great. But the ferrets did better than expected. The Tribe stayed in the ferret recovery business, but one stumbling block we faced was in 2011 and 2013 due to plague. We went from 6,000 acres with 60 ferrets to 600 acres of just prairie dogs. Then we found one ferret and a litter.

In 2017, we released more ferrets—but the prairie dog habitat has still not recovered. Shortly after our reintroduction projects started, I went back to school and did my dissertation on the ecological relationship of ferrets to prairie dogs and badgers. Today we work to reestablish colonies and manage plague. We were the only tribal site to participate in a large, range-wide study to evaluate the use of a plague vaccine bait. Today we use products to control fleas and keep plague out of the system. So prairie dogs are a big part of what we do.

Editors: What other types of projects do you manage?

Shaun: We conduct research on pronghorn, and we're trying to figure out why the populations are declining. We captured and collared them, and deployed radio collars on fawns in the spring. We now have five years of data. We found that coyote predation on pronghorns before they could reach adulthood was contributing to the decline. We sterilized coyotes in the project area, and initial results were promising, but keeping this project going is challenging because we have a very checkerboarded boundary pattern. South Dakota Game and Fish also does intense control in our neighboring areas. And because pronghorn sometimes move, many sometimes shift outside of the areas that have been treated.

Editors: Your program is very innovative with primary research. What helps you be successful?

Shaun: It started internally but also through collaborations with the USGS, nonprofits, and environmental groups. Over the years we've expanded our list of partners. We're fortunate to have a lot of habitat, and we're able to do a lot of work on the ground. I like to think of doing management as a pyramid. Before, hunting was at the top, and everything else was under it. Now, wildlife conservation is at the top. The research component to our program is more for conservation—managing threats rather than the wildlife itself. Some species are in conservation status, and others—like game—are in a sustained yield model.

Editors: What other opportunities do you manage for your tribal membership?

Shaun: We have a recreational fishery. The Missouri River eats away at our shoreline regularly. We've lost hundreds of acres over the years. The Tribe has even had to move playground equipment in some areas. We're creating artificial islands and breakwaters near the shoreline. In theory, this helps stop or slow down ice action, slows erosion, and creates some habitat. We supplement that with riparian vegetation plantings. Shoreline erosion has threatened our sewage lagoons, so the Tribe worked with the corps to armor the shoreline adjacent to the lagoons. We worked with some of our congressional delegation to change some laws so that our breakwater projects could be adequately funded. Now a breakwater has been created in the Missouri River so the water is along a softer edge. This creates swimming places too and more habitat. We also closely monitor for chronic wasting disease.

Editors: What improvements would you like to see in your department?

Shaun: I think the bridge between us and the community could be improved. Luckily, we don't have a lot of conflict, but sometimes there are disagreements. We don't have a good mechanism to get community input and opinions. Our goal would be to create a way to communicate better where we listen to our tribal membership regularly.

Sandhill Sky

above us

each body   a   crossed   t t t   t

oh, flock

indoodem

f o r m i n g   and   re   f o r m i n g

sky w r i t i n g

with   wings

and stick-straight crane legs trailing

out behind and behind lengthening beyond belief

no,

we are not trained to read sky

but our resistance   yields

to throaty   r o l l of song

the ancient   filling day and   echoing

a sonorous chorus

encircles us

we wrapped in sound—enraptured

ah

the s w o o p of cranes, the s i e g e

their motion

a choreography s p e l l i n g and re-spelling

member ing   an alphabet

ajijaakwag

a pause   before   text

a vision   airy   and sweet

a blue harmonics

this ether

essence of

emptiness   now filled.

Chapter 22

Reclaiming Ancestral Lands and Relationships

| *An Interview with* CHIEF JAMES R. FLOYD *by* JANISSE RAY

For a time, the heart of the Muscogee (Creek) Nation was a place called Ocmulgee, near what is now Macon, Georgia. The Creek were mound builders, and at Ocmulgee they built many, including Great Temple Mound, which rises out of the floodplain of the roiling Ocmulgee River. In 1934 the National Park Service (NPS) designated approximately 700 acres of this homeland as the Ocmulgee National Monument. Evidence of 17,000 years of human habitation has been found there. Then, in the early 2000s, John Wilson, a Georgia environmental activist, proposed to recognize the homeland of the Creek people with full national park status, protecting an entire archaeological landscape in the process.

In 2019 President Donald Trump signed into law the Dingell Act, which included the Ocmulgee. The park's designation was changed to "national historical park," a nod toward recognition that the area is significant beyond the mounds themselves. In addition, the park quadrupled in size, to more than 3,000 acres. Most important, the bill authorized the NPS to conduct a much-needed special resources study, which is looking at the national significance of the area.

Going further and establishing Ocmulgee National Park will be an act of forgiveness and restitution, truth and reconciliation, part of our nation's acknowledgment of past wrongdoings and a step toward moving forward into a new and restorative future. Chief James R. Floyd has been very instrumental in the development of the national park at Ocmulgee. Soon after he got involved in the initiative, the Muscogee (Creek) Nation, under his governance and guidance, was presented with the opportunity to purchase 125 acres of the area. They made the purchase and now again own a part of their homeland, the place where the Creek Nation first "sat down."

In January and July 2021, author Janisse Ray interviewed Chief Floyd about the movement to create Ocmulgee National Park and reclaim the Muscogee (Creek) homeland.

Ray: Chief Floyd, to get us started, would you talk about the Muscogee relationship to land along the Ocmulgee River in Georgia?

Chief Floyd: That was our original homeland. It covered parts of 11 states, including Alabama, Georgia, and Florida, and we lived there prior to removal. I can remember from the earliest times my parents teaching us who we were and where we came from, and that always had a deep meaning for me. It helped shape who I am and how my life has worked. What I believe—and I speak for a number of other tribal members—is that our past has taught us to be resilient to the present, and it will help us to be strong in the future if we only learn about our past and the significance of it. The things we have experienced, the history of our Tribe, a lot of that has been tough and sad and very hard to understand. If we could understand our history—of the families, of the places—and utilize that in our lives and in the government of our Tribe, then it makes us stronger. When we look at Ocmulgee, there's a story there and a meaning there, and it's for us to embrace that story, to learn as much as we can about it, and to use that to guide us in what we do.

Part of our tradition is that we really don't revisit the past, particularly when it [comes] to the Trail of Tears. We had to pull up and leave the place that we loved, to come to a place that was foreign to us. When I was the Principal Chief, elders and traditional people in the Tribe would tell me, "We're not supposed to talk about the past, and we're not supposed to revisit or go back there." At the same time, there is a growing interest from our members to know and understand our past. To be able to go back and touch it and be there is very meaningful to our people.

Wikipedia gets our history wrong. It can be put online and be completely different from reality. We've seen that. I've seen that personally. The knowledge we can gather can give us the strength we need to be able to tell our own story, and doing that strengthens us and makes us a fuller population of people that have the knowledge that we've gained from ourselves about ourselves.

Ray: I love that you used the word "touch." That's a verb that not many people would use, because you're not using it in the normal sense.

Chief Floyd: Right. Our whole existence comes from the Earth and from the ground, and so it has a deep meaning to us. It's very much about being attached. That's one of the best things about going back, say to Ocmulgee. We're able to be a part of what made us who we are now, to see it firsthand—the trees, the surroundings—and smell it and touch it. It's a very real thing for us.

Ray: Ocmulgee Mounds was a vibrant, thriving, active place—a town, a community. Why did the Creek people leave? Where did they go?

Chief Floyd: One thing I don't have the full answer for is what happened. But we do know that for over 600 years of consecutive time our ancestors lived there. That's one of the strongest statements we can make, that we know we were there—100% documented—for 600 years. We can't say that about other sites.

I've been reading about the period of the 1600s and 1700s. About the time of the revolution, 1776, we began to be a confederacy. We took in the Seminole, Yuchi, Coushatta, and some smaller tribes that had been in that area and that had been almost destroyed. We gave up life at Ocmulgee Mounds and became more dissipated throughout the Southeast, living in little communities. We still carried on the everyday practices and hierarchies, including a central figure that led spiritually and in governance of the people, as well as the square (the way we assembled in groups around a central location). But we gave up mound building.

Ray: John Wilson has devoted years of study to the intellectual significance of mound building. Is there a story that was passed to you that may explain more?

Chief Floyd: A mound was almost like a major city for us, the place you went for your culture, government, religion, and burials.

Ray: In Oklahoma there's a city called Ocmulgee. I went to Wikipedia, and it said the name means "boiling waters." I thought that was so strange, that your people came from Ocmulgee in Georgia and named their new town "Ocmulgee," but that's not explained or acknowledged in Wikipedia.

Chief Floyd: When we were forced to move to Oklahoma and relocated here, we kind of aligned ourselves similarly to how we were back east. The northeast section of the Creek Nation is Broken Arrow, now a suburb of Tulsa. "Broken Arrow" is the English name of one of our tribal towns back east. I'm sure people who live in Broken Arrow don't know the significance of it.

Ray: How old were you the first time you went back to Ocmulgee Old Fields?

Chief Floyd: My mom and dad both had gone back to the East. When I was growing up, I didn't comprehend everything that meant or could mean. It wasn't until I was Chief that I actually set foot at Ocmulgee Mounds.

It was a very comforting feeling [being] back there the first time. You could see why we loved the area so much, why there was so strong an attachment there. It was so comforting. It was almost like you got to go back to that time when we were there. To have it preserved there at Ocmulgee was so unique, emotional, and powerful. It brought back memories, but it also brought to the surface feelings that I haven't quite had in my life. It was like going back and being baptized in that era. Seeing how much it could mean and experiencing that firsthand strengthened me. I thought, "We've got to do something here. This is who we are, and we have to do all we can do to preserve it. Because it will be useful to us in the future." Everything made sense. It was like a puzzle, and all the pieces fit together.

Ray: I hear you. It's a little hard for me to understand it on a gut level, Chief Floyd. As a white person, maybe I'd have to go back a thousand years to get to my ancestral land, and I've never been to Great Britain. There must be a deep sense that you're home.

Chief Floyd: Very much so. It's a very comforting feeling. I've heard so many other people describe it that way. [They] go back there, and they say, "This is our home." That's why so many people go back, because it gives us so much satisfaction and strength. I cannot believe that it would do anything negative to us. It affirms everything we've been taught.

Ray: When you used the word "baptized," namely going back is "like being baptized," you said it brought back memories before memories.

Chief Floyd: We're an Eastern Woodland Tribe. We're from the Southeast. On my mom's side I'm six generations removed from the East, and my dad five or six. I go there and think, "Now I know why we were here." I can look down and see the river, and you can see what the river meant to us. I let it all wash over me. I take it all in. It's a tremendous feeling.

Ray: When I was visiting Brown's Mount [in Georgia], John Wilson brought me up the back way. We kept climbing and finally pulled ourselves up some rocks, and we came upon a pestle, a hole in a rock, perhaps where grain was pounded.

Chief Floyd: I know the place, up on a hill. We stood at the edge of a cliff and looked at that. We have a responsibility to preserve and protect our past. If we don't, no one will understand it, it will go by the wayside, and we will lose a significant piece of our history. I took several council members down there with me so they would understand it. So when [the land] came available, it was a no-brainer for us to buy it. We needed to have it so we can preserve it and protect it. Maybe I won't understand that in my entire life. We might at a later time learn more about the significance of that area.

Ray: Are you saying that you have purchased a tract of land in the area, that the Nation has bought back some of its ancestral homeland?

Chief Floyd: Yes. We purchased about 125 acres in 2019.

Ray: I'm floored and moved by that. Congratulations!

Chief Floyd: Thank you.

Ray: I'd like to talk to you about wildlife in the Ocmulgee region, but maybe we could start with how the Nation manages wildlife on Oklahoma lands.

Chief Floyd: The authorities that I had as Principal Chief in terms of land, wildlife management, and agricultural management were one commission and several programs directly under me. On land privately owned by tribal members (in trust status) and on tribal-owned land, our primary objective then and now was how to maintain the lands the best way possible. Our concerns were primarily agriculture, secondarily any wildlife. We needed the ground in the best possible shape for cattle and wildlife. We did have some poorly managed lands that were overgrazed. We'd take measures to remediate problems we had.

One animal we had to control was wild hogs. We had so many that we knew that each week we'd have to trap around 50 of them, because they repopulate so quickly. They could tear up five acres a night if you let them. I don't eat them personally, and we don't promote eating them because they've been known to carry diseases. Plus they're not native to us.

Ray: What about hunting?

Chief Floyd: Deer is the animal most commonly hunted, and we have a good supply throughout the Creek Nation. We have an extended season that includes archery, black powder, and guns. We've now opened up some of the tribal land for deer hunting. You can go to the ranger's office and obtain a game permit to hunt them.

We're beginning to see a comeback of people wanting to hunt squirrels, rabbits, and turkey in season. Sometimes we have meals of traditional foods. Squirrels are used in traditional religious ceremonies as well, as squirrel soup. We only kill the ones we really need for the ceremony.

Fish and turtles are sometimes eaten. On some of our ag[ricultural] lands we have stock ponds and small lakes, and if people want to go fishing . . . they can go to the ranger's office and get a permit that will allow them to go on our property and fish.

Ray: You're starting to offer permits for hunting on tribal land. Why is that just now happening?

Chief Floyd: A couple of reasons. We have more land available. We can spread out cattle and not have overworked land. Also, the deer population is on the increase in Oklahoma like in other states, and we want to control that and not get out of hand. Also, some of our lands are desirable hunting lands, and we have had problems with people trespassing. This way, if we issue permits, we can regulate who hunts where. We don't want any poaching and no hunting after dark. We want to control that behavior.

Ray: Do you follow US regulations?

Chief Floyd: We do comply with federal laws. We do not shoot any endangered species or anything protected by federal law. Usually what we do is follow the state requirements as well. I think in the future things are going to be evolving.

It could be that the state and Tribe negotiate, or the state accepts our requirements.

Ray: Do you hunt?

Chief Floyd: I'm an outdoors person and have been since I was young. I love to hunt and be outside. Growing up, I used to go hunting every day. If I find a good spot, I'll probably get a permit and hunt deer this year.

Ray: Are there endangered species on tribal lands in Oklahoma?

Chief Floyd: Anything that is endangered, we would not hunt, and I have not seen any abuse of that.

Ray: In our conversation we've talked a lot about your ancestral homeland. Have any decisions yet been made by the Nation about your property there or the entire Ocmulgee Old Fields?

Chief Floyd: Our main concentration effort is the corridor, in particular the subject of the NPS study, essentially a 50-mile range along the Ocmulgee River. Historically we lived and subsisted and traded along that river route. As we purchased the land, we recognized nearby landowners, the Fish and Wildlife Service, the state of Georgia. We have neighbors who are good caretakers for the land. The study proposed multiple uses, including hunting, fishing, hiking, biking. We certainly don't want to be the one that says, "Everyone but us." When I was in office, our objective was to coexist with others. I would recommend a cooperative agreement in the future. We would want to be part of the bigger plan of the area, and that would be the approach I suggest we take.

Ray: As you know, there is a population of black bear in the Ocmulgee National Park study area. It's one of only three populations left in Georgia. When you talk about recollecting your lives as tribal people, you talk about the animals that were traditionally part of the greater community.

Chief Floyd: Muscogee people were known for our clans—Raccoon Clan, Alligator Clan, Deer Clan. We get our clans from the maternal side of our family. My clan is Wind. My dad was Bear Clan. Those people revere those species. In a sense, through our own culture and history, we regulate ourselves in terms of wildlife, because even though I'm Wind Clan, I respect the Bear Clan as well.

By the same token, some aspects of animals are used in our ceremonies, like turtle shells are used by the ladies who lead the rhythm of some of our songs. When I grew up, we used the shells of dead turtles. You didn't just kill something for the value of the shell. They lived their life, and you respected it. There are plenty of feathers laying out in the woods. I am not aware of any disrespect in this way, and I am hoping not because it goes against the morals we have.

Ray: Were people from different geographical locations associated with different clans?

Chief Floyd: Maybe at some point, but I think it's mostly family. You need different clans to support a robust and diverse society. I have a book that lists the clans, and we had around 40. Presently we have 23 clans recognized. Some are extinct. Our fear is that more are becoming obsolete.

Ray: Does a line just run out?

Chief Floyd: In the same way you could lose a family name, you could lose your clan.

Ray: It would be so sad if an extinct clan were also an extinct animal. [Note: I later learned that the Panther Clan is extinct, and panthers in the wild in the East are down to 120 or 130 in southern Florida.] Bison existed in the Ocmulgee Old Fields when your people were there, and they're gone now.

Chief Floyd: Also eastern elk. But I'm not aware of an Elk Clan or a Buffalo Clan.

Ray: I'm remembering that a lot of images pressed on Creek pottery were of animals. It's such a shame that the apartheid happened, because there was such a rich culture, and layers and layers of stories have been lost.

Chief Floyd: Removal had a more negative impact on us than just the physical trauma. We think about the walking, the boat, winter . . . but we were away from the things we used. We're an Eastern Woodland Tribe. The things we used back there—the moss, the medicinal plants, some of the animals—they don't exist so much here in Oklahoma.

Ray: I noticed some plants while I was wandering around [in Georgia], and I thought, "I don't think these plants grew this way." I interviewed a botanist who studies in the area, and I asked her, "Have you ever been out in the woods and seen some remnant community of plants that the Native people might have tended?" She thought not. I've done enough study to know that we have lost a tremendous amount of place-based knowledge. When you were there, did you see any lasting botanical handprints?

Chief Floyd: I did. Once I went out with John Wilson southeast of the mounds. We didn't have GPS or anything. There were markers that our people followed to get where we needed to be. We could go from Savannah, Georgia, to Montgomery, Alabama, and know how to get there. How did we mark our path so that we would know, so we could tell somebody else? We used bent trees. Young sapling trees were bent a certain way so they'd stay in a bent shape. Other people coming through might not have a clue what it meant. Many have been bulldozed, died, burned, cut down, who knows—they don't exist. Walking with John, we found one! That was one of the strongest indicators I got. You line these things up, and they will show you how to get from one place to another.

We know some of the plants—cypress, for example—were used to make objects for our ceremonies or religious needs. It's all been described to me so many times, how we used that setting for our religious needs. Sometimes it can be almost too much to take in at one time.

There's an effort under way to get some of our heritage plants back. There's one called an Indian peach. I remember eating those when I was little. They were small and not so sweet. Those orchards are gone.

There's also a movement to cook with some of those heritage corns and squashes, and with the wild plants we used to have in our diet.

That land in Georgia that we own, maybe we can find some uses for it, not only for cultur[al] and historical aspects, but also for the plants and animals that are along the corridor. That trend would help build our spiritual strength in addition to physical strength. We need those things. They played an important part in our life in the past, and they can again in the future.

Ray: I know you're a busy man and have many more things to do today. I want to express my deep gratitude to you for your time and for talking to me. Also, it's not my place to say I'm sorry, but I really *am* sorry for the cultural and generational and personal trauma that has been placed on your people and all Native peoples. It's horrifying. And thank you for all the work you put into making Ocmulgee a national park. It's a brilliant idea.

Chief Floyd: Thank you so much for calling me. I hope that COVID will pass soon, and we can be more frequent visitors. We'd like to be. We'll look forward to seeing you.

Ray: As will I, Chief Floyd. Thank you again.

We Feel Our Place in Our Soul

Perspectives from a Fond du Lac Elder

| *An Interview with* **VERN NORTHRUP**

Vern Northrup, a retired wildland fire operations specialist for the Bureau of Indian Affairs, is a visual storyteller who has published multiple books to teach the next generation of Native people. Northrup is Lake Superior Chippewa from the Fond du Lac Reservation and currently resides in Sawyer (Gwabi'iganing), Minnesota. He is Bear Clan from different relatives, especially his grandfather on his mother's side. He learned traditional stories from long ago and the meaning behind them. Editor Serra Hoagland conducted this interview with Northrup on 7 December 2020.

Serra: Your photography documents what I like to communicate as "magical" and what the Creator allows us to see. How important is it that we describe these phenomena to non-Native people? Tell us the "why" behind your photos.

Vern: I see photography as a way of passing on our tradition and our language and our philosophies and our spiritualities, especially to our own children, but also to the greater community, so they can begin to understand our culture.

Serra: You are very open about sharing your knowledge with the next generation through art and your teachings. Where did this come from, and how do you balance information sharing with preservation and cultural sustainability?

Vern: Well, it came from suppression [that I experienced] where I had to become assimilated to become successful. I had to move away from the Reservation and do a lot of things to be successful in my career. Even when I went away, I always had the yearning to come back, and it was natural for me to absorb those teachings again. Some of the things I learned out there raised old memories that I had forgotten, and it brought them back to life again. The language is so moving and describes so much in one word. It's so much a part of us and everything that is out there that has been named. Great care and thought were put into those things that they named.

Serra: Speaking of language, what does the term "wildlife" mean to you?

Vern: It means home. It means this is where I live amongst them. We are part of all those things out there. Everything surrounding us—everything—has a spirit, we are just one thing. For Anishinaabe, there are four of everything out there. It grows in four different conditions, and it has four different meanings. We consider everything that grows, and from that we get medicine, the medicine and the strength of the Earth (Mashkiki).

Serra: Can you tell us a little bit about your career and your family's relationship to fire?

Vern: My grandfather was a fire warden on the Reservation, and he was supposed to keep track of the loggers and how they were disposing of their slash. Their operations tended to start a lot of small fires. In 1918, when he was in the National Guard, a fire burned 250,000 acres in one day. It burned into the city of Duluth down to the lakeshore. He was assigned to fight the fire outside Duluth, and he evacuated a boarding house. He was cited for bravery from his National Guard unit. I didn't find this story out until about three years ago. We've lived [with] and used fire for thousands of years. We would get up and move out of the way, and the fires would stretch for miles. When it came through my village, it was a ground fire—but other places in the pines, it could take off.

Serra: You mentioned burning for food plots to provide foraging for deer and other wildlife. How else is fire used for wildlife management?

Vern: In the lake states, when you burn somewhere it changes that spot, and when other things are emerging in the spring, this spot greens up about two weeks before that, and you get different animals using it. Deer like feeding on things at different stages, and the birds like it too. We use fire to stimulate blueberries to grow all over the state and the Reservation. As one of the burn bosses, I would have a module of firefighters and the engines and do prescribed fires for two weeks.

Serra: Can you share some wildlife terms with me in your language?

Vern: My next book is going to be named *Wisdom of the Trees* in English. It will have the Ojibwe word *metigo nawaakon*, which really describes the composition, structure, color, and nature of the forest. It will have 120–130 photos, and we'll go through each species, name them, and describe what their uses are. Then the next series of photos will be from the roots of the trees and name all of the parts of the trees in both languages, even including the mycelium. Then there will be another series of about 35 photos of the different conditions of the forest around there, spruce or ash or hardwood forest. But for deer, the name is really describing how its eyes shine, its antlers point forward, and how its tail flashes as it runs away—*wahwiskhi*. And since you like owls, the Ojibwe name is really just describing the sound it makes. Really, you can build any word to describe what you are seeing. It's [still] an evolving language.

23.1. *Iskode* (Fire). PHOTO BY VERN NORTHRUP

Serra: Do you recall hearing stories from your parents or grandparents about harvesting wildlife that differ from your practices today? Or are there any traditions associated with hunting or harvesting an animal? You have mentioned the practice of giving away your first. Any others?

Vern: Yes, I once watched my grandfather snare a deer. He didn't use a gun. He bent some material over, and the deer tried to go through—it was hazel brush, which is really strong. He caught the deer, and we whacked it in the head with the ax. A lot of people trap around here, and they have a healthy wolf population, and they like the beavers. Moose hunting is currently not allowed on our Reservation, but they do hunt it on the ceded territories, and they hunt in the Boundary Waters Canoe Area. We get 25 permits a year, and there has to be four family members on the permit. We retain hunting and fishing rights in the ceded territories, which extend far north to the Canadian border. We are signatories on two or three treaties, but in each of the treaties we retained gathering rights. All Chippewa tribes in Wisconsin do. This went to the Supreme Court. We can put nets in Lake Superior or go to Wisconsin and hunt elk.

Serra: What concerns do you have about wildlife or wildlife management today?

Vern: One of the things that concerns me here in Minnesota is the state forestry [department]. About 10 years ago, they went into an accelerated cut rate. They are cutting trees that are only 45 years old. They used to keep the cuts to 40 acres, and they changed that practice. It's creating a lot of regeneration, but it will be even-aged stands over a bigger area now, which makes them more susceptible to disease and fire.

Also, the falling bird population is a concern. It's a noticeable drop just from 10 years ago. If there are fewer birds, there will be an increase in bugs. This is just the little birds; I am not sure what the raptors are doing. But this dropping population is really shifting and changing.

Also, right now we should be under one foot of snow, but we only have an inch on the ground. Temperatures should be below zero. We came out of a dry summer, [and] now it is a dry winter. We need to make up a lot of snowfall. The first day after snow leaves, and then we have the 45-day spring fire season. With the newer clear-cuts, fire can go through those areas with all the slash. One week ago, on December 15, people were getting ticks on their dogs. Animals are crowding in population centers, and we cannot find them in the forests.

Serra: What is the biggest misconception about Indian people or our relationship to place?

Vern: We feel our place in our soul. The other race thinks of it as a possession. For us it is part of our body. It is not that we own it, it is part of who we are. Without it we won't be. We have to take care of it and humble ourselves. *Akiwenzii* means the "Earth caretaker" and in English means "older man," and it is quite an honor to be called this.

Chapter 24

Partnerships Are the Key to Conservation

An Interview with **MITZI REED**

Mitzi Reed is the director and biologist for the Choctaw Wildlife and Parks Department in Choctaw, Mississippi. This interview was conducted by the editors, Serra Hoagland and Steven Albert, by email and Zoom in the spring of 2022.

Editors: Tell us a bit about the Choctaw Reservation and its wildlife.

Mitzi: The tribal lands of the Mississippi Band of Choctaw Indians encompass 10 counties in Mississippi and one in Tennessee. In total there are over 40,000 acres of noncontiguous lands in the east-central portion of the state of Mississippi. Once agricultural grounds, the most common ecosystems are now bottomland forests, wetlands, and pine thickets. Choctaw tribal recreational areas account for approximately one-third of tribal lands that are used for hunting, fishing, and other activities. The wildlife in these areas consists of large game species, like white-tailed deer and turkey, and small game, like rabbit and squirrel, which are popular game species for us. Invasive and nuisance wildlife, like feral hogs, nutria, and beaver, add to the mix of wildlife, as well as predator species, like coyotes and bobcats. Species found on tribal lands are the typical species found in Mississippi as . . . the tribal lands are mere pinholes within the state boundaries.

Editors: What is your approach to conservation issues at Choctaw?

Mitzi: Partnerships are the key to conservation for us. Conservation programs in the Southeast are forced to be dependent on local, state, tribal, and federal partnerships. Many Southeast tribes have established environmental-based programs that incorporate some wildlife aspects, but not many programs are devoted solely to wildlife and natural resources conservation. Southeast tribes are often responsible for several acres of tribal lands; however, established programs are often geared towards environmental concerns dealing with land, water, and air. . . . Specific wildlife programs are few and far between in this region, and many opportunities are overlooked or missed due to other tribal priorities. Fortunately, we have both well-established environmental [programs] and wildlife programs that work for the protection of tribal natural resources, while working together to strengthen their deficiencies.

Editors: When was your program established?

Mitzi: The Choctaw Wildlife and Parks [Department] was established in 1980 as an organization solely dedicated to the regulation of hunting and fishing activities. It was not until 2006 with the hiring of a tribal biologist that more concentration on conservation of natural resources began. Partnerships established since 1980 have been helping sustain the Choctaw Wildlife and Parks Department with assistance ranging from technical support to manpower. With that, we address issues related to management, animal damage and conflicts, and remediation. Agreements and memoranda of understanding are necessary parts of these partnerships. These agreements and understandings are based on meaningful, common goals between two or more partners and outline what services each partner is going to provide. The willingness for all parties to invest in these agreements [is] the driving force of a successful and productive partnership. The challenge obviously is that each party has their own priority objectives to fulfill so time constraints, scheduling conflicts, and lack of available personnel often delay response[s] in an activity. Despite the challenges, partnerships provide added capabilities and benefits for all parties involved.

Editors: Who are some of the partners you work with most?

Mitzi: Besides our intertribal programs and agencies, two of our original partners are the Mississippi USDA Wildlife Services office and the Mississippi Department of Wildlife, Fisheries, and Parks. We've also built numerous partnerships with other agencies and programs in Mississippi. These partnerships have worked as extensions of our department to build knowledge and capacity on the Choctaw Indian Reservation.

Editors: What are some specific wildlife projects that have been successful, and why do you think they were successful?

Mitzi: Feral hog and beaver trapping and remediation activities are two of the most notable projects that have been successful. Through the knowledge and resources gained by our partnerships, we have been able to manage these two species to a controllable level on tribal lands. The main reason [for] success is the availability of our partners in the aspects of resources, manpower, and conservation outreach.

Editors: What benefits has the Tribe gotten from these partnerships, and what benefits do you think your state and federal partners have gotten?

Mitzi: Often, these projects are time sensitive, so prompt assistance when a situation arises is one benefit. We are also able to get extra bodies and manpower for boots-on-the-ground work, such as assistance in actual trapping and remediation events. I recall a situation where beavers had dammed up an auxiliary water opening in a utility access hole off a retention pond. The flooding and pressure of the retention pond caused [the] undercutting of the piping under a main transportation road near the Geyser Falls Water Park. This undercutting caused a substantial sinkhole that would have been detrimental had the asphalt of the road given way. Our tribal roads and engineering office and the Mississippi USDA Wildlife Services office provided labor to manually lower the water level and remove a seven-foot beaver-built staircase inside the utility access hole structure, [which] was preventing the water from draining normally. Sticks and debris from this staircase were removed by hand, piece by piece. Without the extra manpower, the . . . remediation of this area would have been [a] tremendous [challenge].

In addition, the data collected are shared by all parties, so that benefits all managers working to protect the same species, as we work to build long-lasting relationships with our neighbors and colleagues. And of course, the hands-on, in-person training opportunities are great for both tribal members and other parties. Invitations to assist state-organized activities, like big game herd health checks, for example, have helped benefit all parties with the biological data collected. Without access to factual biological data from the Mississippi Department of Wildlife, Fisheries, and Parks, the annual hunting and fishing seasons for the Mississippi Band of Choctaw Indians would be blindly difficult to set each year.

Editors: What do you think you've learned from the partnerships you've developed?

Mitzi: Beginning a partnership shouldn't be taken lightly. It's first necessary to determine if a partnership is even necessary. Selecting a partner should be based on rapport and common objectives, and something that should benefit all parties. Careful planning and negotiation are needed to ensure that all commitments are clearly understood and documented. Finally, managing and monitoring the partnership needs to be an ongoing process to determine if changes are needed or to adapt to the situations that arise. Once a true partnership is established, it is important to bring other contacts that [are] relevant to the work to ensure that the full benefit is met throughout the proposed project and beyond.

Burmese Python Impacts and Management on the Miccosukee Reservation, Florida

| CRAIG VAN DER HEIDEN *and* WILLIAM OSCEOLA

INTRODUCTION

The Miccosukee Tribe of Indians of Florida live in the heart of the Florida Everglades, with tribal lands spanning six counties and nearly 300,000 acres. Their timeless culture and way of life are dependent on a healthy and restored Everglades ecosystem. Unfortunately, Florida's mild tropical climate and environment are conducive to the establishment of several hundred invasive species. Though invasive exotic fish outcompete native fish for spawning sites, and vegetation such as Brazilian pepper (*Schinus terebinthifolius*) engulfs large areas of land, none of these pests are more devastating to the environment and local native fauna than the Burmese python (*Python bivittatus*). The proliferation of Burmese pythons has coincided with drastic reductions in many native mammal species, such as raccoons (*Procyon lotor*), Virginia opossums (*Didelphis virginiana*), white-tailed deer (*Odocoileus virginianus*), and bobcats (*Lynx rufus*) (Dorcas et al. 2012). These population declines have affected the Miccosukees' way of life by reducing the availability of game species in Miccosukee lands.

The Burmese python is a large, nonvenomous constrictor snake that is one of the most infamous invasive species in Florida. The python's rapid and widespread invasion has been facilitated by aspects of its natural history, such as diverse habitat usage, broad dietary preferences, long life-span, high reproductive output, and ability to move long distances (Meshaka et al. 2000, Harvey et al. 2008, Willson et al. 2011). Burmese pythons are prevalent throughout the Miccosukee Tribe of Indians of Florida Reservation.

Mating occurs from December through April as males follow the pheromone trails that females leave behind. Following the mating season, females lay eggs in May and June, and hatchlings emerge in July and August (Harvey et al. 2008). The size of the population is not known, but rough estimates have 100,000 Burmese pythons in Florida (Harvey et al. 2008, Willson et al. 2011,

Avery et al. 2014). The Burmese python is known to consume a wide range of prey, including mammals, birds, and reptiles (Reed 2005, Dorcas et al. 2012). In 2017, we began monitoring wildlife populations on tree islands, we have noted a low abundance of small and medium-sized mammals, which may correlate with a high density of pythons. Despite their large size, pythons are very cryptic, and it has become clear that they spend much of their time hidden in habitats that mitigate human interaction (Avery et al. 2014, Dorcas and Willson 2013).

Their ambush predation methods, coupled with their ability to hide, have prevented many removal methods that have been used effectively with reptiles elsewhere (Reed and Rodda 2009). Some techniques, such as environmental DNA and infrared detectors, are being developed. Avery et al. (2014) demonstrated a 30% increase in success with the use of detection dogs compared to human search teams. These results are similar to data we have collected. In addition, the dogs performed searches 2.5 times faster than human teams alone.

With the goal of increasing the detectability of pythons, the Miccosukee Fish and Wildlife Department utilizes several combined detection strategies to increase capture rates. Since 2019, one of our primary capture techniques has been the use of our python detector dog, Shato (the Shona-language word for "python"). Shato was trained at J&K Canine Academy to use his sense of smell to locate Burmese pythons, a technique similar to that used to train police drug dogs. Shato is trained to indicate (by sitting) when he finds a threshold of python smell. With the vast majority of human surveys being conducted on levees, leaving most of the Everglades unsearched, we added the dog to the team with the goal of searching areas that are harder to reach in the interior of the Everglades, such as tree islands. Through the use of our detector dog, python capture rates have significantly improved. We also search using vehicles and night surveys to locate and remove pythons.

METHODS

Routine python surveys are conducted around the Triangle, cattle pastures, and North Grass levees as well as on tree islands. Python surveys are conducted all year but capture rates are higher during the breeding season and when high water levels and cold fronts cause pythons to utilize high ground, such as levees and tree islands. Python surveys are conducted with and without python search dogs. The standard operating protocol is to survey the levees from an elevated surface, such as a truck bed, or to walk into denser areas, like tree islands, with the dog. From a vehicle, two surveyors search both sides of the levee from the crest of the levee down to the water's edge, actively listening for the slow movement of a snake, which resembles crackling leaves. The sound of a moving python will often prompt further investigation. When surveying with Shato, staff will walk him for 15- to 45-minute periods, depending on the weather conditions. After every search, Shato is allowed to rest for 15–30 minutes. Surveying continues until a python is detected and captured, or until the entire planned survey area has been searched and no pythons were detected.

When Shato does find a python, the dog is placed in a safe location, and the surveyors identify the position of the snake and approach it slowly and quietly. The surveyors capture the python by hand and humanely euthanize the snake. Data collected from the site include the date, time, and exact location; the surveyor's names; the behavior of the snake; a description of the vegetation and microhabitat; the temperature of both the microhabitat and the overall site; cloud cover; and the water depth or distance to water. Dead pythons are transported in the truck for further data collection at the Aquatic Repopulation Center, our field station and laboratory. At the field station, we conduct a necropsy on the python, which is weighed, measured (snout-vent length, total length), and sexed by probing the cloacal opening. With adult male snakes, a probe will enter 10–16 scales, with adult females, 3–5 scales (De Vosjoli and Klingenberg 2012). The snake is then analyzed for marks, scars, or recent wounds, and an incision is made on the ventral portion along the length of the body. Additional incisions are made on the left and right lungs in order to check for reptilian invasive pentastomes (*Raillietiella orientalis*). We then examine the reproductive organs of the snake. The testes of a male snake are analyzed to determine if it is turgid; female snakes are carefully searched for egg follicles or developed eggs, and if they are present, they are measurement and counted. We collect the stomach contents of the snake and freeze them for further investigation, when records are made of identifiable contents. Upon completion of the necropsy, the carcass is properly disposed of.

RESULTS

The surveys for 2018–2021 were conducted predominantly by the Miccosukee Fish and Wildlife Department. Some data were collected by the Miccosukee Police Department, conservation law enforcement officers, tribal members, and staff from other departments. Since 2018, we have documented a 226% increase in total number of pythons captured. The largest python caught in Florida was over 20 feet in length, and the snake was found with the aid of the dog.

We have documented the capture of nesting females with the use of Shato. Without the use of the detection dog, nesting females remain hardly detectable. Our crew had never found them before working with the dog. Overall, male pythons have been captured at a higher rate than females. The winter months (December–February) have resulted in the highest capture rates. Since 2018, the pythons we captured have been mainly adults, although we also documented an increase in juveniles in 2021. Python captures were primarily on the L-28 and L-28 interceptor levees, and the snakes were found to primarily be basking. From 2018 to 2021, the stomach contents of pythons analyzed during necropsy consisted of 49.4% mammals, 22.4% birds, and the remainder undetermined or amphibians/reptiles.

Mammal abundance was documented using game cameras. Miccosukee Ranch North contained the highest number of small mammals ($N = 454$), while the cattle pastures were shown to have the lowest occurrences ($N = 32$). The Triangle was shown to have an increased python population with decreased small mammal occurrences.

DISCUSSION

Our data show a significant increase in python occurrences since 2018, though some of the increase may be due to increased surveying efforts and surveyor experience. Increased water levels in the Triangle during the dry season may have been a factor as well, driving pythons to higher land where they are more easily detected. The data show that pythons were found to be primarily basking, which might inform our future protocols about the timing, temperature, and methods for surveying. Cold fronts are common during the winter months, and they can produce overcast cloud cover with heavy rain. The clear skies that follow the fronts provide excellent conditions for basking by pythons—and detection by surveyors. There can often be a delay, however, in finding pythons after a cold front because the shallow Everglades water takes a few days for the temperature to drop. Reduced water temperatures force the pythons out of the water to bask and thermoregulate on higher ground.

The use of detector dogs is a novel tool in the removal of pythons. In an experimental setting, the detection of pythons by experienced herpetologists is less than 1% (Dorcas and Willson 2013). The use of detection dogs increases the capture rates of pythons. Detection dogs are particularly important in the removal of female pythons and their nests. Removing the female and her eggs is not only removing a single individual but also removing the future potential destruction of the python hatchlings.

The python mating season occurs between December and April, when males follow the pheromone trails left by females. We have captured substantially higher rates of males during the mating season when they are on the move, searching for females, and several males have been found courting a single female in a mating ball.

Analyses of stomach contents show that the primary prey for pythons in this area are small mammals, though large snakes (> 14 feet) eat white-tailed deer. The majority of the pythons are being captured in the Triangle area, which also holds some of the lowest small mammal occurrences throughout tribal lands.

Future surveys in combination with qualitative data collection will increase the knowledge of the Miccosukee Fish and Wildlife Department on the python population and will better assist us in determing the population size and impact of the Burmese python on Miccosukee tribal lands. This will ultimately lead to better python detection and removal and restoring the Miccosukee homelands.

LITERATURE CITED

Avery, M. L., J. S. Humphrey, K. L. Keacher, and W. E. Bruce. 2014. Detection and removal of invasive Burmese pythons: methods development update. Proceedings of the Vertebrate Pest Conference 26(26):145–48.

De Vosjoli, P., and R. Klingenberg. 2005. Burmese pythons plus reticulated pythons and related species. Advanced Vivarium Systems, Irvine, CA

Dorcas, M. E., and J. D. Willson. 2013. Hidden giants: problems associated with studying secretive invasive pythons. Pages 367–85 in W. Lutterschmidt, editor. Reptiles in research: investigations of ecology, physiology, and behavior from desert to sea. Nova Science Publishers, Hauppauge, New York.

Dorcas, M. E., J. D. Willson, R. N. Reed, R. W. Snow, M. R. Rochford, M. A. Miller, W. E Meshaka Jr., P. T. Andreadis, F. J. Mazzotti, C. M. Romagosa, and K. M. Hart. 2012. Severe mammal declines coincide with proliferation of invasive Burmese pythons in Everglades National Park. Proceedings of the National Academy of Sciences 109(7):2418–22.

Harvey, R. G., M. L. Brien, M. S. Cherkiss, M. Dorcas, M. Rochford, R. W. Snow, and F. J. Mazzotti. 2008. Burmese pythons in South Florida: scientific support for invasive species management. EDIS 2008. 10.32473/edis-uw286-2008. Accessed 1 August 2022.

Meshaka, W. E., Jr., W. F. Loftus, and T. Steiner. 2000. The herpetofauna of Everglades National Park. Florida Scientist 63:84–103.

Reed, R. N. 2005. An ecological risk assessment of nonnative boas and pythons as potentially invasive species in the United States. Risk Analysis: An International Journal 25(3):753–66.

Reed, R. N., and G. H. Rodda. 2009. Giant constrictors: biological and management profiles and an establishment risk assessment for nine large species of pythons, anacondas, and the boa constrictor. US Geological Survey Open-File Report 2009-1202. US Geological Survey, Reston, Virginia.

Willson, J. D., M. E. Dorcas, and R. W. Snow. 2011. Identifying plausible scenarios for the establishment of invasive Burmese pythons (*Python molurus*) in southern Florida. Biological Invasions 13(7):1493–1504.

Chapter 26

So Many Things That Humble Me

| An *Interview* with JOHN SEWELL

John Sewell is the wildlife biologist for the Passamaquoddy Tribe at Indian Township, Maine. This interview was conducted by Serra J. Hoagland and Steven Albert on 16 May 2022 and has been lightly edited.

Serra: John, tell us just a little bit about yourself and your position. You were originally hired under an EPA [Environmental Protection Agency] grant?

John: In 2002, I was hired by the Tribe under their Environmental Department. They had had a wildlife biologist for about two years before I started, so that was really the first time the Tribe had their own fish and wildlife staff member. I had been traveling the country, doing wildlife work. I got an undergrad degree from University of Maine at Orono. After that, I worked in Maine for a couple of years for our state wildlife department. Then I traveled from Wyoming to California to Alaska doing seasonal wildlife work and guiding a little bit, and when the opportunity came about to come back to Maine, I interviewed for the position and got the job.

    I started out under an EPA grant, and once we got a grant through the US Fish and Wildlife Service, I switched over to that funding source, and I've been on soft money for the last 18 out of 20 years. That takes up a lot of my time, trying to secure funding to keep me and a technician going and do some meaningful wildlife work. But we have to go where the money is, whether it's to work on northern long-eared bats, or it's migrating alewives (the anadromous fish species here on our home river), or whether it's Canada lynx. It's mostly species that are threatened or endangered, or species of significance not just to the tribe, but [to] the country and the federal government. I think this process ignores a lot of species of tribal significance, so that's something where we struggle a little bit—finding funding for nuts-and-bolts management of tribal sustenance species. That's really a funding challenge for us.

Steve: That gets to an important point, doesn't it, the struggle that tribes have to just fund their basic programs?

John: It is. We chase money for the species du jour, where the funding might be. Not that it's not meaningful research or meaningful management; it's just that sometimes it's not what we would like to concentrate on. But we have to go where the funding takes us. We create a little revenue here with the Tribe. We have a spring bear hunt, which we started in 2007. It was the first spring bear hunt in Maine since the practice was outlawed in the '80s. The Passamaquoddy Tribe has about 130,000 acres of federal trust reservation land, that we have complete jurisdiction over. The spring bear hunt revenue helps fund a tribal game warden, and we also have a revenue moose hunt, where the Tribe sells and auctions off 14 moose permits, all guided by tribal members. That [money] goes to our warden department, but those are one of a kind—the only ones of their kind for the Northeast Region. These are tribal-guided hunts on trust lands.

Our sustenance resources are deer, moose, and bear, with moose being the most significant. Every tribal household gets one moose permit, so it's a big deal for tribal members to get out and harvest that animal to feed the community for the winter. We have a robust moose population, things are going well there. Trying to manage the revenue hunt and manage the sustenance hunt and avoid conflicts while still creating a little revenue—that's all unfunded management. Anything I do with those large sustenance species is really stuff I don't have specific soft money for.

Serra: What types of management-related issues do you face as a Northeast Tribe that is distinct from other tribes? Passamaquoddy has a pretty large land base.

John: The Passamaquoddy Tribe has two distinct Reservations and two separate tribal governments, so that's unique. We have Indian Township tribal government, and we have Pleasant Point tribal government. Saltwater and freshwater is how they describe it. The tribes in the Northeast were always migratory tribes, they moved with the seasons. Saltwater sustenance was a huge part of our Tribe, and so there [are] two Reservations. There's one here on the St. Croix River, which is inland, where I am, and there's one near Eastport, Maine, which is the farthest east you can go in the United States. That's the saltwater component. So we have two separate tribal governments that come together for a joint Tribal Council to make decisions that affect the whole Tribe.

We apply for all our own things through Indian Township tribal government, and Pleasant Point applies for all their stuff on their own, and we come together for large decisions regarding sustenance hunting or anything to do with our forestry practices, and we have a joint Tribal Council.

Steve: What kinds of challenges does that present?

John: It presents a lot of challenges. One is getting the right number of people from each Reservation for the joint Tribal Council to have a quorum when we have a big topic to discuss. If we can't get a quorum, it gets kicked down the road until we get together again. And COVID has just made it harder. It's very hard for us to get together, and some of these decisions are getting put on the back burner. I think in general the tribal communities don't want to meet virtually a lot. They like meeting face-to-face, to be in a big room. I [sometimes] say "we"—[but] I'm not invited to a joint Tribal Council meeting unless I'm summoned. I'm not a tribal member.

Another challenge is that we have a lot of trust land on the western side of Maine, and it takes five and a half hours to get there. So we have 130,000 acres, but it's broken up in small pieces throughout the state, so that creates management challenges, travel challenges. . . . [We] have these islands surrounded by state-owned land—it's actually timber company–owned land. It's mostly private land around us, but it's large landowners. We're lucky in Maine: hunters, trappers, [and] fishermen have free rein in this area.

Serra: And you said you only have one conservation officer for all that? Or do you have one per township?

John: We currently have three conservation officers to cover 130,000 acres, just got two in the last month. We had just one for about two years. We have tribal members as game wardens, but the number is always in flux.

Steve: Being surrounded by private or state lands—does that create any conflicts with respect to sovereignty or trespass?

John: Yes, it does. The state and the tribes have had an okay relationship. As far as fish and wildlife management and sustenance hunting, though, there have been times where that relationship has been strained. I came from a background of knowing a lot of state biologists. I worked for Inland Fisheries and Wildlife before I traveled, and I rekindled a lot of those relationships when I started here. I think we've had a much better relationship in the last 10 years.

Sovereignty rights come up regarding all of those species and fisheries too, so there's a rub there at times on many issues. But the working relationship has been okay for the last 10 or 15 years, much better than in the past, but there are still some things that are coming to a head. Wildlife-wise, not really. I mean, everybody that knows about fish and wildlife management in Maine knows that most tribal land is a source population for most wildlife.

Serra: I was looking on your website, and one of the stated missions is about providing economic opportunities for the community by expanding existing businesses and fostering economic growth.

John: Revenue-wise, we have that revenue moose hunt and the spring bear hunt, [but] the license fees are just a tiny portion of what that brings to our community. . . . Every spring bear hunt outfitter and guide company is run by tribal members. We have four tribal-owned guiding businesses that run the spring bear hunt. In fact, today is the first day of the spring bear season. So I'll be tagging bears tonight and talking to hunters from Texas and New Jersey and you name it. These businesses are all owned by tribal members. They employ tribal guides and tribal helpers, so the influx of revenue to the Tribe and surrounding areas is also paying for lodging and food. People here really know what it means to our area. Washington County, Maine, is one of the poorest counties in Maine and has the highest unemployment.

So these are a big deal for us, and they're the right thing management-wise for black bears, which are very predatory in the spring. Male bears kill sows, dominant bears kill calf moose and deer fawns. So we think by managing bears in the spring, we can have a positive impact on our moose and deer populations. There's no hound hunting for bears in the spring, it's all over baits. Sows and cubs aren't killed, so we're having an impact not just on our own sustenance species—moose and deer—but the surrounding landowners love the spring bear hunt too, because they love the idea of us saving calves and fawns.

Serra: You also do some interesting work with trying to boost loon productivity.

John: We work with the Biodiversity Research Institute. We had gotten some oil-spill mitigation money, and BRI came to the Tribe—people know that we're a large landowner, and we're interested in the St. Croix watershed—so BRI worked with us deploying 22 rafts out on the St. Croix River. A co-worker is going out on the water today to check some of the rafts we deployed last week, to try to supplement our loon nest success. This is paid contract work, and this goes back to always looking for funding. So this is $15,000 through BRI to pay us to help out with loons for the summer. I got another little contract for some international waterway work regarding alewife monitoring. These are all the little pieces of funding that we have to keep coming in for us to have a department and a technician.

Steve: Do you conduct any research?

John: We do quite a bit of research. We just got done with a large deer-collaring study. Our deer are very different from most anywhere in the country. We have depressed deer populations. We want more deer. We have populations of from 2 to 4 deer per square mile. We're trying to increase our white-tailed deer populations to 8, 10, [or] 12 deer per square mile. We had that in the past, but our deer are dependent on what's called a deer wintering area, a DWA, a deer yard. From January through March, all the deer migrate to a certain habitat to spend the winter. It's the only way they can survive. These deer wintering areas have great snow intercept because they have hemlocks and cedar, so the snow depth on the ground is low, and they have thermal cover [since] they're usually related to a riparian area. All the deer in north[ern] and eastern Maine migrate to these DWAs. So when you factor in spruce budworm, we lost most of the softwood deer wintering areas.

And now we have a predator that was never in Maine, the coyote. They showed up here in the 1960s. The deer heyday was in the '40s, '50s, and '60s, and then in the early '70s, spruce budworm came along and the coyote became established. We had to cut down our deer yards [to slow the budworm infestation]. We have a new predator that replaced some old predators, and the deer populations plummeted. It's a big emphasis for tribal members, local guides, and all the communities—deer populations are a huge thing.

We have the largest deer wintering area in eastern Maine on the Indian Township Reservation. It's a 1,200-acre deer wintering area, and we know it houses between 300 and 700 deer per year, depending on the year and severity of the winter. We always took a proactive approach for managing predators in the deer yard, but we didn't know what kind of draw this deer yard had. We knew that deer traveled here, we saw the migration corridors, we knew that we were protecting deer from around [the Reservation], but we didn't know how far or anything like that. We got a Tribal Wildlife Grant to collar deer, track migrations, and track fawn survival and adult doe survival, and to see if it was feasible with GPS collars to capture fawns and do a fawn mortality study. That just wrapped up. Some very interesting things came from that study. We collared 50 deer over three years. The average travel distance to our deer yard was 17 miles. We have some deer that travel[ed] 120 miles, straight line, to get to our deer yard. That's some cutting-edge research that we had done that nobody in eastern Maine knew much about.

We did some fawn mortality work, we also did some adult doe mortality work, and [we] figured out causes of death for these collared does. We also shed some light on fawning habitat selection, where these doe deer are going to have fawns, where they're protected.

We also have a coyote-collaring project we started this fall, where we collared 15 coyotes. It's the first time that's been done in Maine. We're trying to assess territory size, and see if the predator-management program in our deer wintering area is helping our deer population.

Serra: What are some of the other challenges you're facing?

John: Funding is a huge one. I have to chase grants constantly to keep us moving. I just had a presentation last Monday to the BIA Northeast Region. There's some chatter about dedicated funding for fish and wildlife programs, and BIA wanted to hear similar things from us: What are your challenges? What have you done with this money that you got from us for the past few years? I emphasized the same things to them: we really need dedicated funding for fish and wildlife programs, to fund enforcing our ordinances, [to] work on our sustenance species, and to fund research. Last week I did hunter exit interviews after the moose hunt. That took me a week just to get an idea for the hunting effort that's been put into our moose hunt. All those things are unfunded, along with inland fisheries work we'd like to do. Funding for infrastructure would be great too.

Cooperation has been much better with the federal government and the state. I think it's come a long way. I don't know if that's the case for other tribes, but for us, it's been much better the last 10 years, it really has.

Steve: Do you have any written agreements with the feds or the state?

John: We do, with almost every agency, whether it's BIA or Fish and Wildlife Service, USDA. That's been reaffirmed recently, the trust responsibility to tribes. Even my biologist friends in the federal government have said, "That's a priority from our bosses, to talk to you tribes and see what your challenges are." So I can call up my buddies and say, "Hey, we need some help reading otolith samples from white perch, [but] I don't have the funds for it," and they'll say, "Oh yeah, we can take them. We can send them to our lab." EPA has been great at reading our fish tissue samples lately for mercury, so cooperation's been really good, which is a big change from the past.

Serra: My last question is, what do you love about your job?

John: The diversity—the diversity is phenomenal. It's the perfect job for me. It could [be] loons this week; last week, it [was] alewives. [One day,] I could be collaring a coyote or going in a coyote den, [and] the following week, we could be checking on wood duck boxes or American marten nesting boxes that we've deployed. I could be doing animal damage control. I could be electro-fishing for eastern brook trout, [or] running acoustic bat monitors.

I love dealing with tribal members. I love the fact that I can work and manage for what tribal members find significant. We have other interest groups, but working for tribal members is what I love. It makes the job easier in a way and harder in a way too. There are some significant cultural aspects. Me growing up as a hunter and a fisherman and a trapper, coming from a wildlife background, coming from a sustenance, utilitarian background from wildlife—they put things into fish and game that opens my eyes even 20 years later. I'm in awe when I talk to our tribal historian about muskrat harvest and what it meant to the Tribe when times were tough. There are so many things that humble me when it comes to working with tribal members.

Swamp Boy's Pet and Field Guide (after Aimee Nezhukumatathil)

| *A Memoir by* **CHIP LIVINGSTON**

Hickory, Catalpa, Cypress, Pine, Chaste Tree, Cedar, Dogwood, Elm, Sparkleberry, Pawpaw, Maple, Blackhaw, Bay, Hornbeam, Maidenhair, Pecan, Redbud, Birch, Sugarberry, Sweetgum, Sycamore, Hawthorn, Blueberry, Loquat, Guava, Silverbell, Witch Hazel, Juniper, Magnolia, Tupelo, Devilwood, Sweet Olive, Hophornbeam, Sourwood, Plum, Cherry, Laurel, Peach, Kumquat, Nectarine, Holly, Willow, Camelia, Buttonbush, Persimmon, Buckeye, Palm, Fringetree

My life as a young naturalist, junior herpetologist, wannabe zoologist began when I got prescription eyeglasses at eight years old. I suddenly saw the magnificent details of a startlingly defined world. My mother recalls her horror at not knowing how poor my vision was until we drove home from the optician's. With my wire frames too thin for their Coke-bottle lenses, I announced with glee every noticed new thing I could see on our familiar ride through our neighborhood. "Look at the leaves!" I shouted, like an autumn leafer, startled at how individual each creation was. My watercolor imagery transformed with dimension and accuracy into a landscape. "That tree is different from that one! And that one! Look at all the different shapes of leaves!" The leaves, like what one heard about snowflakes, no two exactly alike. Astonishing to me, although I'd never seen snow growing up in northwest Florida.

Live Oak, White Oak, Water Oak, Myrtle Oak, Chinkapin Oak, Bluff Oak, Swamp Oak, Chestnut Oak, Shumard Oak, Willow Oak, Blackjack Oak, Bluejack Oak, Turkey Oak, Post Oak, Red Oak, Unthought-of Oak, Overcup Oak, Sand Post Oak, Sand Live Oak

Oak leaves can be rounded or pointed, oval or lobed or star-shaped, serrated or smooth, with or without bristles. Oak trees are hermaphrodites, producing both male flowers and female flowers. The male ones, wormlike catkins, would litter our windshield and driveway. Its fruit or nuts, called acorns, grow in cuplike structures known as cupules, and turned upside down, look like little round faces wearing tiny ski caps. More than 450 oak species are recorded worldwide, with at least 19 commonly found in Florida.

Four oak varieties and a catalpa tree grew in my mom and stepdad's front yard, and my sister and I needled their acorns onto fishing line we strung around our necks like strands of beads, varying the patterns of the tiny scrub oak acorns with the green live oak berries and the redder nuts from the laurel oak.

The towering pecan trees on my grandparents' land were strung with seasonal bagworms, and for the first time I saw they came from actual baggy nests. Before glasses, I had thought they came out of the high Spanish moss, the shaggy gray beards that hung from so many live oak and cypress trees along the river. The webworms, larvae of the white bagmoth, threatened our harvest—the pecans we gathered and sold to Renfroe's Nut Company for Christmas money. Thick blooming camellias and sprawling white cedars lined the narrow lane between my pawpaw's and my great-grandparents' house, where sweet-sour scuppernongs stretched from the grape arbor to the gate. I chewed the bitter ends of bahiagrass and drew sweet nectar from the honeysuckle flowers.

Twenty-twenty vision on the river was what it must have been like when the People emerged from the fog, an old Creek story of how the clans were formed. As the sun burned the morning mist off the water, I saw my kinsmen clearly in the swampy sloughs: crawfish and water spiders, cranes and herons, bald cypress knees, and palmetto fans. Everything my pawpaw or dad drew my attention to, I now saw, and they had seen all along what I eagerly pointed out. I could even see my bobber move as the bluegill bumped the bait and no longer had to wait for the distant orange blur to disappear beneath the murky water. I could actually see where I cast my rod or pole, the shadows where my dad told me the big fish were hiding. I could see layers of distance and dimension through the thick vines and leafy flora stretching back from the inlet banks, could see the kinds of places my pawpaw pointed out where his grandparents hid to avoid the march to Oklahoma.

My improved eyesight also affected my reading, and I devoured anything related to nature and animals. World Book encyclopedias, my father's *Field and Stream*, every animal book in the school library and the county library's branch near the supermarket. My bedroom wallpaper was an ABC of animals, from aardvark to zebra, and I started a life list of wild "pets" and sightings, imagining my life as a future veterinarian or zookeeper.

Animalia, Chordata, Reptilia, Squamata, Serpentes, Colubridae, *Opheodrys vernalis*
SMOOTH GREEN SNAKE

My first "pet" snake was purchased in a pet shop, a smooth green grass snake docile to hold and an easy feeder. Crickets were plentiful to catch or buy cheap at any local bait store I could ride my bike to. I was eight, a determined young herpetologist, already familiar with the local species of every reptile in my Florida wildlife field guides. While I loved to observe the animals in their natural settings, now that I could see them, although feeling a tinge of guilt in their caging, I fell prey to my personal desires to observe them more closely, to touch them.

I knew the smooth green snake I named Oliver was far from her original habitat, ranging from Toronto and Quebec down through Illinois and Virginia, yet I kept her in a 10-gallon aquarium in my small bedroom. I studied Oliver, was able to gender her by the shape and number of scales around her cloaca, and measured her, 18 inches, an average length for a mature adult, slender, relaxed, graceful, and barely a handful. I'd carefully prepared a terrarium with gravel and dirt, a wide length of bark, and rocks and leafy branches, doing my best to mimic the wild. A water bowl she sometimes coiled into. I carried her to the yard and let her bask in the sun and explore the grass, placed her in the lowest branches of the catalpa tree, and once watched her capture and eat a catalpa worm, larva of the catalpa sphinx moth (*Ceratomia catalpae*) that I'd usually pluck into paper bags to fish with.

Animalia, Chordata, Mammalia, Rodentia, Castoridae, *Castor canadensis*
NORTH AMERICAN BEAVER

The North American beaver is the second largest rodent in the world with a habitat throughout the United States and Canada. With a large head and massive, tree-felling teeth, the beaver is a prime example of evolutionary adaptation for a life around and under water. A beaver can hold its breath up to 15 minutes. Its back feet are webbed, ducklike, for swimming, while its front paws resemble the dexterous hands of a squirrel or raccoon; it can carry logs while walking upright on land. A membrane over its eyes, and valves in its ears and nostrils seal closed for waterproofing. The beaver's long flat tail is scaly, reptilian, but shaped like the bill of an Australian platypus (*Ornithorhynchus anatinus*). It's used as a paddle and rudder, as well as a tool for social warning. Beavers will defend their territories and family members, and will alert others of encroaching danger with a loud, echoing smack of their flat tails against the water's surface.

My dad and stepmom brought home two baby beavers from DeFuniak Springs. They'd heard a hunter brag at the boat launch that he'd shot a big lactating female, and my dad had planned to fish near the dam where the jerk said he killed her. There was only one active beaver lodge on the east side of Juniper Lake, and there, from the bass boat, my stepmom heard the cry of hungry babies. My dad removed a few logs and found two orphaned newborns. They wrapped the kits in a towel, returned to the boat ramp, and bought a doll-sized plastic bottle to feed them a homemade mix of mammal formula. My stepmom prepared a cardboard box for them with towels and water and strips of bark and newspaper, but the kits soon grew teeth and chewed through the box. The babies swam back and forth in our bathtub. Within six weeks, however, my sister's and my favorite, the runt we named Squeaky because of its constant chirring, succumbed to a breathing infection. Its hardier sibling my dad donated to the Swamparium in Cantonment, where the beaver grew up in a fenced-in pool enclosure with various slider turtles, a pair of wood ducks, and an injured ring-billed seagull.

Animalia, Chordata, Mammalia, Carnivora, Procyonidae, *Procyon lotor*
RACCOON

My stepmom had a trained raccoon that she walked on a leash around their trailer park. The raccoon, I think, saw me as a sibling, a new stepsibling, and clearly the trailer was his territory.

"Coon hunting" in my family existed only in spotting the masked ringtails, thankfully not in shooting them. We'd take the boat out after a fish and squirrel fry, shine flashlights into the trees until we found their glowing eyes. I can close my eyes now, 35 years since I've been there, and still recall with distinct clarity the particular smell of the Apalachicola River around midnight.

Animalia, Chordata, Reptilia, Squamata, Serpentes, Colubridae, *Heterodon platirhinos*
EASTERN HOGNOSE SNAKE

Eight months after I purchased the green snake, I traded up, at least in terms of retail price at the pet store, and traded her in for a local species supposedly plentiful but one I hadn't seen alive in the wild, an eastern hognose snake. I wanted a hognose to watch it display the curious traits I had read about. When frightened, the hognose will flatten its head like a cobra, hiss, and strike, usually without even biting, in a head-butting way I imagined comical. But its flipside and far more common form of defense, I found, was to play dead, turning over on its back and shitting a sticky, smelly goo that even a swamp boy didn't like getting on him or cleaning. Piggy wasn't a fun snake, his reticulated scales not as "soft" to pet as Oliver's was, when he let me without going limp. Piggy was practically unholdable, though interesting to watch in his terrarium, the way he'd strike and swallow the oak toads and eastern cricket frogs that I caught on my front porch and around the yard. But usually, as soon as I opened the top of the cage, if I wasn't quiet when I fed him, Piggy'd play dead and poop on his rocks or gravel; his tank began to stink no matter how often I cleaned it.

A few years later, I finally found an eastern hognose in the wild, and when I lifted the rock it hid under, it spread its snouty head flat and hissed like a cobra. I tried to lift it with my snake-catching golf putter, and it flipped over "dead," that familiar stink reminding me it is a reptile best left alone in its natural habitat.

Animalia, Chordata, Reptilia, Squamata, Serpentes, Colubridae, Coluber, *Constrictor priapus*
SOUTHERN BLACK RACER (BLACKSNAKE)

My dad knew how much I wanted a wild snake, a vine snake, the skinny rough green snake (*Opheodrys aestivus*) that grew to be much longer than my smooth green snake ever would have, and Dad said he frequently saw them in the cypress limbs overhanging the river. I'd never seen one, except in pictures, not even a tame one in the pet store, but Dad promised to catch me a local rough green snake. One afternoon he showed up at my mom's with an angry pillowcase tied in a knot and turbulent. "I caught you some beauties," he said. "Two of them so you can start your own ark." I was so excited.

My dad opened the pillowcase and dumped out two big blacksnakes, so pissed off they stood straight up on their tails and hissed at us. Forty-five years later my mom confirmed the memory. "First time in my life I ever saw a snake do that—though I'd heard they could and would if they got mad enough—and not one but two at the same time," she told me on the phone. They stood up, like malevolent vipers in cartoons, then fell to the grass and raced off in different directions. One escaped into the woods behind the house; the other I caught behind the head in the flower bed, with the help of my dad and with my stepdad's golf club. After several stinging bites that were bearable but drew blood, I placed the snake back in the pillowcase. Later I put the black racer in the tank with the hognose, with no immediate trouble. But the first time I tried to feed them, they both chose the same toad, struck, and latched on, and I thought, until Piggy finally let go, the racer was sure to swallow him. I let the blacksnake go in our backyard.

My pawpaw never messed with snakes, nor really any wild animal other than to appreciate them in their natural settings. Black racers, he said, I could let loose any time in his vegetable gardens to keep them clear of rodents. Rattlesnakes, water moccasins, copperheads obviously were best chased from the yards where we played, but a reptile as a pet in a cage—he let me know without so many words that he didn't approve. The other racer common in his fields was the whip snake, also known as the brown racer, red racer, coachwhip, or hoop snake.

Animalia, Chordata, Reptilia, Squamata, Serpentes, Colubridae,
Masticophis flagellum
COACHWHIP

As far as I know, my pawpaw never lied to me, but all the snake books led me to believe that what he told me can't be true. Pawpaw said when he was a teenager, a coachwhip chased him like a loosed bicycle wheel across a field. He swore the snake grabbed its tail, made a hoop, and rolled after him down the lane where he ran from the chicken coop.

The second longest native North American snake, coachwhips are thin, large-eyed, and vary in color depending on their habitats. The mottled variety in northwest Florida was fast. I never caught one, and I never saw one grab its tail, though they often raised their heads above the rows of squash and beans, similar to those angry black racers my dad had caught for me.

Animalia, Chordata, Reptilia, Squamata, Serpentes, Colubridae,
Pantherophis guttatus
CORN SNAKE (RED RAT SNAKE)

The hognose I traded in for a fascinating young red rat snake I named Charlotte. Charlotte was the beginning of a lifetime affection for rat snakes, the large subfamily Colubrinae that includes kingsnakes, milk snakes, chicken snakes, vine snakes, and indigo snakes. Charlotte was a constrictor, a hugger, a climber, an escape artist. I'd buy her live white feeder mice (*Mus musculus*) one at a time, which she'd stalk in her 20-gallon horizontal terrarium, flicking her tongue as she followed it and finally struck, teeth gaining hold and then a quick twist as her whole body wrapped around it, and around herself, as prey and snake formed a scaly coral and copper ball, the stripes on top alternating with the black and white checkerboard of her belly. When the mouse was dead, Charlotte would uncoil, expand her mouth to swallow it, then stretch out. I watched the lump squeeze through her jaws and move through her neck and body while slowly being digested.

Mr. Nowak had a big gray rat snake (*Pantherophis spiloides*) at the Swamparium. He had a black and white ringed kingsnake (*Lampropeltis getula*), a diamondback rattlesnake (*Crotalus adamanteus*), and two pygmy rattlesnakes (*Sistrurus miliarius*). He also had a six-foot eastern indigo (*Drymarchon couperi*), which he took out and held once in a while to lecture on the longest snake in North America, which is an endangered species. His indigo snake was blind, likely poisoned with gasoline poured into a gopher hole to bring out the rattlesnakes, and had been brought to the Swamparium after some rancher found it dehydrated and dying on the highway.

If we weren't out on the Perdido River, or on the Conecuh (Escambia) River, in its bay, or down at Juniper Lake in DeFuniak Springs, camping out along the Apalachicola, or up at the hunting camp in Burnt Corn, Mr. Nowak's Swamparium in Cantonment was my favorite place to go on weekends with my dad. Mr. Nowak's backyard zoo had no admission charge. He had the snakes I mentioned, several big alligators, otters, our former pet beaver, red and gray foxes, a bobcat, and the world's largest recorded alligator snapping turtle, which he named Big Jim.

Animalia, Chordata, Reptilia, Testudines, Cryptodira, Chelydridae, *Macrochelys temminckii*
ALLIGATOR SNAPPING TURTLE (LOGGERHEAD SNAPPER)

Mr. Nowak's Big Jim was estimated to be more than 300 years old. He weighed 307 pounds in 1971 when officially measured for the *Pensacola News Journal* and was over 4 feet long and 12 inches wide across the head. He was the scariest-looking animal I ever saw alive in a pen, or dead on a roadway, and the alligator snappers that occasionally hooked onto my father's catfish trotlines were the most frightening things I've ever seen in freshwater. Mr. Nowak caught Big Jim in 1961, before I was born, on the end of a trotline set in the Perdido River, but Big Jim lived the rest of his life in captivity at the Swamparium created to showcase him.

I remember Big Jim's monstrous open mouth and the dangling red attachment at the back of his throat that looked like the wiggler worms we fished with; it was the bait the turtle had hunted with once, from the bottom of the muddy river. Sometimes Big Jim's mouth would hang open in his pen, and Mr. Nowak would go in with a trimmed stick and demonstrate that, when anything touched or triggered that vermiform tongue appendage, the pointed snout would snap shut like a sprung mousetrap, or like a guillotine, easily removing fingers or breaking a broomstick. Big Jim's dragonlike spikes had eroded over the centuries, but his carapace was still ridged like a dinosaur's, and green-gray algae grew across his shell. One of Big Jim's 13 shell plates had been pierced with a spearhead, and his bony carapace had grown around the broken flint, which was used by scientists to estimate the turtle's age.

I didn't linger at Big Jim's cage. Everything about him reminded me of my dad's scary fishing buddy, Mr. Reeves, whose hawkish and almost toothless mouth hacked a ragged cough I thought he'd surely drop dead of. The three false teeth Mr. Reeves did have could be removed with a lick of his tongue, and he flicked at them endlessly, reminding me of that twitching appendage in the alligator snapping turtle's throat. I ate my first snapping turtle soup at Mr. Reeves's house, cooked up by his wife and served with hot biscuits and honey. To me, the soup tasted like river water, but when fried, snapping turtle tastes like alligator, like rattlesnake, like squirrel, and I guess, if we consider just the crunch of thick batter, a bit like chicken.

Animalia, Chordata, Reptilia, Crocodilia, Alligatoridae, *Alligator mississippiensis*
AMERICAN ALLIGATOR

Once I could see at a distance, I could distinguish an alligator from a log in the water or find it along the shores of the rivers and lakes of our common hunting grounds, and I could now identify different snakes in the wild, usually water moccasins (*Agkistrodon piscivorus*) and banded water snakes (*Nerodia fasciata*). The American alligator is designated the state reptile of Florida (as well as Louisiana and Mississippi), and like coyotes (*Canis latrans*), they can be found almost anywhere in Florida. Male alligators often grow to 15 feet and up to 1,000 pounds.

The only time I saw my father truly frightened, and probably the fastest I've ever seen him move before or since, was when I once threw an empty beer bottle, picked up from the shore of Juniper Lake where we were camping, and aimed it at a large basking alligator. Perhaps thanks to those eyeglasses, or stupid bad luck, the bottle connected with the gator's hard snout with surprising accuracy. The glass burst, and the reptile rose on its strong, short legs, sprinting along the shore before careening with an angry splash back into the lake. My dad had scooped me up in one arm and run in the opposite direction. He ran for some 30 yards toward the road, and then he beat me breathless, and deservedly, for putting us both in such danger. My pawpaw also might have spanked me for disrespecting that formidable predator. I felt terrible for what I'd done that day, and I probably hurt the alligator, but I hadn't been exactly scared. Mr. Nowak climbed in and out of his alligators' pens like they were giant pets, though forever after that day beside the lake, whenever Mr. Nowak entered his fenced-in ponds at the Swamparium, I held my breath for him.

Animalia, Chordata, Reptilia, Squamata, Serpentes, Colubridae,
Thamnophis saurita
EASTERN RIBBON SNAKE

All the Florida reptile books mentioned how abundant and common were garter snakes (*Thamnophis sirtalis*), the most common snake species in North America, but despite my daily outdoor adventures, I'd never come across a wild one. During those almost daily searches in the woods, I always imagined I might see a Florida panther (*Puma concolor couguar*), no matter how rare and unlikely. I trained my eyes for any glimpse of new species and was able to add various animals to my notebooks via my after-school excursions hunting garter snakes.

The closest species I found to the wild garter snake were two ribbon snakes for sale at the pet shop. My corn snake was getting too big, almost three feet long including her tail, and her diet had grown more difficult to maintain. She ate twice a week, mice I had to buy in two separate visits to the pet store, and my mom wouldn't let me bring the mice into the house. Mice are my mother's phobia and where she drew the line. Even in the car on the ride home, each mouse had to be double-bagged and locked in a box inside the trunk. I had to take Charlotte's terrarium outside just to feed her. So after her trip to my elementary school for fifth-grade show-and-tell, I traded in the corn snake for two slender ribbon snakes.

For the ribbon snakes I gave the terrarium a makeover, attempting to improve the resemblance to their outdoor world, my outdoor world indoors. My stepdad helped me convert a corner of the tank into a swimming area, where I'd release minnows and small bream netted from nearby Sugar Creek for them to catch and eat, as well as small frogs and big tadpoles. The ribbon snakes were pretty, skinny cousins to the garter; they had black and yellow stripes with pale beige bellies. They never bit me, seemed not to mind being handled, and often fell asleep or into long trancelike pauses stretched on my forearm from fingers to elbow.

Animalia, Chordata, Mammalia, Carnivora, Musteloidea, Mephitidae,
Mephitis mephitis
STRIPED SKUNK

There are two species of skunk found in Florida, the common striped version and the rarer eastern spotted. Three striped skunk babies my dad saved from the roadside, their mother and siblings flattened by autos, were caged in the backyard at his and my stepmom's new house. (The raccoon walking had resulted in them getting asked to leave the trailer park.) The young skunks' glands were beginning to mature, and my dad took them to a veterinarian to be descented. They still smelled pretty musky, but I didn't mind. I had always liked the pungent smell of the fox pens at the Swamparium. But while funny to watch walk around with their waddle, the baby skunks got mean as adolescents, or perhaps in response to the intrusive surgeries that rendered them naturally defenseless, and those three southern "polecats" were eventually donated to Mr. Nowak's Swamparium.

The skunks got mean, the little beaver died, the raccoon eventually escaped. I was gaining a little perspective on the idea of raising wild animals as so-called pets. My sister got a kitten. My dad and stepmom began to breed Labrador retrievers. I still had my ribbon snakes in a glass cage in my bedroom where I lived with my mom and stepdad. We'd also moved, into a new neighborhood in a supposedly better school district. The subdivision attempted to tame the wild it encroached upon, but there were still rampant creeks and woods all around us. There was even a beaver colony in an estuary half a mile behind our house, and in my Christmas gift of a little-used johnboat, I could paddle up close to the wild beavers while I fished after school with my cane pole.

Animalia, Chordata, Mammalia, Rodentia, Sciuridae, *Sciurus carolinensis*
EASTERN GRAY SQUIRREL

Gray squirrels are commonplace in Florida, the eastern species ranging west to Texas and north to Quebec and Manitoba. My stepmom had sworn off bottle-feeding wild babies, but the gray squirrel fallen from its nest in the pine tree in their new yard seemed to be orphaned, was just growing fur, and had its eyes stuck closed, and it struck a chord for her. The new pet grew up faster than my baby half sister, and learned to leap from shoulder to shoulder, from Dad's lounge chair to the sofa. The squirrel would climb up the curtains and run across the top of the curtain rods, and he began to venture to the porch and then into the yard. He'd adventure and return to sleep inside—until instinct took over. Then he slept outside, chirping to us from the pine tree, scampering back from time to time, and climbing up our jeans to take the apple slices and pumpkin seeds we offered.

My mom preached "leave wild babies alone," haunted from her teenage experience of trying to save a fallen blue jay. She'd climbed the tree to replace it in its nest, only to be "speed-bombed," according to Mom, by the blue jay's angry and protective parents.

Animalia, Chordata, Aves, Passeriformes, Corvidae, *Cyanocitta cristata*
BLUE JAY

The blue jay is a migratory songbird, noted for its frosty blue wings and crest, white face and breast, a striking black mask, and black collar and wing tips. A talkative bird that is bold and aggressive, it is known to attack fearlessly both humans and other species who near its nests.

Animalia, Chordata, Mammalia, Carnivora, Canidae, *Vulpes vulpes*
RED FOX

The backyard of my mom's new house bordered a forest area quickly becoming deforested. The subdivision being developed was called Fox Run, and one of the red foxes began to appear in the back of our yard every morning. For almost a year I would go out and watch it run back and forth as it watched me from the border of the trees. The last time I saw the fox, it had ventured into the front yard. It followed me into the street and halfway to the bus stop.

My new school had specialized studies programs, and I selected a science class in herpetology. There was a kingsnake in the classroom. There were garter snakes, a rough green snake, salamanders and blue-tailed skinks and anole lizards, a box turtle. We first studied the poisonous species found in Florida, a prerequisite precaution for field trips.

Animalia, Chordata, Reptilia, Squamata, Serpentes, Viperidae,
Agkistrodon piscivorus
COTTONMOUTH (WATER MOCCASIN, SWAMP MOCCASIN, VIPER)

Animalia, Chordata, Reptilia, Squamata, Serpentes, Colubridae, *Nerodia sipedon*
COMMON WATER SNAKE (BANDED WATER SNAKE, STREAKED SNAKE, WATER PILOT, WATER ADDER)

Water moccasins are often confused with a nonvenomous relative, the common water snake (banded water snake, water adder, water pilot, or streaked snake), and the woods and creeks where I played and fished were filled with both species. The cottonmouth is thick- or fat-bodied, squat in comparison to the slenderer water snake. The moccasin, a pit viper, has a blunt triangular head and pits behind its nostrils; its eyes have split pupils. The water snake has a narrow, oval head and is round-pupiled. Vertical stripes run down the harmless species' jaw, whereas the cottonmouth has horizontal lines that stretch from its eyes to the back of its head.

Babies of both species are similar in coloration, dimension, and overall characteristics, and I once misidentified a capture I brought to a public speaking class as a visual aid. During a dramatic pause when I was about to open the glass gallon jar and pour the small snake out onto the lectern, I held it up in the classroom's fluorescent lights. It had vertical eyes. It had a horizontal stripe beneath its mouth. I averted a disaster, but it resulted in the banning of live animals being part of any future student presentation.

The mistaken identity was a good wake-up call: I was no expert. An amateur aficionado, I released the baby cottonmouth in the creek where I had captured it. Catch and delayed release had been my initial plan, to only hold local species long enough to observe them for a few weeks. But I was young, impatient, and loved the adventure of the search as much as anything, and no matter how well I tried to take care of my creatures, I became a small-time accomplice in the wildlife trade.

Although I continued to keep snakes as "pets" all through college and drove to reptile shows throughout Florida and Alabama, I became more and more aware of the problem of this pet trade, as well as more critical of any kind of so-called zoo, including the Swamparium. Like my vision had sharpened when I put on corrective lenses, my joy at first seeing the live wild creature was changed, quickly replaced with a sorrow if not repulsion at seeing more clearly the animals' limited living conditions. Caged, or leashed, or aquariumed, I couldn't witness it any longer, couldn't contribute to the disruption of their removal.

I reframed my obsessions to touch or get close to wildlife to appreciation and observation, and I tried to control that sense of possession that lingered to pick up and hold any nonvenomous snake I might come across.

In my 30s, I had the opportunity to caretake a rural property in Vermont in the summers, and the Calais Mountain farmstead was a paradise of garden fauna. Garter snakes were everywhere, on the steps to the front and back porches, on the gravel drive, in the grass, often even climbing the screen doors, not to mention among the water lilies and in the iris beds. There were two ponds on the property, various springs and crossing streams, turtles and beavers and a muskrat, eagles and owls. A blue heron. Red-winged blackbirds in the cattails. Chipmunks. Deer. Once a moose. Once a black bear.

After the first few years, I made a deal with the snakes. I'd only pick up one a season (though moving a snake from inside the house or an outbuilding didn't count as a "capture"). And five minutes was the longest I would let myself hold one. The rationale was they didn't enjoy my fascinated intrusion. It scared them. And it wasn't necessary. Initially after I made this deal, I'd often pick up one of the first garters of the summer, feeling the excitement of the coming abundance and just to get the anticipation over with. Eventually I learned to wait, and in what became my ninth and last August, that patience was rewarded with a foot-long ring-necked snake, something I'd never seen in the wild, never touched or held.

Animalia, Chordata, Reptilia, Squamata, Serpentes, Colubridae,
Diadophis punctatus
RING-NECKED SNAKE

Although ring-necked snakes are believed to be abundant in their range from southeastern Canada to central Mexico, their secretive and nocturnal nature leaves the science on the species unclear. They are slightly venomous, but rear-fanged and docile, no threat to humans. The last snake I held, I picked up in the morning dew, crossing the grass near a maple tree. It was brown, shiny, wet, and smooth, its collar and belly orange-red. It curled in my palm as I carried it into the outbuilding to photograph it. Such a tiny constriction as it wrapped through my fingers. I rubbed my thumb under its chin and then released it in the nearby flower bed.

Part IV

Traditional Ecological Knowledge

Traditional Ecological Knowledge (TEK), also called Indigenous knowledge, has been defined as the evolving universe of information acquired by Indigenous and local peoples over hundreds or thousands of years through direct contact with their local environments. In the past, TEK was viewed as incompatible with Western science, though consensus has now built that these approaches can work together to strengthen resource protection and management. The chapters in this part provide a small window into several approaches taken by tribes today.

Chapter 28

Talutsa
Weaving a Cherokee Future

| ANDREA L. ROGERS

Weaving a basket is an exercise in maintaining balance, in unremitting symmetry. The weavers are woven horizontally around vertical spokes. The basketmaker must keep the spokes straight, upright, and evenly spaced all the way around, even as they weave the reed around the basket, repeatedly adjusting that spacing to draw the sides of the basket in or expand them outward. While shaping the walls, the weave can't be too tight or too loose. To create a double-wall basket, the weaver then reverses the process, pulling and pushing to create an object of beauty, usefulness, and symmetry.

For a long time, I have understood that the primary philosophical principle of traditional Cherokee culture is the importance of balance. It is exemplified in our language, our justice systems, our clan systems, and our familial relationships. It is illustrated in our baskets and our oldest stories.

Tom Belt and Heidi M. Altman write in "Tōhi: The Cherokee Concept of Well-Being" that the worldview of the Cherokee hinges on the importance of balance, within both the world and the individual. "Taken together, these concepts—tōhi: and osi—indicate that there is a normal state of being for the Cherokee world and for the Cherokee individual."[1] When the world is out of balance, the community is at risk, and vice versa.

Ingredients for a balanced Cherokee community include valuing and caring for community members, preserving culture and history, working toward tribal food sovereignty, and supporting language revitalization. Few things have the potential to impact this balance as powerfully as effective land management by the Cherokee Nation.

I first thought of writing about river cane baskets and the tribal land management practices being put into place for the benefit of growing river cane. When I interviewed Feather Smith, ethnobiology manager of the Cherokee Nation Seed Bank (and a basketmaker), she said, "A lot of European accounts talk about Cherokee basket weavers being some of the best basket weavers in the world because of how intricate those double-wall baskets were."[2]

In Cherokee, the word for "basket" is *talutsa*. Many Cherokee noun words that came into the language later are descriptive, but according to Brad

Montgomery-Anderson's book, *Cherokee Reference Grammar*, *talutsa* is a base word.[3] It just means basket. I learned to make Cherokee double-wall baskets from a book while working as a teacher of urban Native children. Armed with Asian honeysuckle and photocopied instructions, I was as wowed as the kids when I turned the soaked reed into a useful and beautiful container. Since then, I have dreamed of utilizing traditional materials, namely river cane.

My introduction to my people's relationship to river cane came after I had learned to weave baskets during a visit to the Cherokee Nation's native plant site. In a presentation, Pat Gwin, senior director of the Cherokee Nation's Environment Resources Group, spoke of the Tribe's efforts, in conjunction with Cherokee elders, to increase the limited patch of river cane that had been relocated to an area creekside at the native plant site. Eventually, that cane will be harvested for use by Cherokees. It can take 20 years for river cane to reach the diameter needed to make the kinds of baskets Cherokees were known for prior to removal.

Various factors have caused the Nation's cane to struggle. According to an article from the US Department of Agriculture on research partnerships with Native communities, at least 98% of canebrakes of once vast amounts of river cane are gone.[4] Michelle Baumflek of the USDA, who is working with tribes on a river cane initiative, said, "River cane is a culturally important plant species

28.1. *Left*, Southeastern river cane basket (date and maker unknown) in author's collection; *right*, basket made by the author. PHOTO COURTESY OF THE AUTHOR

for most of the tribes in the Southeast. . . . It has a lot of potential to help with sediment runoff and bank stabilization, among many other things. It's been basically eradicated. You can still find small stands of river cane in lots of places, but the ecosystem has basically been destroyed."[5]

River cane was and is important to the culture of many southeastern tribes for basketmaking, furniture, weapons, and possibly, food. *Arundinaria gigantea* is also called giant cane and is a type of bamboo. For basketmaking, it must be split to create a flat reed. The splitting and dying of river cane is a time-consuming, difficult process. River cane edges are sharp and tough on bare hands. Artists who make these baskets deserve to be paid well for their knowledge, craft, and labor.

The more I read and listened, the clearer it became that the Cherokee Nation's tribal land management work and the people involved in the many efforts were doing much more than fighting the extinction of river cane. In managing tribal lands for the benefit of Cherokee people, the Cherokee Nation has tried several different approaches. Much like supporting the walls of a basket, the spokes needed to support living Cherokee culture are many.

Among these efforts are the establishment of a secretary of natural resources, creating a seed bank to research and preserve heirloom and native plants and provide them for Cherokee citizens, opening an art gallery where culturally

28.2. The author teaching basket-making online for the Fort Worth Independent School District's American Indian Summer Camp. PHOTO COURTESY OF THE AUTHOR

relevant crafts from native materials are taught, and supporting individuals who are keeping the art, language, and culture alive through their crafts and work. The Cherokee Nation has set aside land for the conservation of native plants, pays for hunting and fishing licenses for tribal citizens, and annually honors the work of numerous national treasures. In addition, the Cherokee language department supports all these efforts, utilizing the latest technology. The department constantly offers language classes (in person and online); creates useful language resources, such as posters, many focusing on natural resources; and produces signage, booklets, and packaging for plants distributed by the Cherokee Nation Seed Bank and labels for plants growing in the native plant site. University professor and Cherokee researcher Clint Carroll has worked with the Tribe and tribal elders in a program called the Medicine Keepers, which supports Cherokee elder speakers in meeting regularly, locates important medicinal and heirloom plants, documents plant knowledge, and then pairs elders with young people in order to pass on cultural knowledge about Cherokee plants, foods, and medicine.[6]

As Feather Smith said in an interview with OsiyoTV, "Everything is sort of interconnected with Cherokees. . . . [Y]ou can't be Cherokee without our language, we can't be Cherokee without our plants, we can't be Cherokee without our medicine. It was really important to preserve these not only for Cherokee people, it was important to have these for our next generation."[7]

Managing land for the good of future generations means much more than managing resources. Cherokees are actively locating and identifying resources and knowledge in danger of extinction. It's difficult to be all things to all people, but that is the balancing act the tribal government does to maintain the Nation's culture and relevance. Realistically, it takes many people using their unique talents to revive, restore, preserve, and share Cherokee culture. The Cherokee Nation is made up of urban Cherokees and rural Cherokees. There are Cherokees who hunt and practice traditional medicine, and Cherokees who have never been out of the city. Cherokee culture, plants, language, and medicine are the spokes around which the Cherokee community is woven. As Smith pointed out in her OsiyoTV appearance, "[O]ur culture and our plants almost go hand in hand and if we don't have any of these native plants that we were using for food and medicines and for crafts, what really are we?"[8]

Tribal policies, land management initiatives, and individual Cherokees shape the world, all the while weaving a strong foundation for a Cherokee future.

NOTES

1. Quoted in L. J. Lefler, *Under the Rattlesnake: Cherokee Health and Resiliency* (Tuscaloosa: University of Alabama Press, 2009), 22.
2. Interview of Feather Smith by author at Cherokee Nation native plant site, 12 May 2021.
3. B. Montgomery-Anderson, *Cherokee Reference Grammar* (Norman: University of Oklahoma Press, 2015).
4. Sarah Farmer, "Research Partnerships with Native American Communities," USDA CompassLive, 8 October 2020, https://www.srs.fs.usda.gov/compass/2020/10/08/research-partnerships-with-native-american-communities/. "The projects focus on sustainability of botanical species that are important to [I]ndigenous communities."
5. Farmer, "Research Partnerships."
6. Lisa H. Schwartz, "Clint Carroll: Rooting Research in Tribal Partnerships," Office for Outreach and Engagement, University of Colorado, Boulder, 5 January 2018, https://www.colorado.edu/outreach/ooe/clint-carroll-rooting-research-tribal-partnerships.
7. OsiyoTV, *Seeding the Future*, 21 January 2021, YouTube, https://www.youtube.com/watch?v=_UyuyrXaOfc.
8. OsiyoTV, *Seeding the Future*.

Chapter 29

A Traditional Strategy to Promote Ecosystem Balance and Cultural Well-Being Utilizing the Values, Philosophies, and Knowledge Systems of Indigenous Peoples

RICHARD T. SHERMAN *and* MICHAEL BRYDGE

The following stories and insights are based on my experiences as a Lakȟóta man, taught by my elders and my surroundings of the Lakȟóta, and later trained in the Western scientific method.[1] If I possess any knowledge at all, it began as a hunter and gatherer of food and medicinal plants. As I later became knowledgeable of the Western method of wildlife biology, I combined the two waves of thought. I, like "[s]ome Indigenous peoples . . . surmise that natural resource 'management' assumes an unequal power dynamic among plants and animals and thus a relationship based on stewardship, or the mutual coexistence of humans and animals, results in more sustainable environments, economies, and communities" (Glenn 2012:1, R. T. Sherman, unpub. data).[2]

Though my thoughts here are a mixture of my lived and learned experiences, I want to emphasize the Lakȟóta notion of not speaking prior to learning from the knowledge of those who are older and wiser than myself. Because of our way, my stories are introduced by and intertwined with the thoughts of my great-uncle Luther Standing Bear (1988, 2006), brother to my great-grandmother Lucy. The world-renowned visionary Black Elk (2008), who lived in Manderson, South Dakota, 20 miles from where I was born and raised, also inspires some of my thoughts. These relatives act as my guides.

The term "stewardship" suggests an innovative way to achieve integrated management with connotations of personal investment and reciprocal relationships, as opposed to the Western management methods, which suggest a more linear approach. Often, society rejects management systems that run counter to its philosophies. The solution may be to create a lingua franca by which both may communicate. The final product will utilize elements of both systems.

THE LOCAL COMMUNITY-BASED ECONOMY AND STEWARDSHIP

The majority of tribal residents on the Pine Ridge Reservation engage regularly in trade and bartering (Sherman 1988). Many of us are experts at trading, bartering, and householding, having participated in intra- and intercommunity commerce and in active regional, cross-continental, and global economies and trade networks for centuries before and after European-American contact. Our economy was sustainable, and when resources became scarce, our people moved, allowing the environment to replenish itself for future seasons and generations. Traditionally, and still today, the Lakȟóta economy and the ways people process "making a living" go beyond calculating cash value. Even land is viewed "not as a utility with which to exploit maximum monetary benefit, but as an area with multiple benefits which cannot be limited merely to dollar representation" (Brydge 2010b:90). Our economy is social and imbued by our ways of viewing the interconnectedness of the world around us (Pickering 2000). Many people hunt and gather. Elders are given free hunting permits. Often, there is a household member or extended family member who provides game meat to those in need. The reservation system sought to undermine our economy and ways of life. Today, it is our lifeways that help dampen the negative effects of the world economy (e.g., the creation of poverty and food deserts) in our communities (Pickering 2000).

There are inherent elements in the Lakȟóta way of thinking and approach to economics that guide our interactions with the living and nonliving world. Our belief that everything is connected is the essence of our being and kinship. Lakȟóta pray before every function, customarily beginning each prayer with the phrase *mi'takuye oyás'iŋ*, which translates as "all my relatives, all my relations." We are of a common spirit with all things animate and inanimate.

The philosophical implications of interconnectivity have garnered support from the mainstream scientific and lay communities in recent years. While Western scientific discourse did not thoroughly accept ecosystem interconnections until the mid- to late 20th century, Indigenous nations have understood ecosystem connections and passed on stories of ecological knowledge since time immemorial (Brydge 2010b). Today, there is a growing acceptance of these attitudes and practices: primary schools emphasize complex pattern recognition and problem solving, universities offer more interdisciplinary education and training (as opposed to the isolated approaches characteristic of Western problem solving), prison systems recognize the need for cleansing ceremonies such as *inipi* (sweat lodge), and health insurance companies are more receptive of and willing to pay for osteopathic or whole body treatments, such as acupuncture and gut health, otherized and long snubbed as "alternative" practices.

The Indigenous view of the economy is holistic and incorporates respect for the Two-Legged, Four-Legged, Winged, and Plant Nations. Animals and plants are known as members of specific nations, such as the Buffalo Nation, Mouse Nation, and Plant Nation (St. Pierre and Long Soldier 1995, anonymous Lakȟóta buffalo rancher, pers. comm., 2007).[3] A 2008 study comparing reactions of Lakȟóta farmers, ranchers, and buffalo caretakers to non-Indian farmers and ranchers in the northern Great Plains illuminated the roles of generosity and recognition of the agency of plants and animals in Lakȟóta culture. Hearing the statement, *I believe we should try to leave as much land aside as possible for other animals, birds, and insects to use*, a Lakȟóta cattle rancher replied, "We love animals, there's a reason for 'em bein' here" (Brydge 2010b:90). This example reiterates the collective sentiment of Lakȟóta agricultural households, of which 72% believe in the maximization of land for other animals, birds, and insects, while just over 31% of non-Indian agricultural operators held the same belief.

SPIRITUALITY AND THE INTERCONNECTEDNESS OF PLANTS, ANIMALS, AND ALL BEINGS

As Lakȟóta people, we see the spirituality all around us and believe that everything has a spirit. When we see a tree, we see the spirit and poetry of it, as opposed to board feet. In the approach we take, we look at the interconnected spirit of nature. We have a medicine, a healing plant called *siŋkpȟéthawote*—or muskrat food (*Acorus calamus*, sweet flag).[4] It is good for colds and the types of infection that may arise in the winter months. And this may seem strange to proponents of a Western model, but a medicine man communicated with a muskrat (*siŋkpȟé*, *Ondatra zibethicus*), and the muskrat informed him of a plant that grows in his habitat. That medicine, brought forth to our people by a muskrat, provides a cure for our people. When I was a young boy, this root was found in most reservation homes. We are innately aware of the connections—our health, well-being, spirituality, and lifeways—to the land.

There are times in our interactions with the Four-Legged and Winged Nations that reciprocity comes in the ultimate sacrifice, that of one's life. Several years ago, while we were conducting mapping exercises and a household survey, a Sičháŋǧu Lakȟóta, just east of the Oglala Oyáte, asked if he could "double-dip" while conducting research. He further explained, "I shot one coyote [šuŋgmánitu, *Canis latrans*] yesterday and one bobcat [igmúgleza, *Lynx rufus*] today. If I get one more bobcat fur, I'll be able to [better support my family]. Then, I'll get this home mapping completed."

This gentleman had very little cash reserves to support his family. He was part-time employed but relied heavily on subsistence living to provide for his family. Through legal animal harvests, he was able to trade and barter. In our way of thinking, an animal sacrifices itself to you, when the time is right. There is an understanding that in times of need, while you are hunting for food or furs, for example, an animal will present itself to you. Our view asserts that animal nations too have agency, and that agency is fulfilled when reciprocity occurs between humans and animal nations. When our people adhere to these concepts—bartering in times of need and understanding an animal will sacrifice itself only when it is supposed to (perhaps the ultimate barter)—there is an inherent respect for the life of wild animals. It is for these reasons, among many others, that in the traditional Lakȟóta way, animals should not be managed but stewarded.

STEWARDSHIP AS OPPOSED TO MANAGEMENT

The traditional Lakȟóta view of animals and plants as revered entities, coequal to humans, shapes human environmental interactions in Lakȟóta communities and creates an alternative understanding of the values of lands, including those not deemed fit for agriculture by Western standards. According to one study, 93% of Lakȟóta agricultural households agreed that plants and animals have as much right to exist as humans, compared to 56% of non-Indian agricultural households (Brydge 2010a). This viewpoint largely explains the absence of large-scale, intensive agricultural operations among the Lakȟóta (Pickering and Jewell 2008). Instead, the maximization of social and environmental benefits is more highly valued.

One buffalo rancher who plants alfalfa detested the idea of turning his property into a commodity agribusiness and refrained from the large-scale monocropping associated with such ventures because "the land needs to be [there] for many plants and animals" (Brydge 2010a:195). In addition, a Lakȟóta buffalo caretaker with the largest private herd on the Reservation equated large-scale agribusiness with ecological disruption, explaining, "We decided a long time ago we would try to bring the buffalo [ptéȟčaka, *Bison bison*] back, the buffalo are more friendly to the land" (Brydge 2010a:195). These attitudes demonstrate generosity for all living beings, a philosophy that perpetuates healthy lifeways and ecosystems, which is echoed through the phrase mi'takuye oyás'iŋ—all my relations (St. Pierre and Long Soldier 1995).

One Lakȟóta caretaker, embodying the attitude of holism and sustainability, described conservation as important " 'so we could have it [the environment] all the time and better-quality bucks and better does and stuff like that for the next generation,' [further] reiterating the popular Lakȟóta thought of responsibility toward wildlife due to the reciprocal relationships between humans and animals." An even more holistic answer came from another Lakȟóta buffalo caretaker: "I believe that those [nonagricultural] plants are natural, you know, and if you disturb them, you disturb the ecosystem. If you disturb the ecosystem, you get the birds and insects [overpopulating,] you know." He continued, "[I]t's just good to keep your natural habitat, you know, the best you could" (Brydge 2010b:90–91).

Leaving land aside for animals unassociated with financial gains exemplifies Indigenous peoples' adherence to traditional values. Dominant society challenges our traditional stewardship practices and values. For instance, acreage utilized for nonagricultural purposes requires a financial sacrifice from Lakȟóta agricultural households. Land maximization increases farm capacity, which in turn provides greater opportunities for harvestable yields and greater access to grants and programs provided by the US Department of Agriculture (e.g., Environmental Quality Incentives Program) and other entities. Setting land aside, however, fulfills two critical components of sustainable resource management in academic (and global) contexts: perpetuating animal and plant habitats, and demonstrating forethought about ecosystem services and provisions from an Indigenous perspective (Brydge 2010b).

Spiritual components and healing processes associated with nonagricultural land are perhaps the most significant to Lakȟóta people. Though Lakȟóta spirituality is woven into our entire value system and environmental ethics, some spiritual practices are completely dependent on the presence of natural beings. For example, St. Pierre and Long Soldier (1995:95) explain the importance of the cottonwood tree (*wáǧačhaŋ*, *Populus deltoides*) during the Sun Dance: "To [the Plains Indians] the cottonwood tree is the cosmic tree, its limbs in the clouds, its spiritual roots spreading deep into Mother Earth and to the four directions. This tree is the center of the sacred universe during the ceremony." In addition, cottonwood trees were home to little people, *Čháŋ othípi* ("they dwell in the trees") whom some believe were teachers of the Sun Dance and keepers of sacred knowledge and others believe were bad spirits (Walker 1980, St. Pierre and Long Soldier 1995). Land along rivers and other water sources must be preserved to ensure the viability of cottonwood trees, other plants, and wildlife habitat. Therefore, both the physical and spiritual natures of beings are necessary for ceremonial purposes and Indigenous continuity. Oglala Lakȟóta elder and my friend from Wounded Knee, Walter Littlemoon, praises the cottonwood and animals (along with humor) for contributing to his path of healing from trauma associated with boarding schools and other harmful life events: "Through this process of remembering, I've become aware that the greatest help for healing has come to me through nature, starting with the cottonwood tree triggering positive memories. . . . I found wisdom from the animals that have helped me to realize and to understand how to use my own sense of hearing, eyesight, touch, taste, and smell. . . . [W]ith help from nature's grandfathers, the rocks, the trees, the wind, I am beginning to see and to understand situations and people just as they are" (Littlemoon and Ridgway 2009:87–88).

To us, nature is not a force to be reckoned with, but a force to coalesce with, a force to understand and learn from. Walter's reconnections with nature were facilitated by elders and memories of what nature taught him as a young boy in regard to "respect[ing] the other forms of life here on earth" (Littlemoon and Ridgway 2009:93, Brydge 2010a:200–201). Thus, for our people and other Indigenous communities around the globe, "a relationship with nature . . . has led to the healing of past wounds, scars left throughout history by westward encroachment and capitalist accumulation of the United States," and though "natural areas do not yield as much economic value in the market economy as arable land, land preservation heightens cultural capital and facilitates" the continuity of healthy Indigenous lifeways much more than increased agricultural yields do (Brydge 2010a:201). Likewise, when environments in which Indigenous communities live, thrive, and belong are affected, either regionally or catastrophically, as the result of damage, degradation, or contamination, so too are their sense of shared identity, traditions, and home (Johnston and Barker 2008, Hoover 2017).

DEVELOPING THE INDIGENOUS STEWARDSHIP MODEL

When I was a boy, we lived a modest (poor by Western standards) but healthy life. I overheard my mom quietly agonize about not having enough for us to eat. We ate everything we had and wished for more. We often did not even have enough scraps to feed the dogs. Walter Littlemoon once explained: "It was never the land of plenty, but it was the land of enough" (Littlemoon and Ridgway 2009:19). The land was healthy, as opposed to being overutilized as it is now. My family depended on four species for subsistence: jackrabbits (*maštíŋska, Lepus californicus*), cottontails (*maštíŋsapela, Sylvilagus* spp.), sharp-tailed grouse (*čháŋšiyo, Tympanuchus phasianellus*), and pheasants (*šiyóša, Phasianus colchicus*).

My father was the hunter for the family until he succumbed to alcoholism, at which time he passed the baton to me. At an early age, I had to learn what kinds of plants we could use for food so we could have a well-rounded meal. We gathered a lot of wild foods, and I learned what types of plants were medicinal. As a young man I tried to steward the resources as best I could, in a way that they would not become depleted: not overharvesting any particular area, moving around to allow game populations to recover and thrive, spreading seed, gathering after plants had properly seeded, and leaving enough behind for plants to maintain themselves. I learned at an early age the importance of providing for my family, but also of the Lakȟóta value of generosity and giving to others. Often my father would give away meat to other families the very night I brought the wild game home.

Entering adulthood, I spent four years in the US Navy. When I was discharged in 1965, I had dreams of managing wildlife for my people. While providing for my family years before, I had realized one thing that was lacking by the tribal government was a management system for resources. So, I attended Utah State University where I studied wildlife management. After earning a bachelor's degree, I served as executive director and then wildlife biologist in the early '90s for the Oglala Sioux Parks and Recreation Authority, managing the fisheries, wildlife, buffalo herd, and ranger program. Five years later, I realized we needed a stewardship model, not a management scheme. After completing a master's degree in regional planning at the University of Massachusetts, Amherst, I returned home to resume my position as a wildlife biologist and put together the Indigenous Stewardship Model.

What I really wanted to do was create a common language that would enable us to better communicate with one another and with proponents of the Western scientific model, so we would not have the same problems over and over. The Indigenous Stewardship Model helps define the conversation, outlines our side of the story, and asks for collaboration—the reciprocity of mutual participation.

Category	Element	Description
ACTIVE INDIGENOUS STEWARDSHIP ON TRIBALLY CONTROLLED LANDS	1	Indigenous Ecology: Land and Habitat Maintenance, Restoration, Preservation
	2	Subsistence Lifestyles: Equal Access to Resources
	3	Promoting Economic Self-Sufficiency: a. Sustainable Harvesting b. Ecologically Sustainable Micro-Enterprise Development
	4	Connections to the Land: Community Monitoring and Reporting
COMMUNITY OUTREACH TO SUPPORT INDIGENOUS STEWARDSHIP	5	Indigenous Ethnobotany: Identifying and Restoring Indigenous Knowledge
	6	Community Input: Synergies between Indigenous Knowledge and Management Systems
	7	Intergenerational Transfer of Indigenous Knowledge
CO-MANAGEMENT Advocating for Indigenous Stewardship on Land Where Authority is Shared or Absent	8	Validating Indigenous Knowledge Systems: Policy, Advocacy, and Reform
	9	Strategies for Genuine Collaboration: Avoiding Appropriation and Dominance
CONSENSUS BUILDING AND CONFLICT MANAGEMENT	10	Indigenous Processes for Decision-Making: Resolving Internal Differences
	11	Indigenous Collaboration in Decision-Making: Resolving External Differences

29.1. The Indigenous Stewardship Model incorporates four main elements.
RICHARD SHERMAN AND MICHAEL BRYDGE

INDIGENOUS STEWARDSHIP MODEL

The Indigenous Stewardship Model was fleshed out over 30 years of listening to my elders, hunting and gathering, studying, and working as a field biologist. The model is a flowing, circular set of actions and suggestions based on what I have been taught from the Lakȟóta way and the Western way. Each element of the model may look a little different through time and for different individuals and groups of people. The combined influence of traditional and Western training on developing the Indigenous Stewardship Model is evident in Ross et al.'s *Indigenous Peoples and the Collaborative Stewardship of Nature: Knowledge Binds and Institutional Conflicts*:

> While recognizing the potential contribution of Western models of resource management, the Indigenous Stewardship Model supports the holistic, locally embedded, and culturally significant role of natural resources in Indigenous worldviews. It promotes sovereignty by explicitly addressing methods for asserting the inherent right of [I]ndigenous peoples to govern themselves with respect to natural resources. By exploring the best elements of both systems, the Indigenous Stewardship Model is intended to establish a method for communication between the Indigenous communities and Western resource managers without the dominance and intrusion that often characterizes relationships between Indigenous peoples and proponents of the Western model. By empowering local people to act as the stewards of their immediate resources the Indigenous Stewardship Model also offers methods for minimizing the costs and maximizing the outcomes of resource monitoring and maintenance. (Ross et al. 2011:260)

This collaborative work introduced 11 elements, some bolstered by thousands of years of field testing. Each element fits within four overarching themes (Ross et al. 2011):

- Active Indigenous stewardship on tribally controlled lands
- Community outreach to support Indigenous stewardship
- Co-management: Advocating for Indigenous stewardship on land where authority is shared or absent
- Consensus building and conflict management

So far, I have emphasized how Indigenous values of stewardship translate into actions that differ from the dominant Western philosophy of management. In the remainder of this chapter, I turn toward the applicability of the model in a variety of contexts, issues, and communities.

OPPORTUNITIES FOR USE OF THE INDIGENOUS STEWARDSHIP MODEL

Tlingit Forestry and "Common" Indigenous Knowledge

Element 1. Indigenous Ecology: Land and Habitat Maintenance, Restoration, and Preservation

Element 8. Validating Indigenous Knowledge Systems: Policy, Advocacy, and Reform

Element 11. Indigenous Collaboration in Decision-Making: Resolving External Differences

In the early 1990s, I spoke to Inuit and Athabascan villagers in Alaska about the knowledge they have and demonstrated the potential for them to create a management plan based on the Indigenous Stewardship Model. State wildlife officials who were present at the meetings were visibly irritated.

While visiting the Tlingit in the fishing village of Hoonah on Chichagof Island, I observed meetings with federal and state officials to discuss the management of forest resources. At that time, the surrounding forest was managed by federal and state officials, and villagers were concerned over the depletion of the forest. The Tlingit suggested that they could create a plan and manage their own resources. A non-Native state manager asked: "How do you propose to do this? Do you have people educated to put this together?" A Tlingit elder stood up in the back of the room and said: "How was it when you got here?" The elder's question was all that was needed to infer that Indigenous knowledge systems, which had maintained the forests for decades, were not being validated. What followed this pointed discussion was a space for Indigenous processes in decision-making to resolve external differences.

Hook-Offs and State-Issued Permits

Element 1. Indigenous Ecology: Land and Habitat Maintenance, Restoration, and Preservation

Element 2. Subsistence Lifestyles: Equal Access to Resources

Element 3. Economic Self-Sufficiency
   a. Sustainable Harvesting
   b. Ecologically Sustainable Micro-Enterprise Development

Element 7. Intergenerational Transfer of Indigenous Knowledge

When I consulted in Southeast Alaska, the Tlingit used hook-offs—areas set aside for each traditional family to use for fishing. In these areas, the seines used to catch salmon took the shape of a broadened U or hook (Langdon 1984). The Tlingit had longhouses where they taught young people to become fishers, and at the end of the training period, the youths would be given their own boats so they could begin stewarding their family's hook-off. Like other Indigenous communities, the Tlingit understood the environmental implications of making a living and engaged in ecologically sustainable micro-enterprise development, as many still do today.

This practice went on for generations, until Alaska began issuing more permits to non-Indians. Overfishing, vandalism, theft, shootings, lack of respect for traditional hook-offs, and even deaths followed. Because of state intervention, a sustainable fishery stewardship management system was nearly ended. Today, the Tlingit are under immense pressure from state and global economic forces, such as sport fishing and large-scale commercial fishing.

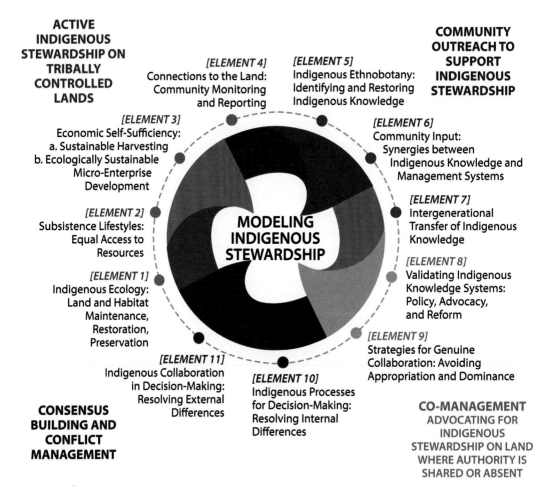

29.2. The circular nature of the Indigenous Stewardship Model. RICHARD SHERMAN AND MICHAEL BRYDGE

Outside Agencies and Tribal Councils as Impediments: Terrestrial and Aerial Counts

Element 4. Connections to the Land: Community Monitoring and Reporting

Element 6. Community Input: Synergies between Indigenous Knowledge and Management Systems

Element 9. Strategies for Genuine Collaboration: Avoiding Appropriation and Dominance

Element 10. Indigenous Processes for Decision-Making: Resolving Internal Differences

Element 11. Indigenous Collaboration in Decision-Making: Resolving External Differences

There is a tendency for outside (state or federal) agencies to try to impose their methods and manage resources on behalf of tribes. That paternalistic behavior has perpetuated the issue of resource appropriation for Indigenous peoples in North America and throughout the world for the past 500 years. I realize, however, that we have adopted some of their methods and that outside agencies are not the only source of frustration for resource stewardship. Often, our own tribal systems are impediments.

When I was growing up, if the Tribal Council deemed animal populations were low, it would issue a moratorium on hunting throughout the Reservation, sometimes creating hardship for the people. Outside agencies would chide us, saying we were not managing wildlife properly. Sometimes the Tribal Council would follow suit. I realized that we had to create something that would accommodate both government(s) and the people, that we would have to create a system to inventory our game animals and set our own hunting seasons and game limits according to our own needs.

The Pine Ridge Reservation was sectioned into 10 stewardship units. My thinking was that after taking inventories and measuring which, if any, units had lower than normal populations of deer (čháŋtȟaȟča, *Odocoileus virginianus*), antelope (ikpísaŋla, *Antilocapra americana*), or other game species, we could restrict hunting in those areas while keeping other units open. With over 2 million acres and surrounding lands providing ample protected habitat, this worked for my people. Survey transects were developed throughout the Reservation, allowing us to do terrestrial counts. One of the first places terrestrial counts were conducted was in the South Unit of Badlands National Park in the northern area of the Reservation.

After the numbers were tallied, a wildlife biologist from the North Unit of Badlands National Park was invited to analyze our methodology and provide his opinion of the results. After comparing our counts to the adjacent state wildlife management unit, we discovered that Reservation land had more game animals per square mile than the state-managed unit. Earlier claims that we were low in white-tailed deer, mule deer (siŋtésapela, *Odocoileus hemionus*), pronghorn, and bighorn sheep (*héčhiŋškayapi, Ovis canadensis*) were false. Outside agencies did not criticize our efforts or our large game counts after that. Ultimately, our counts proved more accurate than prior estimates due to the thoroughness and multipronged approach

It is also instructive to look at Dall sheep counts done in the mid-1990s by the Kluane First Nation in the Yukon. These took several years of a rather confrontational process that was undermined by non-Native wildlife biologists who lacked respect for Kluane ecological knowledge, did not possess the Kluane's deep understanding of the local sheep population's migration habits, and prioritized aerial survey methods over those preferred by the Kluane. To the Canadian government's credit, it involved Kluane elders and hunters in the aerial surveys, though by most accounts, the Kluane were just "along for the ride," and their knowledge and insights were initially regarded as inferior. When on-the-ground sightings and counts by Kluane knowledge-holders ran counter to information collected during aerial counts, the wildlife biologists believed only their own numbers (Nadasdy 2003). Wildlife biologists even began chiding the Kluane as being either ignorant or politically motivated. The Kluane had one main request: "biologists should spend a year or two out on the land—hunting and trapping with a First Nations elder. . . . Perhaps, not surprisingly, biologists never seriously considered this suggestion, much less carried it out" (Nadasdy 2003:205–8).

Agriculture-Based Land Leasing

Element 4. Connections to the Land: Community Monitoring and Reporting

Element 6. Community Input: Synergies between Indigenous Knowledge and Management Systems

Element 8. Validating Indigenous Knowledge Systems: Policy, Advocacy, and Reform

Like most other Indigenous groups worldwide, Lakȟóta people have suffered severe land loss. One result of having our natural resources taken from us is the depletion of our traditional food and medicinal plants and animals. Overgrazing eventually became a palpable problem on Lakȟóta lands. When cattle were removed from the range at the end of the summer, there was neither food nor cover for wildlife, particularly when they needed it the most in order to gain weight and fat for the winter. The inadequate nutrients and caloric intake led to disease and starvation among big game herds. Several times in the early 1990s, I presented the Indigenous Stewardship Model to the Oglala Sioux Tribal Council, and several times I was turned down. According to the Thunder Valley Community Development Corporation (2018):

- Of the 2.78 million acres on the Pine Ridge Reservation, 71% of the land is rangeland and 13% is cropland.

- In 2012, there were only 200 farms owned by American Indians/Alaska Natives.

- From 2007 to 2012 there was a 73% decrease in acreage operated by Native American tenants on the Pine Ridge Reservation.

- The market value of agricultural products sold in 2012 was $87.7 million.

In South Dakota we have some of the best soils west of the Missouri River, yet our people are starving and malnourished, and we live in what is classified as a "food desert." A study conducted by Colorado State University in 2008 revealed that 26% of our household food came from food stamps, 19% came from government-issued commodities, and 6% was received from family, friends, or other donations (Jewel 2008). In addition to unsustainable resource management, the forced reliance on commodity foods and the adoption of nontraditional foodways perpetuates food colonization and contributes to high rates of preventable illness, such as type 2 diabetes, among Indigenous communities (California Newsreel 2008). We have buffalo caretakers with a lifetime of knowledge, and open spaces for traditional food plants and animals to thrive, yet the majority of our lands are used for commodity agriculture. Some models estimate we could feed five times our population from our own lands (Peters et al. 2016).

Organizations run by extended families (*thiyóšpayes* in Lakȟóta) are a common modern way to continue our traditional societies. One is Knife Chief Buffalo Nation Society, a nonprofit organization that manages a cultural buffalo herd for the purpose of supplying buffalo for young men's coming-of-age ceremonies and other ceremonies. Nonprofits such as Tanka Fund (Kyle, South Dakota) are working to protect and revitalize our land base. Makoce Agriculture Development (Porcupine, South Dakota) works to initiate systems-level change to reestablish a local food economy. These and other organizations are combining efforts to dedicate more and more lands for traditional uses, buffalo stewardship, and the foraging and local sales of traditional food sources.

Economic Hardship

Element 3. Economic Self-Sufficiency
 a. Sustainable Harvesting
 b. Ecologically Sustainable Micro-Enterprise Development

Element 8. Validating Indigenous Knowledge Systems: Policy, Advocacy, and Reform

Because of the harsh economic and social environments that many of our Lakȟóta communities live in, decisions are often made in the here and now with short-term thinking. Little regard is given to the seventh-generation thinking embedded in Lakȟóta philosophy, which posits that all our actions and decisions must be made to benefit the child born seven generations after us. An example of disregard for seventh-generation thinking occurred in 1973, when the Oglala Sioux Tribe entered into the Indian Tribal Land Acquisition Program, which provided 40-year loans totaling $19 million at a rate of 5%. According to the *Lakota Times* (2010), from September 2010 onward, the Tribe made annual payments of $870,000 to the USDA Farm Service Agency, which amounted to 73% of the total land lease revenue collected by the Tribe. For several years, lease revenue was actually *less* than the loan amount (Lakota Times 2010). According to a community development corporation focused on Lakȟóta lifeways, health, and well-being, loans such as these "compromise the sovereignty of Tribes and their ability to manage their natural resources for the benefit of their membership, while ensuring the availability of Tribal lands for lease to non-tribal members at low cost" (Thunder Valley Community Development Corporation 2018:89).

In the early 2000s, I co-led a study that revealed "a strong desire and willingness among Lakȟóta households to participate in land stewardship and re-cultivate a practical, physical relationship with the natural environment." In the report, we went on to state: "Nevertheless, tribal, state, and federal bureaucratic institutions fail to recognize such a dwelling approach as a viable strategy to ecological restoration, cultural renewal, and poverty alleviation.... We originally hoped that the results of our research will help policy makers at each of these levels with a means of recognizing the viability of initiating community-based stewardship and resource autonomy.... Principally, this requires reevaluating the merits of the cattle-leasing system, taking into account its ecological costs and its socioeconomic effects on the local communities" (Sherman et al. 2010:516).

Generational Gaps and the Passing of Our Elders

Element 5. Indigenous Ethnobotany: Identifying and Restoring Indigenous Knowledge

Element 7. Intergenerational Transfer of Indigenous Knowledge

Many times, I have been asked, "How would you suppose creating a management system when all the information your people used in the past is gone with the elders?" My answer is, "I don't suppose." All one has to do is consider the values and philosophies of the Lakȟóta people to be the basis for the system, a system inherently of stewardship, not management.

Of the 11 elements of the Indigenous Stewardship Model, "the key component is the transfer of knowledge to the youth" (Ross et al. 2011:253). Programs that facilitate this transfer of knowledge from elders to youth are imperative. In 2012, I partnered with students of cultural anthropology to develop a curriculum and plan for youth ethnobotanical excursions. We spent half a day in the classroom, teaching plant identification and uses to middle school students, and the next morning we went to a buffalo pasture and foraged for traditional food and medicine plants. I emphasized to the students the unique importance of the buffalo pasture because areas where cattle grazed no longer had some of the plants we collected. Elders attended the event and shared stories of doing the same thing during their childhood.

That evening, we gathered at the local elder center in my hometown of Kyle, South Dakota. The tribe donated buffalo meat, and we made a stew with all the plants we had collected. Elders who were incapable of foraging with us arrived and ate too. The meal was prepared and served by the youth, who were able to sit with elders and hear more stories. At the end of the meal, we had a completion ceremony for the youth, and they received certificates. Everyone involved enjoyed the program, and their biggest complaint was they wanted more stew. With adequate funding, time, and prioritization of resources, we can ensure successful programs such as these to facilitate greater knowledge transfer and perpetuate healthy community development through stewardship.

I have conducted ethnobotanical excursions every year since, and I give talks and demonstrate the importance of proper land stewardship for our plant and animal relatives at museums and at Badlands and Wind Cave National Parks. I also participate in buffalo kills, where students will bus in from as far as 90 minutes away to help field dress the animal. Universities and scientists from across the globe incorporate my ideology into their studies to include Indigenous perspectives and Traditional Ecological Knowledge. Many communities have come to realize they already naturally incorporate elements of Indigenous stewardship.

These events are important, but other relationships need to be built as well to pique interest through experiential learning and to develop mentor-apprentice relationships, which are necessary to ensure the continuity of tradition and stewardship. I have built relationships with several individuals and families over the last several decades. One family recently fed me the traditional meal of buffalo meat wrapped in the green husk of the milkweed plant. Through my teachings, the wife now practices sustainable harvesting of mule deer, antelope, and buffalo. Her husband, an avid communicator, writes for the local newspaper. Oftentimes, I see stories and lessons I taught them about food harvesting and preparation in the *Lakota Times*, which is read by thousands of Lakȟóta and other people throughout the world.

LATER AGAIN

The Lakȟóta word for woman—*wíŋyaŋ*—connects the heavens and Earth, the celestial with the terrestrial. *Wí* means "sun," and *íŋyaŋ* means "rock." Thus, the primary power that brings us into this world is connected to both the celestial and the terrestrial. We are nourished by the sun and the Earth, and we will return to the Earth. Even our being is cyclical by nature. Hopefully when we return, we have left it better than we found it.

For the Lakȟóta, the word "goodbye" is not used. A word such as that is terminal, finished, definite. But a circle has no beginning and no end. The thoughts, ideas, and actions of the Indigenous Stewardship Model are continuous, like a circle. The model will mean different things to different people. Certain elements will be more pertinent to some communities and some situations compared to others. But the reciprocal need for collaboration and open communication will always be there. So, not in conclusion but in the Lakȟóta way, I say, *tókša akhé*, or "later again."

ACKNOWLEDGMENTS

We greatly appreciate the stellar efforts of Jordin Simons (independent researcher, environmental studies BA, California State University–Monterey Bay). She assisted with organizing thoughts, adherence to the style guide, developing the animal and plant table, final edits, and more. Seth Allard (Ojibwa, MA, PhD candidate, social work and anthropology, Wayne State University) further assisted with organizing thoughts, additional research, citations, formatting pictures, final edits, and more. His commitment to excellence is admirable. *Wóphila* to you both for your support and encouragement.

TABLE 29.1. Animals, Plants, and Relatives

LAKOTA NAME	ENGLISH TRANSLATION	COMMON ENGLISH NAME	SCIENTIFIC NAME
Čháŋšiyo	Sharp-tailed grouse, fire bird	Sharp-tailed grouse	*Tympanuchus phasianellus*
Cháŋtȟaȟča	White-tailed deer	White-tailed deer	*Odocoileus virginianus*
Čhešlóšlo phežúta	Herb, medicine	Milkweed	*Asclepias syriaca*
Čhetáŋ (generic), čhaŋšká (red-tailed hawk), čhaŋšká šápela (Cooper's hawk)	Hawk	Chicken hawk	*Accipiter striatus*, *Accipiter cooperii*
Héčhiŋškayapi	Bighorn sheep	Bighorn sheep	*Ovis canadensis*
Héčhiŋškayapi ská	White sheep	Dall sheep	*Ovis dalli*
Heȟáka (male), uŋpȟáŋ (female), tȟáhežata (two-year-old)	Elk	Elk	*Cervus elaphus canadensis*
Hoǧáŋša	Red fish	Salmon	*Salmo salar*
Igmúgleza, igmúglezela (variant)	Bobcat	Bobcat	*Lynx rufus*
Ikpísaŋla, niǧésaŋla, tȟáȟčasaŋla, tȟatȟókala (esp. male)	Pronghorn, antelope, domestic goat	Pronghorn, antelope	*Antilocapra americana*
Íŋyaŋ	Stone, rock	Rock	
Kȟaŋǧí, uŋčíšičala	Crow, American crow	Crow	*Corvus brachyrhynchos*
Kȟokȟóyaȟ'aŋla, waȟúpakozA	Chicken, fowl, hen, poultry, winged ones	Fowl	Galloanserae

TABLE 29.1. *(continued)*

LAKOTA NAME	ENGLISH TRANSLATION	COMMON ENGLISH NAME	SCIENTIFIC NAME
Maštíŋsapela	Cottontail rabbit	Cottontails	*Sylvilagus* spp.
Maštíŋska	Jackrabbit, hare, rabbit	Jackrabbit	*Lepus californicus*
Matȟóȟota, šakéhaŋska	Grizzly bear	Grizzly or brown bear	*Ursus arctos horribilis*
Pȟežíȟota	White sage	Sage	*Artemisia ludoviciana*
Ptegléška (southern), ptewániyaŋpi (northern)	Domestic cow, cattle-spotted buffalo, cattle	Cattle	*Bos taurus*
Ptéȟčaka (generic), tȟatȟáŋka (buffalo bull), pté (buffalo cow)	Buffalo	Buffalo	*Bison bison*
Siŋkpȟé	Muskrat	Muskrat	*Ondatra zibethicus*
Siŋkpȟétȟawote	Flagroot, muskrat food	Sweet flag, sway, calamus	*Acorus calamus*
Siŋtésapela	Mule deer, black-tailed deer	Mule deer	*Odocoileus hemionus*
Šiyóka	Greater prairie chicken	Prairie chicken, pin-nated grouse, boomer	*Tympanuchus cupido*
Šiyóša, pȟečhókan háŋska ziŋtkála	Pheasant	Pheasant	*Phasianus colchicus*
Šuŋgmánitu (sometimes applied to wolves), máyašle	Coyote, wilderness dog, coyote that lives in hills	Coyote	*Canis latrans*
Šuŋgmánitu tȟáŋka	Wolf	Wolf	*Canis lupus*
Šúŋka, šuŋgblóka (male), šuŋhpúŋka (dog with puppies), khečhá (shaggy dog)	Dog, horse	Dog	*Canis lupus familiaris*
Šúŋkawakȟáŋ	Horse	Horse	*Equus caballus*
Tȟašíyagmuŋka	Western and perhaps eastern meadowlark	Meadowlark	*Sturnella neglecta*
Tȟaté	Wind	Wind	
Uŋkčékhiȟa, halháta (southern)	Magpie	Magpie	*Pica hudsonia*
Wáǧačhaŋ	Cottonwood	Cottonwood tree	*Populus deltoides*
Waȟpé pȟeží	Leaf/needle grass/hay	Alfalfa, lucerne	*Medicago sativa*
Waŋblí	Golden eagle	Spotted eagle	*Clanga clanga*

NOTES

1. This article is written in the first person to represent the combined personal, academic, professional, and cultural viewpoints, experiences, and beliefs of Richard T. Sherman (the main author). The use of the first person and personal anecdotes is also representative of the common cultural, professional, and academic experiences of Sherman and the autoethnographic method (see Whitinui 2014).

2. While I speak from a Lakȟóta cultural background and experience, references to "Indigenous," "tribal," "Native American," or similarly broad terms, especially when coupled with inclusive pronouns and adjectives (i.e., "us," "we," "our"), can be interpreted as descriptive of common global histories, experiences, viewpoints, and ecological-economic relationships of Indigenous peoples, unless referring to specific tribal groups.

3. Throughout this chapter, the term "buffalo" is used in lieu of "bison" (*Bison bison*).

4. Unless initially referred to in Lakȟóta, the structure for animal description in the text is as follows: common English name (Lakȟóta name, *scientific name*). See the end of the chapter for a table of some animal and plant names.

LITERATURE CITED

Black Elk, J. G. Neihardt, and R. J. DeMallie. 2008. Black Elk speaks: being the life story of a holy man of the Oglala Sioux. State University of New York Press, Albany.

Brydge, M. 2010a. Ethnicity and agricultural household decision making on the high plains: Lakota cultural continuity and environmental ethics. Furthering Perspectives: Anthropological Views of the World 4:180–207.

Brydge, M. 2010b. Lakota agricultural household decision making and environmental conservation: motives, ethics, and barriers. Honor's thesis, Colorado State University, Fort Collins.

California Newsreel. 2008. Bad sugar, place matters. PBS, 10 April.

Glenn, K. 2012. A multi-generational approach to developing natural and human communities: youth, elders, and Indigenous stewardship on the Pine Ridge Indian Reservation. Center for Collaborative Conservation, Colorado State University, Fort Collins.

Hoover, E. 2017. The river is in us: fighting toxics in a Mohawk community. University of Minnesota Press, Minneapolis.

Jewel, B. 2008. The food that senators don't eat: politics and power: the Pine Ridge food economy. Master's thesis, Colorado State University, Fort Collins.

Johnston, B. R., and H. M. Barker. 2008. Consequential damages of nuclear war: the Rongelap report. Left Coast Press, Walnut Creek, California.

Lakota Times. 2010. OST receives reimbursement check in the amount of $345,168.00. 22 September. https://www.lakotatimes.com/articles/ost-receives-reimbursement-check-in-the-amount-of-345168-00/. Accessed 23 July 2021.

Langdon, S. J. 1984. Adaptation and innovation in Tlingit and Haida salmon fisheries. Smithsonian Folklife Festival, Alaska. https://festival.si.edu/articles/1984/adaptation-and-innovation-in-tlingit-and-haida-salmon-fisheries. Accessed 10 August 2022.

Littlemoon, W., and J. Ridgway. 2009. They called me uncivilized: the memoir of an everyday Lakota man from Wounded Knee. iUniverse, Bloomington, Indiana.

Nadasdy, P. 2003. Hunters and bureaucrats: power, knowledge, and Aboriginal-state relations in the southwest Yukon. University of British Columbia Press, Vancouver, Canada.

Peters, C. J., J. Picardy, A. F. Darrouzet-Nardi, J. L. Wilkins, T. S. Griffin, and G. W. Fick. 2016. Carrying capacity of US agricultural land: ten diet scenarios. Elementa: Science of the Anthropocene 4. doi:10.12952/journal.elementa.000116.

Pickering, K. A. 2000. Lakota culture, world economy. University of Nebraska Press, Lincoln.

Pickering, K. A., and B. Jewell. 2008. Nature is relative: religious affiliation, environmental attitudes, and political constraints on the Pine Ridge Indian Reservation. Journal for the Study of Religion, Nature and Culture 2:136–58.

Ross, A., K. P. Sherman, J. G. Snodgrass, H. D. Delcore, and R. Sherman. 2011. Indigenous peoples and the collaborative stewardship of nature: knowledge binds and institutional conflicts. Left Coast Press, Walnut Creek, California.

Sherman, K. P., J. Van Lanen, and R. T. Sherman. 2010. Practical environmentalism on the Pine Ridge Reservation: confronting structural constraints to Indigenous stewardship. Human Ecology 38:507–20.

Sherman, R. T. 1988. A study of traditional and informal sector micro-enterprise activity and its impact on the Pine Ridge Indian Reservation economy. Aspen Institute for Humanistic Studies, Washington, DC.

Standing Bear, L. 1988. My Indian boyhood. Bison Books, Lincoln, Nebraska.

Standing Bear, L. 2006. Land of the spotted eagle. Bison Books, Lincoln, Nebraska.

St. Pierre, M., and T. Long Soldier. 1995. Walking in the sacred manner: healers, dancers, and pipe carriers—medicine women of the plains. Simon and Schuster, New York.

Thunder Valley Community Development Corporation. 2018. Wakígnakapi: developing a food hub and grocery store for the Oglala Lakota Oyáte. Porcupine, South Dakota. In author's possession.

Walker, J. R. 1980. I was born a Lakota: Red Cloud's abdication speech, 4 July 1903. Pages 94 and 136–37 *in* Raymond J. DeMallie and Elaine A. Jahner, editors. Lakota belief and ritual. University of Nebraska Press, Lincoln.

Whitinui, P. 2014. Indigenous autoethnography: exploring, engaging, and experiencing "self" as a Native method of inquiry. Journal of Contemporary Ethnography 43(4):456–87.

Chapter 30

The Making and Unmaking of an Indigenous Desert Oasis and Its Avifauna

Historic Declines in Quitobaquito Birds as a Result in Shifts from O'odham Stewardship to Federal Agency Management

GARY PAUL NABHAN, LORRAINE MARQUEZ EILER, R. ROY JOHNSON, AMADEO REA, ERIC MELLINK, and LAWRENCE STEVENS

INTRODUCTION

Many wildlife managers have begun to accept the value of the Traditional Ecological Knowledge of First Nations in public education, policy, and history; far too few acknowledge that Indigenous peoples have actively managed species and their habitats in ways that increased population densities and species richness at the local or landscape scale. While many wildlife managers may accept passive "species preservation" or "habitat protection" as goals, some new to these fields may know little about landscapes where active management or stewardship had positive ecological value for wildlife or people.

This has been the case with the historic management of Quitobaquito Springs, a sacred oasis for the O'odham peoples of Arizona and Sonora, parts of which are now managed by the US National Park Service (NPS) as part of Organ Pipe National Monument, the US Customs and Border Protection, and the Mexican Commission for Protected Natural Areas as part of a UNESCO biosphere reserve for the Sierra del Pinacate and Gran Desierto de Altar, in Sonora. Although Piman-speaking O'odham lived, prayed, and subsisted at this oasis for at least 4,000 years, they were evicted from Quitobaquito by dubiously legal actions of NPS officials in the mid-1950s. Nevertheless, they retained rights afforded to them by President Franklin Roosevelt and subsequent legislation to harvest the fruits of saguaro and other columnar cacti. Nearly every NPS superintendent of Organ Pipe has facilitated access and, in some cases, road closure for other visitors when Hia c-ed O'odham and Tohono O'odham spiritual practitioners observe ceremonies associated with a cemetery at the oasis.

It has been the documented opinion of the NPS that there was no "open water" reservoir at the oasis until after 1956, when NPS administrators removed most remnants of the O'odham village and traditional water control structures, and bulldozed hundreds of years of pre-Invasion archaeology. This pretense has allowed some managers to claim that the oasis and its post-1950s reservoir are little more than "artificial" constructions.

The oral histories of the Hia c-ed O'odham handed down through Lorraine Marquez Eiler testify to the long record of diversity and abundance of birds and other wildlife at the site. Eiler and her cousin Marlene Vasquez visited the springs during their youth and remember stories handed down from their great-grandfather, who resided there.

In this chapter, we evaluate and contest some of the assumptions made by National Park Service wildlife managers that desert dwellers were technologically incapable of constructing and managing open water reservoirs in this desert landscape; that they were incapable of positively affecting wildlife population density and species richness; and that the spiritual and subsistence traditions of the O'odham that guided the stewardship of this and other desert oases added no value to the long-term conservation and management of water, plant cover, and wildlife.

These issues have consequences for the conservation of global biodiversity and in the Aboriginal homelands of First Nations that have been taken away from them by governments or private landowners. According to conservation biologists who surveyed spring-fed oases from the Baja California peninsula to the Grand Canyon, "Desert oases are among the most threatened ecosystems on earth. . . . Given the small land area that these spring-fed ecosystems occupy, their importance as quality habitat, migratory rest-stops, biodiversity, [culturally used] biodiversity, and economic value in arid regions is largely underestimated" (De Grenade and Stevens 2019:45).

Overall, the Quitobaquito Springs biocultural landscape fits well with the following description of spring-fed desert oases offered by Cartwright et al. (2020:245): "Natural springs in water-limited landscapes are biodiversity hotspots and keystone ecosystems that have a disproportionate influence on surrounding landscapes despite their usually small size." The question we are examining here is whether the now-imperiled Quitobaquito and other desert oases at risk were merely biodiverse because of the availability of water, or because of the time-tried management practices of their Indigenous occupants and stewards.

STUDY SITE

The Quitobaquito Springs habitat complex is located both in Pima County, Arizona, and in the Municipio de Plutarco Elías Calles, Sonora, at an elevation of 351 m. The 1853 border delineation follows the Gadsden Purchase, which essentially bisected an ancient Hia c-ed O'odham settlement. We focus on the avifaunal changes on the 5 ha of the habitat complex immediately north of the US-Mexico border where most bird surveyors have concentrated their observations. We urge archaeologists from the Instituto Nacional de Antropología e Historia in Mexico to make a complementary reconstruction of the historic and prehistoric cultural resources of Quitobaquito in Sonora.

The habitat complex lies in Organ Pipe Cactus National Monument, is administered by the US National Park Service, and contains a diverse mosaic of habitats, including open water and wetlands, hydroriparian, and mesoriparian habitats (Stevens et al. 2021).

Ecological interactivity at Quitobaquito Springs is amplified by the lack of other nearby sources of perennial surface water, a hyperaridity that increases subsidy exchange between the springs and surrounding non-springs ecosystems (Stevens 2020). As a result, Quitobaquito Springs provides critical stopover habitat for Nearctic-Neotropical migratory birds (Nabhan 2004). Quitobaquito Springs also supports endemic aquatic species, including the endangered Quitobaquito desert pupfish (*Cyprinodon eremus*), the rare Sonoyta mud turtle (*Kinosternon sonoriense longifemorale*), Father Kino's tryonia springsnail (*Tryonia quitobaquitae*), and a diverse array of dragonflies and other invertebrates (Kingsley 1998).

This federally designated "traditional cultural property" has played and continues to play a critical role in the lives of Indigenous desert peoples, especially the Hia c-ed O'odham, who retain Traditional Ecological Knowledge about its birds and plants (Nabhan 1982).

30.1. View of Quitobaquito Pond looking toward the west, August 1957, NATIONAL PARK SERVICE; PHOTO BY J. M. EDEN

Spring-fed oases can be highly vulnerable to the vagaries of Indigenous human stewardship or the cessation of them by external forces (Nabhan 2003). In this chapter, we document and analyze historic changes in the Quitobaquito avifauna with respect to changes in the management of this habitat mosaic and the supporting aquifer over the past eight decades. The 2019–2020 border wall construction between Arizona and Sonora initiated a new episode of ecological disruptions, with potentially harmful impacts on the avifauna and the springs ecosystem (Repanshek 2020). These disruptions include

- the overdraft of groundwater from the aquifer that supplies water to the springs;
- noise and habitat destruction from dynamiting in the hills above the springs to provide access to wall constructors;
- blading vegetation within an 18 m strip along the wall;
- installing 24/7 lighting within 80 m of the pond and springs; and
- increased vehicular traffic and human presence.

THE "MAKING" OF AN INDIGENOUS DESERT OASIS

The Spanish names Quitobaquito or Quitovaquita were not commonly used by the O'odham nor are they directly derived from the O'odham name (except for one lexeme, *vaak*, "place where water enters"; Winters 2012). In *O'odham Place Names* (2012), Harry Winters Jr. notes that 'A'al Vaupia is the older O'odham name for Quitobaquito. Some Tohono O'odham also call it 'A'al Vaipia or other variants (e.g., A'al Waipia) in different dialects of O'odham *ha-neoki*. These variants all mean Little Wells, but the term "wells" needs further clarification.

It appears a Spaniard thought that by calling it Quitovaquita, meaning Little Quitovac, it could be contrasted with a similar oasis village about 50 km to the southeast. At both of these oases, O'odham habitat stewards developed ponds with open water and aquatic plant cover fed by several artesian springs (Nabhan 1982, 2003, Felger et al. 1992, Winters 2012).

Winters (2012) quotes the Spanish descriptions of this site from the journals of the Jesuit Eusebio Kino and the Spanish captain Diego Carrasco as they saw it for the first time in October 1698, when Kino named it San Sergio (which Carrasco wrote as San Serjio). As Jesuit historian Burrus (1971:89) translates that passage, "It has water which runs in many places, *carrizales* [canebrakes of *Pluchea* spp. or *Phragmites* spp.], *tulares* [semiaquatic vegetation *Typha* spp. and/or *Scirpus* spp.], and ducks and birds from marshes, and excellent pasturage [grasses such as *Distichlis* spp., *Sporobolus* spp., *Muhlenbergia* spp.] for cattle."

This description indicates that there was already enough open water to harbor ducks and other waterfowl, enough submergent vegetation to harbor marsh birds, enough seepage to support arrowweed (*Pluchea sericea*) or native cane, and enough freshwater to support human needs and those of livestock and wildlife. In the 20th century, geographer Carl Lumholtz (1912) noted that the "springs or wells" of 'A'al Vaupia had supported irrigation agriculture and a permanent settlement of Hia c-ed O'odham. Harvard geohydrologist Kirk Bryan (1925) added further detail by noting that two spring openings showed evidence of excavation by the O'odham residents to "increase and concentrate the flow" in reservoir(s) that allowed irrigation of an alfalfa field.

Winters (2012) makes a key point regarding the original O'odham name of the habitat complex, 'A'al Vaupia: "where the water table is not too deep, a hand-dug *vavhia* might be roughly circular in cross-section, vertical, and with a diameter of only a few feet, and penetrate down to a little ways below the water table." With respect to the Pozos de Coyote (Coyote Wells) on the ocean side of the nearby Pinacate, which the Hia c-ed O'odham also frequented, sandy seeps were excavated initially by coyotes, horses, or passing travelers and were gradually amplified into (three 30-yard-wide) open water reservoirs downstream from the springs to hold spring water.

This remains the case at 'A'al Vaupia's sister oasis in Sonora, Quitovac, where there are "five springs . . . and a large pond that [is] a sink for the washes running into it" (Winters 2012). The etymology he describes fits both oases: "in place names, Vaak means a 'sink (in a wash or river)'" or "a place where flowing water stops at a pond and sinks under the ground."

The excavations of the O'odham created watered oases, which were occasionally deepened first by the use of rawhide scoops, then by mechanized metal scoopers invented in 1883 called "Fresno scrapers" in English, *fresnos* in Spanish, and *i:takud* in O'odham. It is erroneous to claim that the Hia c-ed O'odham were incapable of excavating seeps into elaborate wells, deepening hand-dug ditches to make water-delivery canals, or redesigning puddles in low-lying places in watercourses to make them into deeper reservoirs. Once they accomplished these tasks, they created elaborate networks of hundreds of linear meters of acequias or *vaikka* to irrigate fields or orchards, which can still be seen at Quitobaquito more than 70 years after they were initially abandoned.

These forms of spring water concentration, gravity-fed transport, and spreading onto croplands created additional seeps and open water habitats that attracted myriad forms of plants and wildlife. Alkaline grass flats, *carrizales*, *tulares*, riparian bosques, and open waters all supported diverse flora and fauna. The O'odham not only created or enhanced such habitats, their stewardship maintained them over at least three centuries. To call their biocultural landscape either a "natural" or an "artificial" oasis is to succumb to a dualism that fails to comprehend the unique interactions among humans and other species in a place that one O'odham elder described as being "where the birds are our friends" (Nabhan 1982).

THE SHAPING OF AN INDIGENOUS OASIS AVIFAUNA: 1938–1957

The first professionally recorded bird observations at Quitobaquito Springs were made by Laurence M. Huey of the San Diego Natural History Museum in 1939, two years after the establishment of Organ Pipe Cactus National Monument. This was at a time when Hia c-ed O'odham families still lived, drank, irrigated, farmed, foraged, hunted, and prayed at this sacred site, even though it had come under the administrative management of the NPS. Huey's (1942) generalized vertebrate surveys and Hensley's (1954) breeding bird surveys were conducted when Hia c-ed O'odham farmers and foragers still actively managed the habitat complex, and these two studies form a reliable baseline documentation of the birds while the site was still influenced by the TEK of the O'odham. In table 30.1, we characterize the oasis habitats of each time period and correlate them with the bird community assemblage of the same period.

During the period 1938–1957, the initial years after the establishment of the Organ Pipe National Monument, Hia c-ed O'odham farmers, foragers, and stockmen were still in residence and actively managed spring flows for orchards, fields, and livestock watering on both sides of the US-Mexico border. Gould (1938) measured Quitobaquito Springs discharge at 2.7 liters per second (L/s), as reported by Cole and Whiteside (1965). Twenty-three of the bird species documented at Quitobaquito Springs by Huey (1942) and Hensley (1954) during this period were absent during the years of border wall construction.

Reichhardt et al. (1986) inferred from these early observations that the following species of birds occupied bordering riparian habitats, orchards, and fields in the Sonoran Desert: crissal thrasher (*Toxostoma crissale*), northern cardinal (*Cardinalis cardinalis*), pyrrhuloxia (*Cardinalis sinuatus*), and Abert's towhee (*Pipilo aberti*). In addition, Reichhardt et al. (1986) identified the following species occurring in irrigated fields and orchards adjacent to the open surface water of irrigation reservoirs: eared grebe (*Podiceps nigricollis*), great egret (*Ardea alba*), snowy egret (*Egretta thula*), black-crowned night heron (*Nycticorax nycticorax*), white-faced ibis (*Plegadis chihi*), northern pintail (*Anas acuta*), blue-winged teal (*Anas discors*), American wigeon (*Mareca americana*), northern shoveler (*Spatula clypeata*), bufflehead (*Bucephala albeola*), sora (*Porzana carolina*), violet-green swallow (*Tachycineta thalassina*), and tree swallow (*Tachycineta bicolor*). During this same period, the Indigenous settlement at Quitobaquito Springs harbored crissal thrasher, northern cardinal, pyrrhuloxia, black-crowned night heron, white-faced ibis, northern pintail, blue-winged teal, American wigeon, northern shoveler, sora, violet-green swallow, belted kingfisher (*Megaceryle alcyon*), and tree swallow.

Curiously, while Inca dove (*Columbina inca*) was reported during this period, other birds that commonly occur in anthropogenic habitats were not found in the O'odham settlement at Quitobaquito Springs, including Gambel's quail (*Callipepla gambelii*), house sparrow (*Passer domesticus*), and great-tailed grackle (*Quiscalus mexicanus*). Because the pre-NPS reservoirs of the O'odham were rather shallow, American coot (*Fulica americana*) may have been rare or absent during this period (Cole and Whiteside 1965).

The management of water for irrigation supported populations of sora, western sandpiper (*Calidris mauri*), solitary sandpiper (*Tringa solitaria*), greater yellowlegs (*Tringa melanoleuca*), and great blue heron (*Ardea herodias*), which were not found during border wall construction. Birds of fencerows and irrigated orchards, such as black-headed grosbeak (*Pheuticus melanocephalus*), blue grosbeak (*Passerina caerulea*), and lazuli bunting (*Passerina amoena*), were present during Indigenous occupation but not recorded during the years of border wall construction.

By October 1956, the NPS dubiously "acquired" Quitobaquito and its surroundings for a few thousand dollars from a single descendant of a deceased medicine man (*makai*), without consulting other family members or what is now known as the Tohono O'odham Nation Legislative Council. During that month, Natt N. Dodge and Jim Eden photographed the oasis (NPS n.d.:files 13, 518). Buildings and livestock corrals were still intact, and irrigation ditches and ponds excavated by the O'odham were still viable, with emergent aquatic plants such as alkali bullrush (*Scirpus olneyi*) and giant cottonwood trees (*Populus fremontii*). Within two years, much of what was documented in the Dodge and Eden photos had been removed by NPS staff or had degraded.

TABLE 30.1. Riparian and Wetland Birds Recorded by Several Researchers at Quitobaquito Springs

RIPARIAN AND WETLAND BIRDS RECORDED BY GROSCHUPF ET AL. (1988) AND TIBBITTS AND DICKSON (2005)	SCIENTIFIC NAME	O'ODHAM NAME	CULTURAL USES	CONSERVATION STATUS	1939 SURVEY (HUEY 1942) AND 1948–1949 SURVEY (HENSLEY 1954)	1980–1982 (NABHAN ET AL. 1982)	JOHNSON ET AL. (1983)	EBIRD 2016–2019	EBIRD 2019–2020 (FINK ET AL. 2020)
Wood duck	*Aix sponsa*	vaka'ig		SCC					
Blue-winged teal	*Spatula discors* (*Anas discors*)	vaka'ig			X	X		X	
Cinnamon teal	*Spatula cyanoptera*	vaka'ig			X	X		X	X
Northern shoveler	*Spatula clypeata*	vaka'ig			X			X	
Gadwall	*Mareca strepera*	vaka'ig						X	
American wigeon	*Mareca americana*	vaka'ig			X				X
Mallard/Mexican duck	*Anas platyrhynchos*	vaka'ig						X	
Northern pintail	*Anas acuta*	vaka'ig		SCC	X			X	
Green-winged teal	*Anas crecca*	vaka'ig			X	X			X
Canvasback	*Aythya valisineria*	vaka'ig							

TABLE 30.1. (continued)

RIPARIAN AND WETLAND BIRDS RECORDED BY GROSCHUPF ET AL. (1988) AND TIBBITTS AND DICKSON (2005)	SCIENTIFIC NAME	O'ODHAM NAME	CULTURAL USES	CONSERVATION STATUS	1939 SURVEY (HUEY 1942) AND 1948–1949 SURVEY (HENSLEY 1954)	1980–1982 (NABHAN ET AL. 1982)	JOHNSON ET AL. (1983)	EBIRD 2016–2019	EBIRD 2019–2020 (FINK ET AL. 2020)
Redhead	*Aythya americana*	*vaka'ig*			X				X
Ring-necked duck	*Aythya collaris*	*vaka'ig*						X	
Lesser scaup	*Aythya affinis*	*vaka'ig*						X	
Bufflehead	*Bucephala albeola*	*vaka'ig*							
Common goldeneye	*Bucephala clangula*	*vaka'ig*							
Hooded merganser	*Lophodytes cucullatus*	*vaka'ig*						X	
Common merganser	*Mergus merganser*	*vaka'ig*							
Red-breasted merganser	*Mergus serrator*	*vaka'ig*							
Ruddy duck	*Oxyura jamaicensis*	*vaka'ig*						X	
Grebes									
Least grebe	*Tachybaptus dominicus*				X				X
Pied billed grebe	*Podilymbus podiceps*					X		X	X
Eared grebe	*Podiceps nigricollis*								
Western grebe	*Aechmophorus occidentalis*								
Doves									
Inca dove	*Columbina inca*	*gúgu*			X			X	
White-winged dove	*Zenaida asiatica*	*okokoi*	T, F		X	X	X	X	X
Mourning dove	*Zenaida macroura*	*ho:hi*	F		X	X	X	X	X
Swifts									
Vaux's swift	*Chaetura vauxi*				X			X	
White-throated swift	*Aeronautes saxatalis*				X	X		X	X
Hummingbirds									
Costa's hummingbird	*Calypte costae*	*wipismal*	T	SCC		X	X	X	X
Rufous hummingbird	*Selasphorus rufus*	*wipismal*	T	SCC		X		X	
Allen's hummingbird	*Selasphorus sasin*	*wipismal*	T			X			
Broad-billed hummingbird	*Cynanthus latirostris*	*wipismal*	T	SCC			X	X	

(continued)

TABLE 30.1. (continued)

RIPARIAN AND WETLAND BIRDS RECORDED BY GROSCHUPF ET AL. (1988) AND TIBBITTS AND DICKSON (2005)	SCIENTIFIC NAME	O'ODHAM NAME	CULTURAL USES	CONSERVATION STATUS	1939 SURVEY (HUEY 1942) AND 1948–1949 SURVEY (HENSLEY 1954)	1980–1982 (NABHAN ET AL. 1982)	JOHNSON ET AL. (1983)	EBIRD 2016–2019	EBIRD 2019–2020 (FINK ET AL. 2020)
Rails, Gallinules, Coots									
Virginia rail	Rallus limicola								
Sora	Porzana carolina				X				
Common gallinule	Gallinula galeata								
American coot	Fulica americana	vachpik	F		X	X		X	X
Black rail	Laterallus jamaicensis			SCC					
Stilts, Avocets, Plovers									
Black-necked stilt	Himantopus mexicanus	chevel u'uwhig			X				X
American avocet	Recurvirostra americana	haiñ uushab						X	X
Killdeer	Charadrius vociferus	chivi-chu:ch							
Sandpipers, Phalarpopes, and Allies									
Least sandpiper	Calidris minutilla	shu:dagi mamad				X			X
Western sandpiper	Calidris mauri	shu:dagi mamad							
Long-billed dowitcher	Limnodromus scolopaceus	shu:dagi mamad							
Wilson's snipe	Gallinago delicata	shu:dagi mamad						X	
Spotted sandpiper	Actitis macularius	shu:dagi mamad			X	X	X	X	X
Solitary sandpiper	Tringa solitaria	shu:dagi mamad						X	
Lesser yellowlegs	Tringa flavipes	shu:dagi mamad							
Willet	Tringa semipalmata	shu:dagi mamad							X
Greater yellowlegs	Tringa melanoleuca	shu:dagi mamad			X			X	
Wilson's phalarope	Phalaropus tricolor	shu:dagi mamad							
Red phalarope	Phalaropus fulicarius	shu:dagi mamad							

TABLE 30.1. (continued)

RIPARIAN AND WETLAND BIRDS RECORDED BY GROSCHUPF ET AL. (1988) AND TIBBITTS AND DICKSON (2005)	SCIENTIFIC NAME	O'ODHAM NAME	CULTURAL USES	CONSERVATION STATUS	1939 SURVEY (HUEY 1942) AND 1948–1949 SURVEY (HENSLEY 1954)	1980–1982 (NABHAN ET AL. 1982)	JOHNSON ET AL. (1983)	EBIRD 2016–2019	EBIRD 2019–2020 (FINK ET AL. 2020)
Gulls, Terns									
Heermann's gull	*Larus heermanni*	ka:chk ba:sho u'uhig							
Forster's tern	*Sterna forsteri*	ka:chk ba:sho u'uhig							
Least tern	*Sternula antillarum*	ka:chk ba:sho u'uhig							
Black tern	*Chlidonias niger*	ka:chk ba:sho u'uhig							
Storks, Cormorants, Pelicans									
Wood stork	*Mycteria americana*								
Brown pelican	*Pelecanus occidentalis*	chu'aggiakam vakoañ		SCC					
Double-crested cormorant	*Nannopterum auritum*								X
Bitterns, Herons, Ibises									
Least bittern	*Ixobrychus exilis*	vakoañ							
Great blue heron	*Ardea herodias*	ko:magi vakoañ			X	X		X	
Great egret	*Ardea alba*	toha vakoañ							X
Snowy egret	*Egretta thula*	toha vakoañ		SCC					X
Cattle egret	*Bubulcus ibis*	toha che:nam vakoañ						X	
Green heron	*Butorides virescens*	tash che:nam vakoañ				X			X
Black-crowned night-heron	*Nycticorax nycticorax*	vakoañ			X			X	
White-faced ibis	*Plegadis chihi*			SCC	X				X
Hawks									
Osprey	*Pandion haliaetus*	vakoañ ba'ag		SCC				X	X
Sharp-shinned hawk	*Accipiter striatus*				X		X	X	
Cooper's hawk	*Accipiter cooperii*				X	X		X	X

(continued)

TABLE 30.1. (continued)

RIPARIAN AND WETLAND BIRDS RECORDED BY GROSCHUPF ET AL. (1988) AND TIBBITTS AND DICKSON (2005)	SCIENTIFIC NAME	O'ODHAM NAME	CULTURAL USES	CONSERVATION STATUS	1939 SURVEY (HUEY 1942) AND 1948–1949 SURVEY (HENSLEY 1954)	1980–1982 (NABHAN ET AL. 1982)	JOHNSON ET AL. (1983)	EBIRD 2016–2019	EBIRD 2019–2020 (FINK ET AL. 2020)
Common black hawk	*Buteogallus anthracinus*								
Harris's hawk	*Parabuteo unicinctus*	tobavi							X
Swainson's hawk	*Buteo swainsoni*							X	
Red-tailed hawk	*Buteo jamaicensis*	haupal			X	X		X	X
Kingfisher									
Belted kingfisher	*Megaceryle alcyon*	ba'ivchul		SCC	X				
Woodpeckers									
Gila woodpecker	*Melanerpes uropygialis*	hikvig		SCC	X		X		
Gilded flicker	*Colaptes chrysoides*	kudat		SCC			X		
Caracaras, Falcons									
Crested caracara	*Caracara plancus*	kusijim		SCC					X
American kestrel	*Falco sparverius*	sisiki				X		X	X
Merlin	*Falco columbarius*							X	
Flycatcher									
Ash-throated flycatcher	*Myiarchus cinerascens*		T					X	
Brown-crested flycatcher	*Myiarchus tyrannulus*							X	X
Tropical kingbird	*Tyrannus melancholicus*	wa'akek							
Cassin's kingbird	*Tyrannus vociferans*	wa'akek							
Thick-billed kingbird	*Tyrannus crassirostris*	wa'akek		SCC					
Western kingbird	*Contopus verticalis*	chukukmal			X	X	X		X
Western wood pewee	*Contopus sordidulus*					X	X	X	X
Southwestern willow flycatcher	*Empidonax traillii*			FE, SCC		X	X	X	
Gray flycatcher	*Empidonax wrightii*							X	
Black phoebe	*Sayornis nigricans*				X	X	X	X	X
Eastern phoebe	*Sayornis phoebe*								
Say's phoebe	*Sayornis saya*	hewel mo:s	T					X	X
Vermilion flycatcher	*Pyrocephalus rubinus*				X	X	X	X	X

TABLE 30.1. (continued)

RIPARIAN AND WETLAND BIRDS RECORDED BY GROSCHUPF ET AL. (1988) AND TIBBITTS AND DICKSON (2005)	SCIENTIFIC NAME	O'ODHAM NAME	CULTURAL USES	CONSERVATION STATUS	1939 SURVEY (HUEY 1942) AND 1948–1949 SURVEY (HENSLEY 1954)	1980–1982 (NABHAN ET AL. 1982)	JOHNSON ET AL. (1983)	EBIRD 2016–2019	EBIRD 2019–2020 (FINK ET AL. 2020)
Vireos									
Bell's vireo	*Vireo bellii*			SCC		X	X	X	X
Hutton's vireo	*Vireo huttoni*							X	
Warbling vireo	*Vireo gilvus*					X	X	X	X
Swallows, Verdin									
Bank swallow	*Riparia riparia*	gi:dval						X	
Tree swallow	*Tachycineta bicolor*	gi:dval			X	X		X	X
Violet-green swallow	*Tachycineta thalassina*	gi:dval			X	X		X	X
Northern rough-winged swallow	*Stelgidopteryx serripennis*	gi:dval			X	X		X	
(Desert) Purple martin	*Progne subis*	gi:dval		SCC	X			X	X
Cliff swallow	*Petrochelidon pyrrhonota*	gi:dval			X	X		X	X
Verdin	*Auriparus flaviceps*	gi:sobi	T		X	X	X	X	X
Wrens									
House wren	*Troglodytes aedon*					X	X	X	X
Marsh wren	*Cistothorus palustris*					X		X	X
Bewick's wren	*Thyromanes bewickii*							X	X
Kinglet, Gnatcatcher									
Black-tailed gnatcatcher	*Polioptila melanura*	s-cuk mo'okam gi:sobi				X	X	X	X
Ruby-crowned kinglet	*Corthylio calendula*								
Thrashers									
Curve-billed thrasher	*Toxostoma curvirostre*	kudvich						X	
Bendire's thrasher	*Toxostoma bendirei*	biit keishnam							X
Crissal thrasher	*Toxostoma crissale*	chev chiñkam				X		X	X
Silky Flycatcher									
Phainopepla	*Phainopepla nitens*	kuigam			X	X	X	X	X

(continued)

TABLE 30.1. (continued)

RIPARIAN AND WETLAND BIRDS RECORDED BY GROSCHUPF ET AL. (1988) AND TIBBITTS AND DICKSON (2005)	SCIENTIFIC NAME	O'ODHAM NAME	CULTURAL USES	CONSERVATION STATUS	1939 SURVEY (HUEY 1942) AND 1948–1949 SURVEY (HENSLEY 1954)	1980–1982 (NABHAN ET AL. 1982)	JOHNSON ET AL. (1983)	EBIRD 2016–2019	EBIRD 2019–2020 (FINK ET AL. 2020)
Finches, Sparrows									
House finch	*Haemorhous mexicanus*	bahidag u'uhig			X	X	X	X	X
Purple finch	*Haemorhous purpureus*								
Pine siskin	*Spinus pinus*								
Lesser goldfinch	*Spinus psaltria*	o'am u'uhig(?)			X	X	X	X	X
Rufous-winged sparrow	*Peucaea carpalis*								
Black-throated sparrow	*Amphispiza bilineata*								
Lark sparrow	*Chondestes grammacus*				X	X	X	X	X
Golden-crowned sparrow	*Zonotrichia atricapilla*								
White-throated sparrow	*Zonotrichia albicollis*								
Song sparrow	*Melospiza melodia*								
Lincoln's sparrow	*Melospiza lincolnii*							X	
Swamp sparrow	*Melospiza georgiana*								
Chat, Blackbirds, Orioles, Cowbirds									
Yellow-breasted chat	*Icteria virens*	o'am u'uhig(?)		SCC		X	X	X	
Yellow-headed blackbird	*Icterus xanthocephalus*	shashañ			X	X			X
Hooded oriole	*Icterus cucullatus*	wakokam			X	X	X	X	
Bullock's oriole	*Icterus bullockii*	wakokam			X	X	X	X	X
Red-winged blackbird	*Agelaius phoeniceus*	shashañ			X	X	X	X	
Bronzed cowbird	*Molothrus aeneus*	shashañ				X	X	X	
Brown-headed cowbird	*Molothrus ater*	shashañ			X	X	X	X	X
Brewer's blackbird	*Euphagus cyanocephalus*	shashañ					X		
Great-tailed grackle	*Quiscalus mexicanus*	shashañ					X		
Warblers									
Lucy's warbler	*Leiothlypis luciae*			SCC		X	X	X	X
MacGillivray's warbler	*Geothlypis tolmiei*					X	X	X	
Common yellowthroat	*Geothlypis trichas*				X	X	X	X	X

TABLE 30.1. (continued)

RIPARIAN AND WETLAND BIRDS RECORDED BY GROSCHUPF ET AL. (1988) AND TIBBITTS AND DICKSON (2005)	SCIENTIFIC NAME	O'ODHAM NAME	CULTURAL USES	CONSERVATION STATUS	1939 SURVEY (HUEY 1942) AND 1948-1949 SURVEY (HENSLEY 1954)	1980-1982 (NABHAN ET AL. 1982)	JOHNSON ET AL. (1983)	EBIRD 2016-2019	EBIRD 2019-2020 (FINK ET AL. 2020)
American redstart	*Setophaga ruticilla*								
Yellow warbler	*Setophaga petechia*			SCC	X	X	X	X	X
Yellow-rumped warbler	*Setophaga coronata*				X	X	X	X	X
Grace's warbler	*Setophaga graciae*			SCC					
Wilson's warbler	*Cardellina pusilla*						X	X	X
Tanagers									
Summer tanager	*Piranga rubra*	he:t kawudk							
Scarlet tanager	*Piranga olivacea*								
Western tanager	*Piranga ludoviciana*						X		
Grosbeaks and Allies									
Northern cardinal	*Cardinalis cardinalis*	sipo:k			X	X		X	
Pyrrhuloxia	*Cardinalis sinuatus*	sipo:k				X		X	
Rose-breasted grosbeak	*Pheucticus ludovicianus*								
Black-headed grosbeak	*Pheuticus melanocephalus*						X		
Blue grosbeak	*Passerina caerulea*	hevachud			X		X	X	
Lazuli bunting	*Passerina amoena*				X		X	X	
Varied bunting	*Passerina versicolor*			SCC					

Cultural uses: F = food, T = taboos.
Conservation status: FE = federally endangered, SCC = species of conservation concern.

THE UNMAKING OF AN INDIGENOUS DESERT OASIS

1958–1975

This period began with the expulsion of Hia c-ed O'odham residents from Quitobaquito Springs. Buildings were razed, fields were abandoned, orchard trees were left unmanaged, the pond was deepened, and livestock were removed. Cole and Whiteside (1965) reported that spring discharge was greatly reduced in 1964, running only 0.85 L/s, while channels from the springs that ran 7.5 cm deep through June 1963 were running only 2 cm deep by May 1964. Deeply rooted pomegranates (*Punica* spp.), figs (*Ficus* spp.), and date palms (*Phoenix* spp.) persisted, offering shade, roosts, and food for both resident and migratory birds. Although there were no comprehensive bird surveys of Quitobaquito Springs during this period, great-tailed grackle was first reported there at this time, and American coot and unidentified grebes were common (Monson and Phillips 1964, Cole and Whiteside 1965).

Reichhardt et al. (1986) suggested that the cool, humid, shady microhabitats provided by three mesquite species (*Prosopis* spp.), graythorn (*Ziziphus obtusifolia*), wolfberry (*Lycium* spp.), ash (*Fraxinus velutina*), and fruit trees at that time would still have provided habitat for *Empidonax* flycatchers, Nashville warbler (*Leiothlypis ruficapilla*), yellow-throated warbler (*Setophaga dominica*), black-throated gray warbler (*Setophaga nigrescens*), blue grosbeak, lazuli bunting, pine siskin (*Spinus pinus*), savannah sparrow (*Passerculus sandwichensis*), Scott's oriole (*Icterus parisorum*), and Lincoln's sparrow (*Melospiza lincolnii*) (table 30.1). Rea's (1997) and Rea and Cassa's (2008) surveys of historic O'odham field and orchard complexes in the desert less than 100 km to the north of Quitobaquito on the Gila River document much of the same songbird diversity, especially the insectivorous birds.

1976–1998

In 1976, the designations of Organ Pipe as a new UNESCO biosphere reserve and of 95% of its lands as a wilderness ushered in a new era of scientific study, including ornithological surveys at Quitobaquito Springs. Thirty-three species of birds were documented that were not observed later during the period of border wall construction. Even though the Hia c-ed O'odham had been evicted from living at the site, their ceremonies and wild plant gathering continued with NPS acquiescence. Quitobaquito Springs was placed on the National Register of Historic Places in 1978.

More rigorous groundwater monitoring, nonnative plant control, and more scientifically rigorous field surveys began during this time. Carruth (1996) noted that Quitobaquito Springs discharge averaged 1.77 L/s between 1981 and 1992. Rainfall at two gauges in the recharge area northeast of Quitobaquito Hills averaged 168 mm/year during an 11-year monitoring program, which ended in September 1992.

Several birds of conservation significance were first observed at Quitobaquito Springs during this time, including the broad-billed hummingbird

(*Cynanthus latirostris*), southwestern willow flycatcher (*Empidonax traillii extimus*, federally endangered), Bell's vireo (*Vireo bellii*), and purple martin (*Progne subis*).

1999–2021

Quitobaquito sustained accelerated impacts from climate change and groundwater table decline, and those impacts were exacerbated by the border wall construction. Forty-three bird species recorded earlier in this period were not observed at Quitobaquito Springs during the first year of border wall construction, although both the COVID-19 pandemic and Puerto Blanco Drive border road closures limited survey access to the habitat complex.

Summarizing the data for springs flow rates from NPS and USGS monitoring stations, Main (2020) reported that the strongest spring remaining on the site flowed below 0.63 L/s for much of the period when groundwater was being pumped for the wall construction, nearly one-third less than the average discharge in the 1980s. Quitobaquito Springs discharge dipped to a low of 0.35 L/s in July 2020, which is even lower than the low discharge records documented in May 1964 (Cole and Whiteside 1965).

Reduction of discharge by at least 30% since March 2019 was a primary factor in the lowering of the Quitobaquito Pond by 30–45 cm, which left 75%–80% of the pond's bottom dry. Former Arizona state meteorologist Paul Brown (pers. comm., 2020) estimated that 2 m/year of evaporation occurred from the surface of the Quitobaquito Pond in 2020.

Some NPS, Customs and Border Protection, and the Army Corps of Engineers staff contest the idea that the border wall construction's use of 1,670,380 liters of groundwater per linear kilometer contributed to declines in springs discharge or to avifaunal assemblage changes, arguing that the pond levels began dropping in late December 2019 before wall construction began due to a leak in the pond lining, which has since been patched. Other NPS officials argue that the 2020 heat wave in Organ Pipe Cactus National Monument, with 151 days over 43°C, probably intensified "temporary" declines in the Quitobaquito Pond; the previous record was 116 days in 2018.

However, University of Arizona hydrologists and biologists are skeptical that the pond liner leakage fully explained the timing and volume of hydrological and ecological changes at Quitobaquito Springs (Main 2020). They observe that the groundwater source of the aquifer feeding Quitobaquito Springs lies entirely within Organ Pipe National Monument and its 20 m wide Roosevelt Reserve (Carruth 1996) where (1) wells used by the army corps for wall foundation construction and dust control operations were based, and (2) dynamiting along a fragile fault line in the Quitobaquito Hills adjacent to the springs occurred during wall construction and the period of rapid reduction in pond stage.

We believe it is most likely that wall construction has a prominent role in explaining the decline or absence of 47 bird species previously recorded at Quitobaquito.

CONCLUSIONS

Johnson et al. (in press) document 210 bird species at the site; 110 of them are "oasis" obligate or preferential birds, including 4 federally listed species of concern and 25 species of conservation concern. Given different survey methodologies, our conclusions must remain tentative. Nevertheless, it is clear that changes in the avifauna between the period of O'odham Indigenous management of wildlife habitat and later NPS management were larger than shifts recorded between any of the other periods when Quitobaquito was under NPS control. In particular, waterbird species richness decreased considerably between the period of O'odham occupation and the 1980–1983 surveys under NPS supervision.

Waterfowl were especially affected, and three marsh bird species, two shorebird species, and two wading bird species disappeared from the site. It appears that species richness of waterbirds was enhanced both by O'odham management prior to 1958 and by ecologically oriented park management immediately following 1980–1983. It may have dipped between the late 1980s and 2010, but it measurably improved in late 2010. Waterbirds decreased again during the wall construction, but shorebirds increased, perhaps due to the larger areas available as mudflats as opposed to open water habitat.

Forty-seven bird species were recorded in the 3.5 years prior to the onset of wall construction that were not detected after, and 12 species were detected during wall construction that had not been there just prior to it. Of the 47 species, 38 (81%) required or heavily utilized oasis habitats because of their dependence on wetland, hydroriparian, and mesoriparian habitats for some portion of their life cycle (e.g., migration, nesting; table 30.2).

TABLE 30.2. List of Oasis Birds That Are Obligate or Preferential Wetland, Hydroriparian, and Mesoriparian Species

Blue-winged teal	Northern shoveler	Lesser scaup
Northern pintail	Hooded merganser	Gambel's quail
Ring-necked duck	Ruddy duck	Rufous hummingbird
Inca dove	Vaux's swift	Solitary sandpiper
Broad-billed hummingbird	Calliope hummingbird	Cattle egret
Greater yellowlegs	Great blue heron	Sharp-shinned hawk
Black-crowned night heron	Swainson's hawk	Willow flycatcher*
Zone-tailed hawk	Yellow-breasted chat	Hooded oriole
Gray flycatcher	Great-tailed grackle	Bronzed cowbird
Red-winged blackbird	Curve-billed thrasher	MacGillivray's warbler
Swainson's thrush	Pyrrhuloxia	Summer tanager
Northern cardinal	Blue grosbeak	Lazuli bunting
Black-headed grosbeak	Gadwall	

Note: These species were recorded during nesting or migration at Quitobaquito in the 3.5 years prior to the September 2019 onset of border wall construction but not recorded there from construction through 2021. Birds are listed in taxonomic order. See table 30.1 for scientific and O'odham names.

*Federally endangered (subspecies *traillii*).

At least five factors, alone or in combination, may account for the putative decline in avian species richness: (1) limited or biased sampling that failed to detect species that were actually present; (2) hotter, drier weather leading to a loss of habitat; (3) interannual variation in the arrival and departure of migratory birds at particular stopover habitats; (4) long-term declines in spring discharge leading to loss of habitat, pond size, and vegetation; and (5) border wall construction impacts, including increased heavy equipment traffic, fugitive dust from bulldozed areas and traffic, noise and habitat destruction from dynamiting, destruction of native vegetation, groundwater overdraft that reduced springs discharge and pond stage, and die-off of riparian trees and wetland vegetation. Although direct impacts of wall construction rapidly and substantially altered this oasis ecosystem, significant changes in both habitats and regional species abundance predate initiation of that construction. Above, we grouped those changes into four time periods to be able to more easily detect them.

Since President Joseph Biden halted wall construction in January 2021, the 60 m by 18 m pond basin at Quitobaquito has refilled to cover the entire surface of its area to a depth of 32–50 cm, far less than its average pre-wall depth. Continuous lighting within 50 m of the pond edge and the wall itself continue to disrupt the nocturnal environment. The NPS and the International Sonoran Desert Alliance have responded to the requests of the O'odham descendants of Quitobaquito settlers and to 30 prominent desert scientists by raising over $150,000 for ecohydrological restoration of Quitobaquito Springs.

There is evidence in the NPS files of legal irregularities in the transfer of Quitobaquito Springs to NPS stewardship with the consent (and payment) of only one of the many O'odham descendants and part-time dwellers tied to the oasis community (Nabhan 2003). It is urgent that the NPS and the Indigenous nation historically affiliated with this unique landscape collaborate further in the management of this traditional cultural property.

LITERATURE CITED

Bryan, K. 1925. The Papago country, Arizona. US Geological Survey, Water Supply Paper 499. US Government Printing Office, Washington, DC.

Burrus, E. J. 1971. Kino and Manje, explorers of Sonora and Arizona: their vision of the future, a study of their expeditions and plans. Jesuit History Institute, St. Louis University, Missouri.

Carruth, R. L. 1996. Hydrogeology of the Quitobaquito Springs and la Abra Plain area, Organ Pipe Cactus National Monument, Arizona, and Sonora, Mexico. US Geological Survey, Water-Resources Investigations Report 95-4295, Tucson, Arizona.

Cartwright, J., K. Dwire, Z. Freed, S. Hammer, B. McLaughlin, L. Misztal, E. Schenk, J. Spence, A. Springer, and L. Stevens. 2020. Oases of the future? springs as potential hydrologic refugia in drying climates. Frontiers in Ecology and the Environment 18:245–53.

Cole, G., and M. Whiteside. 1965. An ecological reconnaissance of Quitobaquito Spring, Arizona. Arizona Academy of Sciences 3:159–63.

De Grenade, R., and L. E. Stevens. 2019. Desert oasis springs: ecohydrology, biocultural diversity, mythology, and societal implications. Encyclopedia of the World's Biomes 36–46.

Felger, R. S., P. L. Warren, S. L. Anderson, and G. P. Nabhan. 1992. Vascular plants of a desert oasis: flora and ethnobotany of Quitobaquito, Organ Pipe Cactus National Monument, Arizona. Proceedings of the San Diego Society of Natural History 8:1–44.

Fink, D., T. Auer, A. Johnston, M. Strimas-Mackey, O. Robinson, S. Ligocki, B. Petersen, I. Wood, C. Davies, B. Sullivan, M. Iliff, and S. Kelling. 2020. eBird status and trends, data version: 2018, released 2020. Cornell Lab of Ornithology, Ithaca, New York. https://doi.org/10.2173/ebirdst.2018.

Gould, C. N., ed. 1938. Geology of Organ Pipe Cactus National Monument. Pages 455–61 in Southwestern monuments monthly report, supplement for June. US National Park Service Southwest Regional Center, Santa Fe, New Mexico.

Groschupf, K. D., B. T. Brown, and R. R. Johnson. 1988. An annotated checklist of the birds of Organ Pipe Cactus National Monument, Arizona. Southwest Parks and Monuments Association, Tucson, Arizona.

Hensley, M. M. 1954. Ecological relationships of the breeding bird population of the desert biome in Arizona. Ecological Monographs 24:185–208.

Huey, L. M. 1942. A vertebrate faunal survey of the Organ Pipe Cactus National Monument, Arizona. Transactions of the San Diego Society of Natural History 32:353–76.

Johnson, R. R., B. T. Brown, and S. Goldwasser. 1983. Avian use of Quitobaquito Springs oasis, Organ Pipe Cactus National Monument, Arizona. Quitobaquito Science Series No. 1. Cooperative National Park Resources Studies Unit, University of Arizona, Tucson.

Kingsley, K. J. 1998. Invertebrates of Organ Pipe National Monument. Technical Report No. 60, University of Arizona Cooperative Ecosystem Studies Unit and Organ Pipe National Monument, Tucson, Arizona.

Lumholtz, C. 1912. New trails in Mexico. Charles Scribner's Sons, New York.

Main, D. 2020. Sacred Arizona spring drying up as border wall construction continues. National Geographic, 20 July. https://www.nationalgeographic.com/science/article/quitobaquito-springs-arizona-drying-up-border-wall. Accessed 5 June 2022.

Monson, G., and A. R. Phillips. 1964. An annotated checklist of the species of birds in Arizona. Part 4 in C. Lowe, editor. The vertebrates of Arizona. University of Arizona Press, Tucson.

Nabhan, G.P. 1982. Papago fields: agroecology and ethnobiology. PhD dissertation, University of Arizona, Tucson.

Nabhan, G. P. 2002. The desert smells like rain: a naturalist in O'odham country. University of Arizona Press, Tucson.

Nabhan, G. P. 2003. Destruction of an ancient Indigenous cultural landscape: an epitaph from Organ Pipe Cactus National Monument. Ecological Restoration 21:290–95.

Nabhan, G. P. 2004. Stresses on pollinators during migration: is nectar availability at stopovers a weak link in plant-pollinator conservation. In G. P. Nabhan, editor. Conservation of migratory pollinators and their nectar corridors in North America, 4–23. University of Arizona Press, Tucson.

Nabhan, G. P., A. M. Rea, K. L. Reichhardt, E. Mellink, and C. F. Hutchinson. 1982. Papago influences on habitat and biotic diversity: Quitovac Oasis ethnoecology. Journal of Ethnobiology 2:124–43.

National Park Service [NPS]. n.d. Western Archaeological Center Special Collections. Tucson, Arizona.

Rea, A. M. 1997. Once a river: bird life and habitat changes on the middle Gila. University of Arizona Press, Tucson.

Rea, A. M., and C. Cassa. 2008. Wings in the desert: a folk ornithology of the northern Pimans. University of Arizona Press, Tucson.

Reichhardt, K. L., E. Mellink, G. P. Nabhan, and A. M. Rea. 1986. Habitat heterogeneity and biodiversity associated with Indigenous agriculture in the Sonoran Desert. Ethnoecology 2:1–17.

Repanshek, K. 2020. Concerns voiced over border wall construction damaging Quitobaquito Springs at Organ Pipe. National Park Traveler, 2 August. https://www.nationalparkstraveler.org/2020/08/concerns-voiced-border-wall-construction-damaging-quitobaquito-springs-organ-pipe. Accessed 5 June 2022.

Stevens, L. E. 2020. The springs biome, with an emphasis on arid regions. Encyclopedia of the World's Biomes 2:354–70.

Stevens, L. E., E. R. Schenk, and A. E. Springer. 2021. Springs ecosystem classification. Ecological Applications 31:e2218.

Tibbitts, T. A., and L. L. Dickson. 2005. An update to "An annotated checklist of the birds of Organ Pipe Cactus National Monument." Unpublished annotated species list for Organ Pipe Cactus National Monument, Ajo, Arizona. In possession of author.

Winters, H., Jr. 2012. O'odham place names: meanings, origins, and histories, Arizona and Sonora. Nighthorse Press, Tucson, Arizona.

Chapter 31

How Traditional Ecological Knowledge Informs the Field of Conservation Biology

SARAH E. RINKEVICH and CRYSTAL LEONETTI

INTRODUCTION

Traditional Ecological Knowledge (TEK) is increasingly being applied in the field of wildlife conservation, yet conceptualizations of what TEK is and views regarding whether TEK is science remain diverse (Ramos 2018). According to Huntington (1998), TEK offers ecological information and insights relevant to ecological management and research that cannot be obtained from other sources. The term has been used to describe the knowledge held by Indigenous cultures about their immediate environment and the cultural practices that build on that information (Menzies and Butler 2006).

For thousands of years, Indigenous peoples have used biological knowledge of their local environment to sustain themselves and to maintain their cultural identity (Johnson 1992). Indigenous peoples around the world possess a broad knowledge base of the complex ecological systems in their own localities (Gadgil et al. 1993). This information functions as the basis for time-tested resource management systems of long-resident peoples. Yet the involvement of Indigenous people in environmental decision-making remains rare, and Western science often overlooks and disparages Indigenous systems and associated TEK (Westley and Miller 2003).

According to Berkes (1993), TEK parallels the scientific discipline of ecology because both TEK and Western science share the practices of observation and description of the empirical world. TEK is conceptually holistic, however, in that Indigenous knowledge systems consider the biotic and abiotic as being connected and not compartmentalized, as they often are in Western science (McGregor 2004, Pierotti 2011). Many examples of holistic approaches that consider ecological relationships exist in Western science, such as community ecology (Krebs 2009), macroecology (Brown 1995), and ecosystem management (Grumbine 1994). Other disciplines outside of ecology recognize the importance of holism because it emphasizes relationships. According to Mayr (1982:67), Einstein based his entire relativity theory on the consideration of relationships. Similarities between Indigenous knowledge and quantum theory, which stresses the irreducible link between the observer and the observed and the basic holism of all phenomena, have been discussed (Peat 1994, 2005).

In this chapter, we present an overview of TEK and its relationship to Western science, examples of convergence with Western science, challenges with using TEK in the context of endangered species management, and its legal basis. Our goal is to present how TEK has enhanced the field of conservation biology. Published literature on TEK is vast (see Inglis 1993, Berkes 1993, Berkes et al. 2000, Huntington 2000, Klubnikin et al. 2000, Nabhan 2000, Pierotti and Wildcat 2000, Menzies 2006, Pierotti 2011, Ramos et al. 2016, Ramos 2018) and extends globally with the majority of case studies from Canada, Australia, South America, and Asia (e.g., Fernadez-Gimenez 2000, Donovan and Puri 2004, Chambers and Fabricius 2007, Wehi 2009, Rist et al. 2010). We focus on literature from North America that involved Native peoples and studies that addressed threatened and endangered species across taxa.

TRADITIONAL ECOLOGICAL KNOWLEDGE

There is no universally accepted definition for the term TEK (Berkes 1993, Johnson 1992). A variety of terms that are analogous to TEK have been used in the past, including Indigenous knowledge (Warren et al. 1995), local knowledge (Berkes 1993), Native science (Cajete 2000), and Aboriginal science (Aikenhead 2006). We use the term TEK because it is the one most widely used in the literature (see Johannes 1989, Berkes 1993, Pierotti and Wildcat 2000, Pierotti 2011).

TEK is not a phrase used in Native American or other Indigenous cultures; the concept would be simply considered "knowledge" (S. Pilsk, pers. comm. 2012). This knowledge encompasses practical and empirical aspects of understanding and is both information itself and a way of knowing (Mailhot 1994, Pierotti 2011). Native peoples further point out that TEK is not so much knowledge as it is a "way of life" (Nadasdy 2003). Importantly, TEK is not a uniform concept among Indigenous peoples (Battiste and Henderson 2000).

TEK includes intimate and detailed knowledge of plants, animals, and natural phenomena, as well as the development and use of appropriate technologies for hunting, fishing, trapping, agriculture, forestry (Berkes 1993), and survival (Huntington 2000). One characterization of TEK is the sum of the data and ideas acquired by a human group about its environment resulting from the group's use and occupation of a specific region over many generations (Mailhot 1994). Berkes et al. (2000) described TEK as a cumulative body of knowledge, practice, and belief—evolving by adaptive processes and handed down through generations by cultural transmission—about the relationship of living beings (humans and nonhumans) with one another and with their environment. TEK is a useful construct that represents knowledge gathered from undertaking different pursuits, such as hunting, medicine collection, preparation for spiritual ceremonies, or maintenance of a household economy. These are generalized activities found in many traditional societies and are some of the ways in which Indigenous people interact with the natural world. These interactions carried out over countless generations are the genesis of TEK (Drew 2005, Drew and Henne 2006).

FOUNDATIONS OF TEK

Indigenous peoples throughout the world have always had "science," defined as a body of practical and empirical knowledge of their environment, because without it a society could not survive (Cajete 2000, Nadasdy 2003). As the Cherokee elder quoted in chapter 17 remarked, "We have always thought like scientists." TEK's foundations include a process of environmental learning in order to survive, and passing that knowledge to the next generation. Indigenous people who have lived for generations in a particular environment develop intimate familiarity with the land. Native people have depended on the animals and plants of these environments for food, clothing, shelter, and companionship and developed strong ties to the local wildlife (Pierotti and Wildcat 1999).

As Native American people observed their fellow beings, they noted that each species had characteristics that set them apart from other species and enhanced their chances of survival (Marshall 1995). The way for humans to survive and prosper was to pay careful attention and learn as much as possible about the strengths and weaknesses of all the other organisms, so that they could take them as food and avoid being taken by them. The body of knowledge acquired through careful observations was passed on to others through detailed conversations and stories, which had to be repeated constantly so that the knowledge would be available to future generations (Pierotti and Wildcat 1997, 1999, Pierotti 2011).

DIFFERENCES BETWEEN TEK AND WESTERN SCIENCE

Western science and TEK fundamentally are based in different worldviews, institutions, and methods (Lertzman 2010). Indigenous people have a considerable body of empirically derived knowledge about the natural world and a philosophical approach that is quite different from that found in any Western philosophical tradition (Mander 1991, Suzuki and Knudtson 1992, Barsh 2000, Pierotti and Wildcat 2000, Pierotti 2011). Western scientific information is highly dynamic and changes rapidly. In contrast, the body of data typical for Indigenous knowledge is based on long-term personal observations and elders' memories, and it usually changes relatively slowly (Krupnik and Ray 2007) though Indigenous people have also at times had to cope with and adapt to environmental variability that may have occurred rapidly (Huntington 2002).

The two bodies of knowledge can thus be thought of as two distinct epistemologies—theories about the origins, limits, and meaning of knowledge (Lertzman 2010) and "how we know what we know" (see Landesman 1997, Audi 1998). TEK, for example, consists of not only ecological "data" but also spirituality, values, normative rules, and cultural practices (Casimirri 2003). In contrast, Western science became detached from religion and spirituality beginning with the Enlightenment in the 18th century (see Mayr 1982).

Another difference between TEK and Western science is that TEK is recorded and transmitted through oral tradition, often through stories, while Western science normally employs written works (Barnhardt and Kawagley 2005). TEK is rooted in a social context that views the world in terms of social and spiritual relations between all life forms. Western science tends to be hierarchical or linearly organized and reduces the environment to conceptually discrete components that are managed separately (Johnson 1992). Native peoples do not subscribe to the dichotomy of nature and culture but view the world as an integrated entity. One gets to know the forest and the plants and animals that dwell therein in the same way that one becomes familiar with other people, by spending time with them, investing in those relationships the same qualities of care, feeling, and attention. Knowledge of the world is gained by moving about in it, exploring it, attending to it, ever alert to the signs by which it is revealed. Native people do not construct the environment, but acquire skills to engage with it (Ingold 1996).

CONVERGENCE OF TEK AND WESTERN SCIENCE

Observation of a pattern (i.e., induction) is the starting point of the scientific method in the field of ecology (Romesburg 1991). In this sense, induction in both Western science and Indigenous knowledge is quite similar. Western science and Indigenous ecological knowledge also share the next step: generating potential explanations and predictions for the pattern observed. Traditional knowledge is based on an accumulation of empirical data (Berkes et al. 2000) resulting from patient observation of the natural world, its patterns (Pierotti 2011), and interrelationships among species of wildlife and plants, a process similar to Western science.

Nabhan (2000) reported on the oral traditions of the O'odham and Comcáac in the southwestern United States, which contain over 20 interspecific relationships encoded in their biosystematic lexicon. For example, a key issue regarding conservation management of the desert tortoise (*Gopherus agassizii*) has been providing protected habitat with sufficiently diverse forage available. Despite 60 years of incidental reports on desert tortoises' feeding behavior, stomach contents, and fecal pellet analysis, Western scientists' knowledge of the species' dietary needs had remained fragmentary (Van Devender and Schwalbe 1998).

But the Comcáac have 4 species of desert plants that are referred to as "desert tortoise's forage" in their Native language.

The Comcáac also associate an ephemeral legume to the endangered Sonoran pronghorn (*Antilocapra americana sonoriensis*), calling the plant "pronghorn—its wild bean." The Comcáac have a name for a wild onion that translates as "desert bighorn eats it" because they have observed that desert bighorn sheep (*Ovis canadensis mexicanus*) consume that plant. These are tangible examples of how Indigenous ecological knowledge can be used to guide Western empirical or experimental studies to learn more about plant-animal interactions. Many experts believe that TEK could inform endangered species recovery efforts and habitat restoration planning (Nabhan 2000).

The depth of Indigenous knowledge rooted in long inhabitation of a particular place can benefit Western scientists (Barnhardt and Kawagley 2005). Ferguson et al. (1998) reported that detailed observations of caribou (*Rangifer tarandus*) had been preserved in Inuit oral traditions and were corroborated by Western scientific written records (Ferguson and Messier 1997, Ferguson et al. 1998). Written records from the 1800s, though limited spatially and temporally, support Inuit knowledge that South Baffin caribou populations follow a regular abundance and movement cycle over periods of 60–80 years.

Huntington (1998) interviewed Inupiaq peoples in Point Lay, Buckland, and Norton Bay, Alaska, to gather ecological information about beluga whales (*Delphinapterus leucas*), including migratory patterns, local movements, feeding behavior and prey patterns, predator avoidance, calving, bathymetry, ecological interactions, and human influences. The descriptions were broadly in accordance with current Western scientific understanding, although the overlap was not complete. Huntington (1999) reported that TEK provided more specifics than published scientific research on beluga whale interactions with humans and anthropogenic influences, such as noise (e.g., Frost et al. 1993).

Fraser et al. (2005) reported that TEK revealed long-term trends on the viability of divergent brook char (*Salvelinus fantinalis*) populations in the region that were not achievable with Western scientific data. Gilchrist et al. (2005) reported that Inuit correctly identified aspects of harlequin duck (*Histrionicus histrionicus*) biology, including preferred habitats and seasonal movements. Krupnik and Ray (2007) reported detailed information about Pacific walrus (*Odobenus rosmarus divergens*) migration patterns from Yup'ik hunters in Alaska. Walrus hunters had knowledge of walrus biology and ecology, including seasonal differences in distribution, separation of different groupings, and two seasonal peaks of abundance around St. Lawrence Island (Krupnik and Ray 2007).

Indigenous peoples are similarly interested in the knowledge of Western scientists and where their understandings overlap. The convergence of TEK and Western science suggests that there may be areas in which TEK can contribute insights and possibly new concepts about and/or connections between species unknown to or unrecognized by Western ecologists (Pierotti and Wildcat 2000). During a study of beluga whales, for example, researchers confused over why belugas no longer entered certain rivers were told by Indigenous people that it was because of beavers (*Castor canadensis*), which build dams that inhibit salmon movements, making certain streams less attractive to the belugas, which feed on salmon (Huntington and Myrmin 1996).

A convergence of TEK and Western science has also led to better population estimates for some species. Huntington (2000) reported that Inupiaq whalers' knowledge of bowhead whale (*Balaena mysticetus*) populations proved more accurate than that of scientists during his study. Combining the data from the Western scientific census and the Inupiaq whalers' TEK of bowhead whales provided a more accurate population estimate.

Indigenous people in Arctic communities who still rely on fish and wildlife for subsistence continue to recognize patterns of nature, such as shifts in wildlife populations. Inuit people from the community of Sanikiluaq on the shore of Hudson Bay, for example, reported that regional common eider (*Somateria mollissima sedentaria*) populations had declined in the early 1990s (Gilchrist et al. 2005). Inuit residents stated that extensive sea ice had formed during the winter of 1991–1992 and limited the locations where eiders could feed in open water, resulting in their mass starvation. Although the ultimate cause of the severe winter ice conditions was unknown to them, the Inuits detected both a change of sea ice conditions and the resulting mass die-off of eiders. Gilchrist et al. (2005) also documented that Inuit residents accurately reported declining populations of ivory gull (*Pagophila eburnean*) and thick-billed murre (*Uria lomvia*). Fully incorporating Indigenous people into research regimes and policy debates would be a major step toward a more equal dialogue and a partnership built on data sharing and mutual respect (Fenge 2001, Berkes 2002).

Information encoded in the lexicons of Native peoples can be difficult and time consuming to understand and interpret. Nabhan (2000) demonstrated that O'odham and Comcáac oral traditions include many ecological insights that can be known only once the language is understood. One Yup'ik elder explained, "[T]hey say all animals have ears through the ground." The Yup'ik language is second only to Navajo in numbers of speakers (19,000 in 2010, according to the US Census), and accordingly it retains many social patterns and a traditional knowledge base. Human-animal relations among Yup'ik are articulated by *qanruyutet* (oral instructions; Fienup-Riordan 2020). In Yup'ik lands, where subsistence hunting and fishing are part of daily life, the attribution to an ancestral voice is critical to understanding animal-human-environment relations. Animals are cohabitants, not resources, are part of the sentient world as nonhuman persons, and like the environment are responsive to human thoughts, words, and deeds (Fienup-Riordan 2020). One Nunivak Island elder shared that the plants, animals, and land understand only the Yup'ik language (pers. comm., 2010).

Understanding a culture and learning an Indigenous language takes years, if not decades. Cultural traditions, such as only being able to share stories or information about a certain species during appropriate seasons or times of the year or among select groups of knowledge-holders, can limit the collection of

TEK. Because the methods for documenting TEK derive from the social sciences (Huntington 2000), conservation biologists seeking to collect this type of information would need to be trained in ethnographic methodologies, or work closely with professionals that are, although even those professionals should proceed with caution. Collecting TEK may not be appropriate in every situation, especially in instances where the information should not be shared with people outside of the culture.

Although an integration of Indigenous and Western scientific ways of knowing and managing wildlife is difficult to achieve (Nakashima 1993), there have been several successful instances we are aware of. Here are a few.

During the 1989 *Exxon Valdez* oil spill disaster in Prince William Sound, Alaska, federal and state agencies recognized the vast traditional knowledge of Native people, who provided detailed information on conditions in the years prior to the spill. Indigenous people had knowledge of the historical population sizes and ranges of many of the species injured by the spill and observations concerning the diet, behavior, and interrelationships of injured species (Huntington 2000). The use of Western scientific data and traditional knowledge, along with increasing the involvement of local communities in cleaning up the oil spill, enhanced the success of the restoration effort (Miraglia 1998).

Alaska Native peoples have been at the forefront in bringing Indigenous perspectives into a variety of policy arenas through a wide range of research and development initiatives (Barnhardt and Kawagley 2005; chapters 13 and 16 of this volume). Alaska Indigenous knowledge, for example, was cited with regard to polar bear (*Ursus maritimus*) population trends (US Fish and Wildlife Service 2008), describing the legal designation of critical habitat for spectacled eider (*Somateria fischeri*; US Fish and Wildlife Service 2001), and subsistence harvest regulations for migratory birds. The National Marine Fisheries Service worked extensively with Native hunters to use traditional knowledge when proposing the Cook Inlet beluga whales (*Delphinapterus leucas*) as a distinct population segment (National Marine Fisheries Service 2007a). The TEK of Alaska Native peoples along with systematic aerial survey data documented a contraction of the summer range of Cook Inlet belugas over the previous two decades (National Marine Fisheries Service 2009a).

The National Marine Fisheries Service also used Alaska Native TEK in developing the stock conservation plan for the eastern Pacific northern fur seal (National Marine Fisheries Service 2007b). Numerous ethnographic studies regarding the First Nations' TEK of eulachon (*Taleichthys pacificus*), a small, anadromous ocean fish, were used to describe the species' historical distribution and abundance in the proposed rule to list the fish (National Marine Fisheries Service 2009b). In addition, federal and state biologists and managers in Alaska collect and use TEK for research and monitoring fish populations under the Federal Subsistence Management Program (see US Fish and Wildlife Service 2010a).

The US Fish and Wildlife Service and National Marine Fisheries Service have referenced TEK in several proposed and final rules concerning threatened and endangered species in Alaska. One example of the federal government referencing TEK in the lower 48 states occurred in the 12-month petition finding for the Sonoran Desert population of bald eagle (*Haliaeetus leucocephalus*). The White Mountain and San Carlos Apache Tribes and the Salt River Pima–Maricopa Indian Community provided the US Fish and Wildlife Service with their knowledge about bald eagle populations and habitat in Arizona (US Fish and Wildlife Service 2010b). More than 200 years of Indigenous knowledge of bald eagles in Arizona supplemented 30 years of service and state monitoring data.

CONCLUSION

Understanding the ecology and population dynamics of key species in particular ecosystems needs to be multidisciplinary due to the complexity and magnitude of the many intersecting problems faced by wildlife and their habitats (Westley and Miller 2003). The need to find integration between social and biological sciences was discussed in detail by Wilson (1998), who argued that sound environmental policy can only be formed at the juncture of ethics, social science, and biology. He used the term "consilience"—the unity of knowledge—for this concept. It means a "jumping together" of knowledge, the linking of facts and fact-based theory across disciplines to create a common groundwork of explanation, following William Whewell in his 1840 synthesis, *The Philosophy of the Inductive Sciences*. Such a merger could draw Indigenous peoples into a dialogue that could generate a deeper understanding of complex ecological processes and concepts. According to Lertzman (2010), TEK and Western science are parallel, potentially complementary knowledge systems. Rather than trying to "integrate" TEK into Western science, TEK could be thought of as a parallel science.

We believe TEK is underutilized in natural resource management, and many professionals in applied ecology and resource management have been slow to embrace it. Recognition of the scientific importance of TEK should lead to more cooperative relations between researchers and local communities in which Indigenous people who are repositories of knowledge and skills become an integral part of a research program as consultants or collaborators rather than merely as guides or assistants (Healey 1993). The incorporation of TEK may enhance Western society's appreciation of the cultures that hold ecological knowledge about specific species, which could result in less controversy and more mutual respect. Ethnographic studies show that by engaging the people who live in areas of interest, resource managers can obtain cultural information for management purposes while building relationships with local communities. Scientific investigations that make use of this knowledge are likely to be particularly successful.

As the pace of ecological changes increases, so too does the need for baseline information with which to direct conservation and restoration activities (Ford and Martinez 2000). Species are becoming extinct, and ecosystems are being dramatically altered (Casey and Myers 1998, Drew 2005). The field of conservation biology is constantly evolving to better address a complex and dynamic suite of threats to biodiversity. The use of TEK is an approach with the potential to greatly augment existing conservation programs and help shape new ones (Drew 2005). Listening to Indigenous peoples as well as respecting and understanding their needs and aspirations will help Western scientists partner more effectively in wildlife and habitat conservation. TEK and Western science offer each other externally derived, independent reference standards that can be a basis for bicultural verification. The two represent complementary knowledge systems with their own methods, philosophical foundations, and communities of respected experts. Indigenous peoples' knowledge could be key to developing a sound basis for conservation planning and action (Westley and Miller 2003). With regard to climate change in the Canadian Arctic, for example, a number of projects have begun to engage northern Aboriginal peoples and focus on Indigenous knowledge of climate processes, changes, and impacts. In order to better understand the complex nature of northern ecosystems, all available knowledge must be considered and valued (Furgal et al. 2002). Indigenous elders have knowledge attained over multiple generations that could likely save Western scientists many years of learning if they are willing to listen.

Public lands are managed by science agencies across the North American continent. These public lands are the homelands of Indigenous peoples, who adapted through many cycles of climate change and who are the most logical stewardship leaders. Western scientists are charged with the protection of marine mammals, migratory birds, and life-giving landscapes, and they should also recognize their responsibility to the people whose lives, ancestors, and futures are inseparable from these aspects of the Earth. Indigenous-led stewardship would exemplify a coexistence of and healthy relationship between land and humans and would be an enormous benefit to biodiversity, wildlife health, and clean water.

LITERATURE CITED

Aikenhead, G. S. 2006. Science education for everyday life: evidence-based practice. Teachers College Press, New York.

Audi, R. 1998. Epistemology: a contemporary introduction to the theory of knowledge. Routledge, New York.

Barnhardt, R., and A. O. Kawagley. 2005. Indigenous knowledge systems and Alaska Native ways of knowing. Anthropology and Education Quarterly 36:8–23.

Barsh, R. L. 2000. Taking Indigenous science seriously. Pages 152–73 in S. A. Bocking, editor. Biodiversity in Canada: ecology, ideas, and action. Broadview Press, Toronto, Canada.

Battiste, M., and Y. Henderson. 2000. Protecting Indigenous knowledge and heritage: a global challenge. Purich Publishing, Saskatoon, Saskatchewan, Canada.

Berkes, F. 1993. TEK in perspective. Pages 1–10 in J. T. Inglis, editor. TEK: concepts and cases. International Program on TEK and International Development Research Centre, Ottawa, Canada.

Berkes, F. 2002. Epilogue: making sense of Arctic environmental change? Pages 335–349 *in* I. Krupnik and D. Jolly, editors. The Earth is faster now: Indigenous observations of Arctic environmental change. Arctic Research Commission of the United States, Fairbanks, Alaska.

Berkes, F., J. Colding, and C. Folke. 2000. Rediscovery of TEK as adaptive management. Ecological Applications 10:1251–62.

Brown, J. H. 1995. Macroecology. University of Chicago Press, Chicago, Illinois.

Cajete, G. 2000. Native science: natural laws of interdependence. Clear Light Publishers, Santa Fe, New Mexico.

Casey, J., and R. A. Myers. 1998. Near extinction of a large, widely distributed fish. Science 281:690.

Casimirri, G. 2003. Problems with integrating TEK into contemporary resource management. Oral presentation at the International Institute for Sustainable Development, 12th World Forestry Congress, 21–28 September 2003, Quebec City, Canada.

Chambers, N., and C. Fabricius. 2007. Expert and generalist local knowledge about land-cover change on South Africa's wild coast: can local ecological knowledge add value to science? Ecology and Society 12:10.

Donovan, D., and R. Puri. 2004. Learning from traditional knowledge of non-timber forest products: Penan Benalui and the autecology of Aquilaria in Indonesian Borneo. Ecology and Society 9:3.

Drew, J. A. 2005. Use of TEK in marine conservation. Conservation Biology 19(4):1286–93.

Drew, J. A., and A. P. Henne. 2006. Conservation biology and TEK: integrating academic disciplines for better conservation practice. Ecology and Society 11:34.

Fenge, T. 2001. Inuit and climate change: perspectives and policy opportunities. Canadian Journal of Policy Research 2:79–85.

Ferguson, M. A. D., and F. Messier. 1997. Collection and analysis of TEK about a population of Arctic tundra caribou. Arctic 50:17–28.

Ferguson, M. A. D., R. C. Williamson, and F. Messier. 1998. Inuit knowledge of long-term changes in a population of Arctic tundra caribou. Arctic 51:201–19.

Fernadez-Gimenez, M. E. 2000. The role of Mongolian nomadic pastoralists' ecological knowledge in rangeland management. Ecological Applications 10:1318–26.

Fienup-Riordan, A. 2020. Nunakun-gguq ciutengqertut: they say they have ears through the ground: animals essays from Southwest Alaska. University of Alaska Press, Fairbanks.

Ford, J., and D. Martinez. 2000. TEK, ecosystem science, and environmental management. Ecological Applications 10:1249–50.

Fraser, D. J., P. Duchesne, and L. Bernatchez. 2005. Migratory charr schools exhibit population and kin associations beyond juvenile stages. Molecular Ecology 14:3133–46.

Frost, K. J., L. F. Lowry, and G. M. Carroll. 1993. Beluga whale and spotted seal use of a coastal lagoon system in the northeastern Chukchi Sea. Arctic 46:8–16.

Furgal, C., D. Martin, and P. Gosselin. 2002. Climate change and health in Nunavik and Labrador: lessons from Inuit knowledge. Pages 266–300 *in* I. Krupnik and D. Jolly, editors. The Earth is faster now: Indigenous observations of Arctic environmental change. Arctic Research Commission of the United States, Fairbanks, Alaska.

Gadgil, M., F. Berkes, and C. Folke. 1993. Indigenous knowledge for biodiversity conservation. Ambio 22:151–56.

Gilchrist, G., M. Mallory, and F. Merkel. 2005. Can local ecological knowledge contribute to wildlife management? case studies of migratory birds. Ecology and Society 10:20.

Grumbine, R. E. 1994. What is ecosystem management. Conservation Biology 8:27–38.

Healey, C. 1993. The significance and application of TEK. Pages 21–26 *in* N. M. Williams and G. Bains, editors. Traditional Ecological Knowledge. Centre for Resource and Environmental Studies, Australian National University, Canberra.

Huntington, H. P. 1998. Observations on the utility of the semi-directive interview for documenting TEK. Arctic 51:237–42.

Huntington, H. P. 1999. Traditional knowledge and ecology of beluga whales (*Delphinapteus leucas*) in the eastern Chukchi and northern Bering Seas, Alaska. Arctic 52:49–61.

Huntington, H. P. 2000. Using TEK in science: methods and applications. Ecological Applications 10:1270–74.

Huntington, H. P. 2002. Human understanding and understanding humans in the Arctic system. Pages xxi–xxvii *in* I. Krupnik and D. Jolly, editors. The Earth is faster now: Indigenous observations of Arctic environmental change. Arctic Research Commission of the United States, Fairbanks, Alaska.

Huntington, H. P., and N. I. Myrmin. 1996. TEK of beluga whales: an Indigenous knowledge pilot project in the Chukchi and northern Bering Seas. Oral presentation at the Inuit Circumpolar Conference, 19–24 May 1996, Anchorage, Alaska.

Inglis, J. T., editor. 1993. TEK: concepts and cases. International Program on TEK and International Development Research Centre, Ottawa, Canada.

Johannes, R. E. 1989. Traditional Ecological Knowledge: a collection of essays. Gland Switzerland, International Conservation Union.

Johnson, M. 1992. Lore: capturing Traditional Environmental Knowledge. Dene Cultural Institute, International Development Research Centre, Ottawa, Canada.

Klubnikin, K., C. Annett, M. Cherkasova, M. Shishin, and I. Fotieva. 2000. The sacred and the scientific: TEK in Siberian River conservation. Ecological Applications 10:1296–1306.

Krebs, C. J. 2009. Ecology: the experimental analysis of distribution and abundance. University of British Columbia Press, Vancouver, Canada.

Krupnik, I., and G. C. Ray. 2007. Pacific walruses, Indigenous hunters, and climate change: bridging scientific and Indigenous knowledge. Deep-Sea Research 2(54):2946–57.

Landesman, C. 1997. An introduction to epistemology. Blackwell, Cambridge, Massachusetts.

Lertzman, D. A. 2010. Best of two worlds: Traditional Ecological Knowledge and Western science in ecosystem-based management. British Columbia Journal of Ecosystems and Management 10:104–26.

Mailhot, J. 1994. TEK: the diversity of knowledge systems and their study. Great Whale Environmental Assessment No. 4. Great Whale Review, Montreal, Canada.

Mander, J. 1991. The absence of the sacred. Sierra Club Books, San Francisco, California.

Marshall, J., III. 1995. On behalf of the wolf and the First Peoples. Red Crane Books, Santa Fe, New Mexico.

Mayr, E. 1982. The growth of biological thought: diversity, evolution, and inheritance. Belknap Press, Cambridge, Massachusetts.

McGregor, D. 2004. TEK and sustainable development: towards coexistence. Pages 72–91 *in* M. Blaser, J. A. Feit, and F. McRae, editors. In the way of development: Indigenous peoples, life projects and globalization. International Development Research Centre, Zed Books, London, England.

Menzies, C. R. 2006. TEK and natural resource management. University of Nebraska Press, Lincoln.

Menzies, C. R., and C. Butler. 2006. Understanding ecological knowledge. Pages 1–20 *in* C. R. Menzies, editor. TEK and natural resource management. University of Nebraska Press, Lincoln.

Miraglia, R. A. 1998. TEK handbook: a training manual and reference guide for designing, conducting, and participating in research projects using TEK. Division of Subsistence, Alaska Department of Fish and Game, Anchorage.

Nabhan, G. P. 2000. Interspecific relationships affecting endangered species recognized by O'odham and Comcáac cultures. Ecological Applications 10:1288–95.

Nadasdy, P. 2003. Hunters and bureaucrats: power, knowledge, and Aboriginal-state relations in the southwest Yukon. University of British Columbia Press, Vancouver, Canada.

Nakashima, D. J. 1993. Astute observation on the sea ice edge: Inuit knowledge as a basis for Arctic co-management. Pages 99–110 in J. T. Inglis, editor. TEK: concepts and cases. International Program on TEK and International Development Research Centre, Ottawa, Canada.

National Marine Fisheries Service. 2007a. Endangered and threatened species: proposed endangered status for the Cook Inlet beluga whale. Federal Register 72:19854–62.

National Marine Fisheries Service. 2007b. Notice of availability of final eastern Pacific northern fur seal stock conservation plan. Federal Register 72:73766–70.

National Marine Fisheries Service. 2009a. Endangered and threatened species: designation of critical habitat for Cook Inlet beluga whale. Federal Register 74:63080–95.

National Marine Fisheries Service. 2009b. Endangered and threatened species: proposed threatened status for southern distinct population segment of eulachon. Federal Register 74:10857–76.

Peat, F. D. 1994. Lighting the seventh fire: the spiritual ways, healing, and science of the Native American. Birch Lane Press, Secaucus, New Jersey.

Peat, F. D. 2005. Blackfoot physics: a journey into the Native American universe. Weiser Books, Boston, Massachusetts.

Pierotti, R. 2011. Indigenous knowledge, ecology, and evolutionary biology. Taylor and Francis, New York.

Pierotti, R., and D. Wildcat. 1997. The science of ecology and Native American tradition. Winds of Change 12:94–98.

Pierotti, R., and D. Wildcat. 1999. Traditional knowledge, culturally based worldviews and Western science. Pages 192–99 in D. Posey, editor. Cultural and spiritual values of biodiversity. UN Environment Program. Intermediate Technology Publications, London, England.

Pierotti, R., and D. Wildcat. 2000. TEK: the third alternative (commentary). Ecological Applications 10:1333–40.

Ramos, S. C. 2018. Considerations for culturally sensitive Traditional Ecological Knowledge research in wildlife conservation. Wildlife Society Bulletin 42(2):358–65.

Ramos, S. C., T. M. Shenk, and K. M. Leong. 2016. Introduction to Traditional Ecological Knowledge in wildlife conservation. Natural Resource Report NPS/NRSS/BRD/NRR-2016/1291. National Park Service, Fort Collins, Colorado.

Rist, L., R. U. Shaanker, E. J. Milner-Gulland, and J. Ghazoul. 2010. The use of TEK in forest management: an example from India. Ecology and Society 15:1. http://www.ecologyandsociety.org/vol15/iss1/art3/. Accessed 6 August 2022.

Romesburg, H. C. 1991. On improving the natural resources and environmental sciences. Journal of Wildlife Management 55:744–56.

Suzuki, D., and P. Knudtson. 1992. Wisdom of the elders: sacred Native stories of nature. Bantam Books, New York.

US Fish and Wildlife Service. 2001. Endangered and threatened wildlife and plants: final determination of critical habitat for the spectacled eider (*Somateria fischeri*). Federal Register 66:9146–85.

US Fish and Wildlife Service. 2008. Endangered and threatened wildlife and plants: determination of threatened status for the polar bear (*Ursus maritimus*) throughout its range: final rule. Federal Register 73:28212–303.

US Fish and Wildlife Service. 2010a. 2010 Fisheries resource monitoring plan. https://www.doi.gov/sites/doi.gov/files/2020-fisheries-resource-monitoring-plan.pdf. Accessed 6 August 2022.

US Fish and Wildlife Service. 2010b. Endangered and threatened wildlife and plants: 12-month finding on a petition to list the Sonoran Desert population of the bald eagle as a threatened or endangered distinct population segment. Federal Register 75:8601–21.

Van Devender, T. R., and C. R. Schwalbe. 1998. Diet of free-ranging desert tortoises (*Gopherus agaaizii*) in the north-eastern Sonoran Desert, Arizona. Final Report No. 195043. Arizona Game and Fish Department, Phoenix.

Warren, D. M., L. J. Slikkerveer, and D. Brokensha. 1995. The cultural dimension of development: Indigenous knowledge systems. Intermediate Technology Publications, London, England.

Wehi, P. M. 2009. Indigenous ancestral sayings contribute to modern conservation partnerships: examples using *Phormium tenax*. Ecological Applications 19(1):267–75.

Westley, F. R., and P. S. Miller. 2003. Experiments in consilience: integrating social and scientific responses to save endangered species. Island Press, Washington, DC.

Wilson, E. O. 1998. Consilience: the unity of knowledge. Knopf, New York.

Chapter 32

Yurok Traditional Ecological Knowledge as Related to Elk Management and Conservation

JULIANA SUZUKAWA, SEAFHA C. RAMOS, and TIANA M. WILLIAMS-CLAUSSEN

With increasing threats toward wildlife arising from factors such as climate change, scholars have advocated for exploring collaborative solutions with Indigenous researchers and communities, which often have deep understandings of and cultural connections to their environments and can provide unique expertise (Vales et al. 2017, Arsenault et al. 2019). This combination of knowledge and connection is often referred to as Traditional Ecological Knowledge (TEK), which has been defined as "the culturally and spiritually based way in which [I]ndigenous peoples relate to their ecosystems" (LaDuke 1994:127). This relationship has been a part of Indigenous peoples' cultures in what is now the United States of America since before the arrival of European settlers and continues today in many communities.

TEK research has been conducted in various wildlife management and conservation aspects (Dowsley and Wenzel 2008, Polfus et al. 2014) and in the human dimensions of wildlife studies (Reo and Whyte 2012). However, TEK and contemporary wildlife management are rooted in different scientific philosophies and epistemologies, leading to diverse interpretations of what TEK is and how it may be applied in wildlife research, management, and conservation (Moller et al. 2004, Ramos 2018, Learn 2020). This is complicated by the fact that much of Indigenous knowledge is passed down through oral tradition (Morris and Eldridge 2020). TEK is place-based, so conducting qualitative research through interviews with knowledgeable individuals integrated within a given system of interest can allow researchers to gain a better understanding of a particular community's cultural lens and its perspectives on wildlife (Ramos et al. 2016, Ramos 2022).

For the Yurok Tribe of northwestern California, TEK can be conceptualized as *hlkelonah 'ue-megetohl*, "a system where Yurok people and wildlife collaboratively strive to create and maintain [the] balance of the Earth via physical and spiritual management in tandem" (Ramos 2019:86). Many wildlife species are important to Yurok TEK, such as the Roosevelt elk (*meyweehl*; *Cervus canadensis roosevelti*), which is a cultural and subsistence resource for the Yurok people. Although several herds of elk are found on Yurok ancestral lands, Yurok people have limited access to them; the population of elk within the Yurok Tribe's jurisdiction, a small proportion of that ancestral territory, is greatly diminished from historic levels. Herds are often clustered on protected federal lands or inaccessible private lands. There is considerable interest in the tribal community and government in restoring the elk population on lands within Yurok jurisdiction (Yurok Tribe 2016).

Elk were described in the scientific literature as early as 1936 (Bailey 1936) and have since been the subject of many studies (Harper et al. 1967, Weckerly 2017). Primary elk management strategies have included population surveys to inform hunting programs, monitoring elk distribution and abundance, and monitoring habitat (Vales et al. 2017, CDFW 2018). To improve long-term elk management in California, we need innovative management actions and greater understanding among all parties interested in elk conservation and management (CDFW 2018). Though some scholars have specifically identified subsistence activities, such as hunting by Indigenous people, in the broader context of TEK and socioecological systems (Reo and Whyte 2012), management approaches have been slow to adapt to the divergent cultural contexts of Indigenous subsistence and ceremonial hunting versus recreational hunting (McCorquodale 1997). Further, studies examining the relationship between Yurok TEK and elk are limited. Such understanding could contribute significantly to elk management across jurisdictional boundaries within Yurok ancestral lands, such as in areas of Redwood National Park where large herds of elk persist. The objective of this research is to gain an understanding of Yurok TEK as related to elk.

STUDY AREA

Yurok ancestral lands cover approximately 400,000 acres in northwestern California, with the Yurok Reservation centered along the Klamath River (figure 32.1; Huntsinger and Diekman 2010). In 1993, the Yurok organized a Tribal Council and formal government to "[p]reserve and promote our culture, language, and religious beliefs and practices" and "restore, enhance, and manage the tribal fishery, tribal water rights, tribal forests, and all other natural resources" (Yurok Tribe 1993). Natural resources, such as salmon (*nepuy*; *Oncorhynchus* spp.), acorns (*woomehl*; *Notholithocarpus densiflorus*), and elk, have had a role in Yurok people's overall health and ceremonies. When the United States began outlawing the practice of Indigenous cultures (Getches et al. 2005), many Yurok people stood up for their rights. Their persistence has allowed them to practice much of their culture today (Lara-Cooper and Lara 2019). Though the Yurok experienced cultural genocide, as did many tribes across the United States, they have been resilient and have maintained their TEK for the purpose of "fixing the Earth" and taking care of natural resources, including wildlife (Ramos 2022).

METHODS

Interview Procedures

We conducted semistructured interviews in 2018–2019 with three adult males who are culturally affiliated with Yurok. Individuals were invited to participate based on a previous tribal wildlife TEK interview project and suggestions from community members. Interviews were audio-recorded and transcribed. We provided participants with a photo of an elk to facilitate discussion of the utility of various parts, as well as a map of Yurok ancestral territory to facilitate discussion of important elk habitat. Interviews included the following topics:

- How land management has impacted elk populations and habitat in Yurok ancestral lands.
- Whether lands managed by the Yurok Tribe can sustainably maintain elk populations.
- Needs of elk over their life cycle.
- Yurok language as related to elk.
- Items that can be made from elk and how these items are part of Yurok culture.

Interview Analysis

We shared and stored all data according to strict cultural and individual data protection protocols. We assigned each participant (P) a unique identifier: P1, P2, and P3. We used qualitative analysis software (NVivo v. 12) to conduct thematic analyses of the interview transcripts (Terry et al. 2017). We coded information from the interview transcripts that related to topics determined a priori, such as elk management and items made from elk. We also analyzed themes that emerged from the process of coding. Because coding took place by two of the authors working independently, we utilized codebooks and code hierarchy charts (NVivo) for interpretation of the results. We created a word cloud of the most frequently used terms (N = 25) in the coded text (NVivo).

In reporting results, we edited quotes for clarity. We also employed member checking: each participant was provided with their interview transcript and a draft of the chapter to verify whether quotes or summaries accurately represented their perspective.

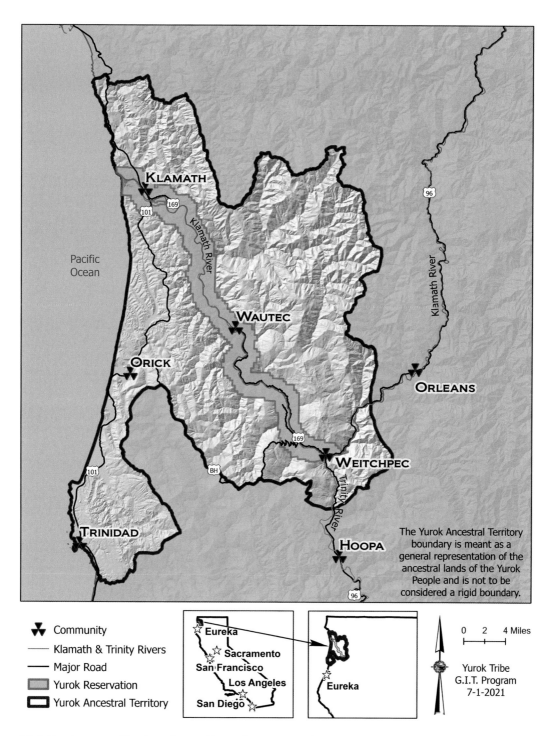

32.1. Yurok ancestral lands are located in northwestern California, with the Yurok Reservation centering on the Klamath River. YUROK TRIBE GIT PROGRAM

RESULTS

Below, we report on five themes: Yurok language, items made from elk, elk in ceremony, elk habitat and management, and continuance of Yurok TEK as related to elk. Our word cloud (figure 32.2) revealed that the three most frequently used terms were "elk," "people," and "cultural." Other words of significance were "Yurok," "knowledge," and (Yurok) "language."

Yurok Language

The Yurok language is important in Yurok TEK (Ramos 2022), and there are ongoing efforts in language revitalization by the Yurok Tribe. P1 and P3 used Yurok language in their interviews. P1 mentioned that the inside of one type of *tekwonekws*, a redwood trunk where valuables are kept, can be shaped with tools made with elk, such as an adz. P3 shared that *'we-s'ech* refers to an elk's antlers. When discussing elk habitat, P1 also mentioned an extensive historical prairie system, *knewsep*, which translates to "long prairie."

The Yurok language has various counting systems with different word structures, depending on what is being counted. For example, there is a different counting system for people and for round things (Garrett 2014). P3 commented that the size of a bull elk is described by the counting of his tines, which are counted using the same system as for tools: *kerhterpee'* (spike or one tool), *ner'erpee'* (two-pointer or two tools), *nerhkserpee'* (three-pointer or three tools), and so on. We believe this research might provide the first documentation of the practice of utilizing the "tool" counting system for describing elk. P3 mentioned he had learned this way of counting bull elk from first-language speakers.

32.2. Word cloud representing the 25 most frequently used terms in interviews about Yurok Traditional Ecological Knowledge as related to elk (*meweehl*). JULIANA SUZUKAWA

A sentiment shared by P3 is that there are no general words for broader taxonomic genera or families in the Yurok language; everything has a name. This reflects the belief that everything that has a life, such as elk, should be respected. P3 stated, "I think that's really important.... [A]s Yurok people, we were really connected to our natural environment.... [T]here's not a single generic word for tree. That's a redwood tree, or that's a Sitka spruce. All the plants, and all our animals, they all have names. Because when you give something a name, you don't generalize it. It's like you're giving it life. And when you have any sort of life, you have a responsibility . . . to manage and care for it."

Items Made from Elk

One of the main themes of the interviews was items made from elk. Participants mentioned 20 specific items (table 32.1) in relation to utilitarian and ceremonial use, and several items were made by one of the interviewees (figure 32.3). Participants identified specific components of elk as providing particular utility. For example, P3 stated, "Elk hides are a lot deeper-based [in sound] than deer hides if you're making drums out of them." P1 made reference to "that particular elk bone.... You can have adz that have stone heads, but their stone don't do this." Referencing the incredible hardness of elk leg bones, P3 affirmed "that bone is really strong."

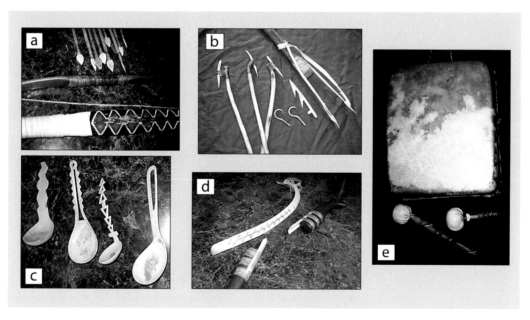

32.3. Items made from elk. (*a*) *Top*, arrow tips; *bottom*, sinew made into bow string; (*b*) *left to right*, eel hooks, fishing hooks, harpoon; (*c*) spoons; (*d*) *top*, towel (rib bone); *bottom*, two adzes; (*e*) drum and drumsticks. ALL ITEMS MADE AND PHOTOGRAPHED BY DALE WEBSTER (YUROK)

TABLE 32.1. Items Made from Elk, as Stated by Interview Participants

ENGLISH TERM	YUROK TERM*	DESCRIPTION, COMMENTS
Adz blade	ner'merwue'	Made with sharpened elk leg bone. An elk adz gives more precision than a stone adz.
Armor		No translation of "armor" was found. However, *skoy* is the Yurok word for "hide." Elk hide can be used to make armor.
Arrow	ner'wkwert	Target arrows are made from elk antler, which can be reused.
Bag	puueweesh	One participant viewed a bag made from elk scrotum in a museum.
Blanket	ka'an	Elk hide can be made into a blanket.
Drum/drumstick	chokchoop'a'r	Elk hide can be made into a drum and sticks.
Fishhook	nerhcherr'	Can be made from elk bone.
Hairpin	nohchuer	Often referred to as a "hair stick"; can be made from elk antler. Contemporary use by women.
Harpoon		Elk parts can be used to make a toggle harpoon head traditionally used to harvest seal and sea lion.
Headband (hookman's)		Part of ceremonial regalia in the White Deerskin Dance. Sea lion teeth can be inserted into carved elk antler to extend the hooks in the headband.
Hook (eel)	lemolohl 'o kwar	Elk bone or antler can be used for the blade or sharp needle of an eel hook.
Knife (women's)	chegeyoh	*Chegeyoh* is translated as "small knife, pocket knife." Participant described small elk antler or bone pocket knives used by women.
Needle	kwar	Can be made from elk bone.
Net shuttle/gauge	pla's	Stick for measuring meshes of a fishing net.
Pressure or percussion flaker/chisel	hokchkehl	*Hokchkehl* is translated as "arrow flaker." Elk bone can be used to carve intricate designs, shape arrows, or chisel out the inside of a redwood box.
Purse	puueweesh	Traditionally, men used carved elk antler to store their *cheeek* (dentalia, traditional money). Commonly referred to as elkhorn purses.
Sinew	hophl	Can be used in a variety of items, such as *rego'* (ceremonial morning feathers), bowstring, and tying arrow parts together.
Spoon	hegon	Elk antler can be made into a spoon.
Towel	chelogehl	*Chelogehl* translates as "rib." An elk rib can be used as a towel to scrape water off the body.
Wedge	s'echoh	*S'echoh* is translated as "horn" or "antler." The same word is used for "wedge" because it is made of antler.

Note: Translations are from the Yurok Language Project dictionary website, Department of Linguistics, University of California, Berkeley. https://linguistics.berkeley.edu/~yurok/index.php.

*Empty cells indicate that a Yurok term was not found.

Many of the utilitarian objects mentioned by participants were tools used for carving and fishing. P2 explained that elk was important in making good fishing tools, including hooks, toggles, harpoons, and net shuttles. Another item mentioned was the elkhorn purses that men traditionally used to carry money (dentalia). Many of these items continue to be used today.

Elk in Ceremony

P1 and P2 expressed that elk are important to several ceremonies. For example, sinew is used to bind together items, such as the ceremonial dance feathers known as "morning feathers." One participant noted that elk sinew is preferred over deer sinew because it is "tighter." He noted that when morning feathers are bound with elk sinew, they have a certain way of bouncing more strongly, whereas when made with deer sinew, they are a bit limper. Elk hide also can be made into medicine pouches.

P3 referenced the spiritual component of elk materials beyond their utilitarian contribution to ceremony, particularly in relation to the requirement to "think only good thoughts" when crafting regalia or drums. This is important "because you're making . . . medicine. . . . [Y]ou have to put that [good] energy into that." P3 also explained that for several types of ceremonies it is important to feed others. This is good medicine, and there are specific protocols to show respect for the animal harvested. He also shared that he has given elk meat to ceremonies, feeling a responsibility to help others in the community.

Elk Habitat and Management

P1 said that elk populations were practically nonexistent a century or more ago. However, he has seen elk numbers increase throughout his lifetime. All the participants were glad that the elk have returned to Yurok ancestral lands, but they noted that their populations have grown much faster outside of the Reservation. P1 and P2 both referenced the benefit provided by wolf (*werhlkereeshneg*; *Canis lupus*) reintroduction to Yellowstone National Park, and P1 proposed that the reintroduction of predators such as wolves could help balance local elk populations on the Reservation in a similar way. P1 and P2 both placed emphasis on the importance of habitat such as prairies to healthy elk populations.

P1 and P2 referenced the contributions humans can make to maintaining elk habitat through traditional prescribed burns. Traditional burning clears brush and increases forage used by elk, a practice that P2 referenced as both cultural and ecological stewardship. P2 confirmed that in areas where traditional prescribed burns are being implemented, elk have been seen by community members. Although positive language, such as "we need to get to burning," was used regarding fire as a tool for elk habitat management, one participant noted that some nonnative grasses that were introduced by European settlers in areas around the Reservation, such as Redwood National Park, respond well to fire. Still, there was a general interest in expanding the reestablishment of traditional prescribed burns.

P1 described the role that elk play in sustaining their own ecosystem, indicating that they move back and forth between grass and brush patches, and there are "a lot of little trees that are getting killed every year . . . so that's almost like brush control." There was also mention of the need for broader cross-jurisdictional elk management between tribes and governmental agencies.

Continuance of Yurok TEK as Related to Elk

All participants described various aspects of TEK through the Yurok cultural lens, explaining how the lands were managed sustainably and responsibly before European settler contact; the Yurok people made sure to not overharvest to preserve the land for future generations. P2 indicated there may be value in continuing to harvest elk and teaching how to create items from elk "so that way we have that depth of . . . understanding and respect for the animal." He referenced the elk as "relations" and described the amount of work and commitment that goes into harvesting, processing, and producing utilitarian or ceremonial items.

P2 expressed that if elk harvest were "more openly allowed . . . then people might take the time to actually do more with it," to learn these skills. P2 and P3 noted that having respect for the elk, especially in relation to hunting, is a traditional perspective they aim to teach the younger people in their families.

DISCUSSION

The elk population in Humboldt and Del Norte Counties (i.e., in and around Yurok ancestral lands) was significantly more abundant prior to European colonization, and it is presumed that tribal subsistence hunting had minimal impact on elk populations. Intense hunting by European settlers began in the mid-1800s, and the elk's range was significantly reduced (NPS 2016). Today, Roosevelt elk in California persist only in Humboldt, Del Norte, Siskiyou, and Mendocino Counties (CDFW 2018, Moran et al. 2020). The participants in our study recognized this decline. That the higher historical abundance of elk provided significantly for the social-cultural evolution of the region is clear in the breadth and scope of the ways that elk was used for both utilitarian and ceremonial purposes. This is further supported by the finding that elk can be counted using the "tools" counting system, which may be indicative of their importance and utility in everyday traditional Yurok life. Elk remain an important part of Yurok cultural vitality: the species continues to contribute to subsistence and ceremonial use, though other utilitarian uses have declined in recent generations, presumably due to lack of access to elk to harvest.

Our results indicate that the Yurok relationship with elk is mutually beneficial. Interviewees described elk management as encompassing sustainable harvest practices, respectful use of the animals harvested, and proactive landscape management (e.g., traditional prescribed burning) to create and maintain habitat beneficial to the species. The interview participants expressed that elk contribute to the maintenance of the landscape through their foraging style and movement, which are distinct from other native (i.e., deer) and nonnative (i.e., feral cattle) ungulates in the region, and by supporting humans in a reciprocal relationship. Yurok people and their animal relations, including elk, mutually contribute to a balanced system by serving their particular functions in Yurok TEK (Ramos 2022).

Because of the diverse ways they contribute to cultural and ecosystem maintenance, elk within the Yurok system are possibly both a cultural keystone species and an ecological keystone species. An ecological keystone species is one "whose impact on its community or ecosystem is large, and disproportionately large relative to its abundance" (Power et al. 1996:609, Kotliar 2000). A cultural keystone species is one of particular importance to the ethnosphere it contributes to, referenced by Garibaldi and Turner (2004) as the "cultural component of the [E]arth's systems." In our study, P2 identified elk as a cultural keystone species for Yurok. Arguably, without elk in their part of the biosphere, the Yurok people would have developed a significantly different regional ethnoecology, making elk integral to the Yurok's cultural evolution.

The human-mediated ecosystem blurs the lines between a cultural and ecological keystone species within this system. Roosevelt elks' ability to help maintain their own ecosystem is part of what contributes to their role as an ecological keystone species. With their reduced presence, the vegetation communities in Yurok territory have lost a significant resource for maintaining the open prairies and oak woodland found pre-European settlement. The distinction between cultural and ecological keystone species becomes less clear due to the synergy of Yurok people and elk as co-managers of their shared ecosystem. For example, Yurok people instituted traditional burning in support of elk. The prohibition of traditional prescribed burning and unnatural fire suppression (Lake 2007) post-European contact have resulted in the reduction of historically maintained prairies and oak woodland to only 1% of their historical extent; the landscape has reverted to a "natural" or unmodified and unmanaged state (J. Hostler, pers. comm.). Elk habitat restoration was prioritized by the interviewees as one of the most important issues to address, often referenced hand in hand with the reinstitution of traditional burning. Restoration of traditional prairies for elk habitat might even mitigate catastrophic wildfires by reducing fuels as prairies expand to a more historical size.

Given the geographically expansive habitat needs of elk, one interviewee suggested the need for broader collaborative management across jurisdictional boundaries. Of particular note were the difficulties associated with the eradication of introduced grass species in Redwood National Park, which encompasses areas of Yurok ancestral lands and has one of the largest herds of elk (NPS 2016). Given the challenges, control, as opposed to eradication, efforts could be a feasible approach to nonnative grass management (K. Grantham, pers. comm., 19 July 2021). One participant suggested that the reintroduction of elk predators, such as wolves, could aid in maintaining ecosystem balance. Yellowstone

National Park reintroduced wolves in 1995, and research has shown the restoration of important ecosystem processes between wolves and elk (Mao et al. 2005). Perhaps there are opportunities for co-management of elk between Redwood National Park, the Yurok, and other local tribes and agencies.

Ultimately, the question arises of how the mutualistic relationship between Yurok and elk can be expressed in contemporary times and in practical ways. Contemporary hunting in Indigenous communities can be shaped by traditional moral codes that guide the community's hunting practices, highlighting the interdependence of epistemological, practical, and ethical dimensions of TEK (Reo and Whyte 2012, Yurok Tribe 2021). P2 noted that the reestablishment of the Yurok relationship with elk might be helped with more legal access. Interviewees also indicated a desire to reclaim the full scope of how elk were traditionally harvested and utilized. Increasing the capacity for hunting through increasing elk populations and access to harvest opportunities can provide complementary opportunities to train Indigenous youth in respectful and sustainable harvest practices, which are important to the continuance of Yurok TEK and a healthy ecosystem. Mechanisms for this reconnection with elk could also include continued revitalization of the Yurok language and more widespread community use. Subsistence hunting of elk also contributes to Yurok ceremonies in various ways, such as meat to feed the people at ceremonies and contribution to regalia. Both P2 and P3 spoke about the responsibility of sharing their harvest with their family and the larger community. Thus, increasing opportunities for harvest contribute to the spiritual vitality of the Yurok people, which in turn engenders respect for elk and a desire to support their continued persistence.

By exploring TEK, holistic solutions for long-term culturally sensitive monitoring, conservation, and management for elk and other wildlife species can potentially be achieved (Popp et al. 2020). This research provides valuable insight into how Yurok TEK can beneficially contribute to elk management restoration planning both within Yurok tribally managed lands and through co-management with relevant entities throughout Yurok ancestral territory. Through these interviews with Yurok-affiliated community members, we were able to identify ways in which elk have historically integrated with the sociocultural landscape in Yurok Country, and how they may do so again.

ACKNOWLEDGMENTS

We are especially thankful to the interview participants: Frank K. Lake, James Gensaw, and Dale Webster. We thank R. Matilton (Hupa/Yurok), wildlife major, Humboldt State University, for assistance with interview transcription. The Yurok Tribe Geospatial Information Technology Program kindly provided the map of Yurok ancestral lands for this chapter. We thank K. Grantham for comments that improved the final manuscript. This research was conducted in part under coperative agreement P17AC01144 through the Californian Cooperative Ecosystem Studies Unit (CESU) between the National Park Service and Humboldt State University Sponsored Programs Foundation. This research was conducted under a memorandum of understanding among the National Park Service, Redwood National Park, Yurok Tribe, and Humboldt State University Sponsored Programs Foundation, as a component of said CESU agreement. The material in this chapter is based in part on work supported by the National Science Foundation Postdoctoral Research Fellowship in Biology (Grant No. 1906338). Any opinions, findings, conclusions, or recommendations expressed in this material are those of the authors and do not necessarily reflect the views of the National Science Foundation.

LITERATURE CITED

Arsenault, R., C. Bourassa, S. Diver, D. McGregor, and A. Witham. 2019. Including Indigenous knowledge systems in environmental assessments: restructuring the process. Global Environmental Politics 19:120–32.

Bailey, V. 1936. The mammals and life zones of Oregon. North American Fauna No. 55. USDA Bureau of Biological Survey, Washington, DC.

California Department of Fish and Wildlife [CDFW]. 2018. Elk conservation and management plan. https://wildlife.ca.gov/Conservation/Mammals/Elk. Accessed 29 June 2021.

Dowsley, M., and G. Wenzel. 2008. "The time of the most polar bears": a co-management conflict in Nunavut. Arctic 61:77–89.

Garibaldi, A., and N. Turner. 2004. Cultural keystone species: implications for ecological conservation and restoration. Ecology and Society 9:3.

Garrett, A. 2014. Basic Yurok: survey of California and other Indian languages. Report No. 16. University of California, Berkeley.

Getches, D. H., C. F. Wilkinson, and R. A. Williams. 2005. Cases and materials on federal Indian law. 5th edition. West, St. Paul, Minnesota.

Harper, J. A., J. H. Harn, W. W. Bentley, and C. F. Yocom. 1967. The status and ecology of the Roosevelt elk in California. Wildlife Monographs 16:3–49.

Huntsinger, L., and L. Diekmann. 2010. The virtual reservation: land distribution, natural resource access, and equity on the Yurok forest. Natural Resources Journal 50:341–69.

Kotliar, N. B. 2000. Application of the new keystone-species concept to prairie dogs: how well does it work? Conservation Biology 14:1715–21.

LaDuke, W. 1994. Traditional Ecological Knowledge and environmental futures. Colorado Journal of Environmental Law and Policy 5:127–48.

Lake, F. K. 2007. Traditional Ecological Knowledge to develop and maintain fire regimes in northwestern California, Klamath-Siskiyou bioregion: management and restoration of culturally significant habitats. PhD dissertation, Oregon State University, Corvallis.

Lara-Cooper, K., and W. J. Lara Sr., editors. 2019. Ka'm-t'em: a journey toward healing. Great Oak Press, Pechanga, California.

Learn, J. 2020. Two-eyed seeing. Wildlife Professional 14:18–26.

Mao, J. S., M. S. Boyce, D. W. Smith, F. J. Singer, D. J. Vales, J. M. Vore, and E. H. Merrill. 2005. Habitat selection by elk before and after wolf reintroduction in Yellowstone National Park. Journal of Wildlife Management 69:1691–1707.

McCorquodale, S. M. 1997. Cultural contexts of recreational hunting and native subsistence and ceremonial hunting: their significance for wildlife management. Wildlife Society Bulletin 25:568–73.

Moller, H., F. Berkes, P. O. Lyver, and M. Kislalioglu. 2004. Combining science and Traditional Ecological Knowledge: monitoring populations for co-management. Ecology and Society 9:2.

Moran, A., K. Morefield, and K. Denryter. 2020. Report on spring aerial surveys of tule elk (*Cervus canadensis nannodes*) in the Mendocino elk management unit. https://nrm.dfg.ca.gov/FileHandler.ashx?DocumentID=181899. Accessed 27 July 2021.

Morris, C. B., and L. A. Eldridge. 2020. The heart of Indigenous research methodologies. Studies in Art Education: A Journal of Issues and Research 61:282–85.

National Park Service [NPS]. 2016. Redwood National and State Parks: 2015 herd unit classification and management of Roosevelt elk. National Park Service, US Department of Interior, Washington, DC.

Polfus, J. L., K. Heinemeyer, M. Hebblewhite, and Taku River Tlingit First Nation. 2014. Comparing Traditional Ecological Knowledge and Western science woodland caribou habitat models. Journal of Wildlife Management 78:112–21.

Popp, J. N., P. Priadka, M. Young, K. Koch, and J. Morgan. 2020. Indigenous guardianship and moose monitoring: weaving Indigenous and Western ways of knowing. Human-Wildlife Interactions 14:296–308.

Power, M. E., D. Tilman, J. A. Estes, B. A. Menge, W. J. Bond, L. S. Mills, G. Daily, J. C. Castilla, J. Lubchenco, and R. T. Paine. 1996. Challenges in the quest for keystones. BioScience 46:609–20.

Ramos, S. C. 2018. Considerations for culturally sensitive Traditional Ecological Knowledge research in wildlife conservation. Wildlife Society Bulletin 42:358–65.

Ramos, S. C. 2019. Sustaining hlkelonah ue meygeytohl in an ever-changing world. Pages 85–93 in K. Lara-Cooper and W. J. Lara, editors. Ka'm-t'em: a journey toward healing. Great Oak Press, Pechanga, California.

Ramos, S. C. 2022. Understanding Yurok Traditional Ecological Knowledge and wildlife conservation. Journal of Wildlife Management 86:1–21.

Ramos, S. C., T. M. Shenk, and K. M. Leong. 2016. Introduction to Traditional Ecological Knowledge in wildlife conservation. Natural Resource Report NPS/NRSS/BRD/NRR-2016/1291. National Park Service, Fort Collins, Colorado.

Reo, N., and K. P. Whyte. 2012. Hunting and morality as elements of Traditional Ecological Knowledge. Human Ecology 40:15–27.

Terry, G., N. Hayfield, V. Clarke, and V. Braun. 2017. Thematic analysis. Pages 17–37 in C. Willig and W. Stainton-Rogers, editors. The SAGE handbook of qualitative research in psychology, 2nd ed. SAGE Publications, Thousand Oaks, California.

Vales, D. J., M. P. Middleton, and M. McDaniel. 2017. A nutrition-based approach for elk habitat management on intensively managed forestlands. Journal of Forestry 115:406–15.

Weckerly, B. 2017. Population ecology of Roosevelt elk: conservation and management in Redwood National and State Parks. University of Nevada Press, Reno.

Yurok Tribe. 1993. Yurok Tribe Constitution. https://yurok.tribal.codes/Constitution. Accessed 10 June 2021.

Yurok Tribe. 2016. Resolution of the Yurok Tribal Council: tribal wildlife grant FY 2017 submission: Yurok Roosevelt elk restoration planning. Resolution No. 16-40. Yurok Tribe, Klamath, California.

Yurok Tribe. 2021. Draft of Yurok Tribe wildlife harvest ordinance. Yurok Tribe, Klamath, California. In possession of author.

Chapter 33

Kue Meyweehl 'esee kue 'Oohl Megetohlkwopew
The Elk and the Yurok People Take Care of Each Other

SEAFHA C. RAMOS *and* **JAMES GENSAW**

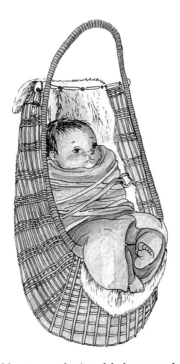

1. Kol' 'we-no'ohl keetee melee' mehl chaanuueks 'o 'Ernerr'.

2. Kue chaanuueks 'we-cheemos keech merwerkseeshon' 'ohlkuemee skewok kue 'we-rorowenek'. 'O ges', "Meyweehl tokseemek', kwesee skuyenee chemeen keetee hohkuemek'."

3. Hehlkeek 'o sepolah ko 'w-ekwsue' kue meyweehl. Kue 'we-too'mar keech 'o le'lohtehl sepolah kohtoh koma loksee'hl mee' kee skuy' soo huenem' 'o wey'. Kee skeyweenepeem' meyweehl.

379

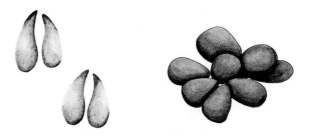

4. Noohl hehlkeek sootok'w 'o laayekw, 'ap newom' meyweehl 'ue-mechkah 'esee 'ue-melox.

5. Kohchew termerpernee meyweehl keech koh, noohl, 'o gem', "Wokhlew' mehl 'ne-kohchewok' k'ee meyweehl."

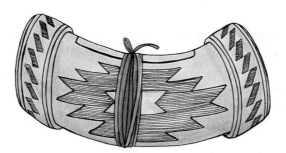

6. Noohl hohkuem' puueweesh, kerhl 'esee nohchuer. Meyweehl 'we-s'ech mehl hohkue'.

7. Noohl 'o ge's, "Kee nahchesek' 'ne-mey' kue kerhl 'esee kue nohchuer."

8. Kuy', kue pegerk 'we-toomar meloo'mehl noohl nahchelehl kue meyweehl 'we-nerperw. Kue we'yon 'we-chek 'o nergery 'ue-pewomek' mehl kue 'ue-meloyek'. Kue puuewomeen 'o gegoyehl, "Pewomekw kue meyweehl 'we-'ahspeyue'r!"

9. Syoolah kue we'yon. Syoohlah mehl muenchehl; hohkue' mehl meyweehl 'we-skoy.

10. Merwer'y noohl weryerkerhl. Meskwee' mehl kue chaanuueks noohl kue we'yon 'o gem', "Nooluechek, kue 'ne-meechos. Kos'ela son' kee k'-ewechek'." Kwesee 'o nepue' kue meyweehl 'w-ahspeyue'r. Kue meyweehl 'esee kue 'oohl megetohlkwopew.

ENGLISH TRANSLATION

1. One time there was going to be a brush dance for a baby at 'Ernerr' [a village at Blue Creek].

2. The baby's uncle became spiritually clean because he wanted to go hunting. He thought, "I respect the elk, so I will make strong medicine [for hunting]."

3. Elk can be found in prairies way up in the mountains. His family had burned a prairie the previous year so that it would grow well there. Then the elk can eat well.

4. Then he went on a trail into the mountains, and then he saw elk tracks and scat.

5. He harvested a six-point elk and he said, "Thank you that I have caught this elk."

6. He then made a purse, a pair of earrings, and a hair stick. The items were made with the elk's antlers.

7. He then thought, "I'll give the earrings and hair stick to my daughter."

8. Later, the man's family went to the brush dance, and they gave the elk meat. The girl's mother helped cook for the dance. The cooks were told, "Cook some elk stew!"

9. The girl danced in the brush dance. She danced with a ceremonial dress; it was made with elk hide.

10. The final round was danced, and it was the end of the brush dance. Medicine had been made for the baby, and the girl said, "I love you, Cousin. I wish that you will live a good life." And then the elk stew was eaten.
The elk and the Yurok people take care of each other.

DRAWING CONNECTIONS / ABOUT THE STORY

In the Yurok worldview, Traditional Ecological Knowledge (TEK) is a system with various components, such as physical management (e.g., hunting or traditional prescribed burning) and spiritual management (e.g., ceremony). Yurok people are a part of this system and, together with wildlife, help balance the world. Although wildlife are taken during hunting, cultural protocols are intended to help people to harvest sustainably (Ramos 2022). This Yurok language story highlights aspects of the relationship between Yurok people and elk (chapter 32 of this volume). Since language is a component of TEK (Ramos 2022), communicating lessons through a story framework facilitates ongoing efforts in TEK and language revitalization in the Yurok community.

The man's family in the story maintains the prairie through traditional burning, taking care of the elk. The girl participates in the brush dance with a dress made with elk hide, and thus the elk's spirit contributes to the continuance of TEK. It is not through his physical action, but through the giving of his life that the elk helps take care of the Yurok people. A moral of the story is that the elk and the Yurok people take care of each other.

Yurok was historically an oral, not a written, language. In the 1950s it was claimed that not more than 20 people had conversational speaking ability beyond words and phrases (Robins 1958). Various writing systems, such as unifon and the New Yurok Alphabet, have been used over the last century to document the language (Robins 1958, Trull 2003, Garrett 2014). To this day, there continue to be different ways of spelling used among the Yurok Tribe, community members, and programs such as the Yurok Language Project at the University of California, Berkeley. At the heart of efforts to create materials, such as this story, is the common goal of community language use and continued language revitalization. We have used a combination of the Yurok Tribe's New Yurok Alphabet and the spelling of the University of California, Berkeley, Yurok Language Project (Garrett 2014).

ACKNOWLEDGMENTS

We are deeply thankful to the Yurok language community, including those who have passed on (*'aawok*) and those who will come, for all the efforts made and love given to keep the language alive. The beautiful illustrations in this story were created by Melitta Jackson (Yurok/Hupa/Karuk/Modoc). We thank A. Garrett for a review that improved the final version of this story. The material in this chapter is based in part on work supported by the National Science Foundation Postdoctoral Research Fellowship in Biology (Grant No. 1906338). Any opinions, findings, conclusions, or recommendations expressed in this material are those of the authors and do not necessarily reflect the views of the National Science Foundation.

LITERATURE CITED

Garrett, A. 2014. Basic Yurok. Survey Report No. 16. Survey of California and other Indian languages. University of California, Berkeley. https://escholarship.org/uc/item/2vw609w4. Accessed 4 August 2022.

Ramos, S. C. 2022. Understanding Yurok Traditional Ecological Knowledge and wildlife conservation. Journal of Wildlife Management 86:1–21.

Robins, R. H. 1958. The Yurok language: grammar, texts, lexicon. University of California Publications in Linguistics No. 15. University of California Press, Berkeley.

Trull, G. 2003. Georgiana Trull's Yurok language conversation book. Center for Indian Community Development, Humboldt State University, Arcata, California. https://linguistics.berkeley.edu/~yurok/web/YLCB.php. Accessed 11 August 2022.

Chapter 34

Power Parade in Pablo, Montana

| SERRA J. HOAGLAND

A coyote walks down the road and sees another coyote. They walk in tandem while sharing whatever it is coyotes talk about.

In the distance, they see a group of kids and two bikes. One young girl has dark, straight hair that hasn't been brushed since before school in the morning and custom self-cut bangs creeping closer to her eyes. Girl 1 wears a purple shirt. Another girl, blonde, who appears to be her best friend, is equal in height to Girl 1, but is walking her bike alongside the group. A much taller girl towers above the others. She's quiet, a little more reserved, but still part of the gang. A young boy rides his bike out in front of the group, with a bag tied under the seat.

The coyotes slow down as they are approached by what is the closest thing to a circus around these parts. Coyote 1 notices Girl 1 has a stick in her hand.

Girl 1 proudly displays the stick in her hand over her head like it's a spirit stick with all the power in the world, and announces to the coyotes, "We are having a power parade!"

"That's some kind of Sherman Alexie type of shit," Coyote 2 mumbles to Coyote 1.

The little girl, unphased by their remarks, not listening to the side conversation, raises her stick higher over her shoulder, so that she is almost standing only on one foot, looking to the Sky.

She repeats, so the whole street can hear her this time without mistake, "We are on a power parade!"

The cadre of children join forces again, circling around both coyotes and continuing their walking-that-is-more-like-dancing. Kids have that ability. The young boy gets off his bike, hands it to the tallest girl, reaches out to pick up some trash from the street, and gently places it into the plastic bag tied to the bike.

The gang continues with their parade, the stick waving back and forth, over their heads, side to side, to the ground, back and forward.

Below their feet, the wounds in the ground heal, the people around Her recover and sing. The entourage dances, and the sounds of drums and singing fill the houses on every street and the infinite space between all the trees on the Reservation.

ACKNOWLEDGMENTS

This story is inspired by and dedicated to Cienna, Chenoa, Keilan, the Spike Ct crew (Abby, Alex, Xavion), and all the youth in Indian Country who are the answers to our ancestors' prayers.

Index

adaptive management, 7
Agdaagux Tribe, of King Cove, AL, 177
agriculture-based land leasing, 324–25
Ahtna Intertribal Resource Commission, 216
Ahtna language, 35
Aikin, Scott, 39–44
Alaska Conservation Foundation, 187
Alaska Department of Fish and Game, 185, 215, 217, 218
Alaska Marine Tissue Archival Project (AMMTAP), 174–75
Alaska Maritime National Wildlife Refuge (AMNWR), 172
Alaska Migratory Bird Co-Management Council (AMBCC), 206, 214, 215–22; Alaska Native Caucus, 215, 220; governance structure and processes, 216–17; Harvest Assessment Program, 219
Alaska National Interest Lands Conservation Act (ANILCA), 183, 209
Alaska Native Claims Settlement Act (ANCSA), 184, 185–86, 208–9
Alaska Native peoples, 12–13, 101, 169–70, 208; federally-recognized villages, 100, 135, 208; Indigenous Sentinels Network, 169–80; legal definition, 214, 218; migratory bird harvesting, 206–22; regional corporations, 135, 138, 184, 185–86, 188, 209; socioeconomic and cultural changes, 208–9; subsistence-cash mixed economy, 209–10
Aleut Communities, of St. Paul and St. George Islands, AL, 169–80
Aleutian Pribilof Islands Association, 216
alewives, 289, 292, 294
algae, 186, 201, 301
alligators, American (*Alligator mississipiensis*), 140, 300, 301
Alutiiq-Sugpiaq, 208
American Eugenics Society, 7
American Indian/Native Alaska Native, pre-Columbian populations, 12–13
American Indian Science and Engineering Society, 35, 241
Anishinaabe people, 45–46, 107–8, 109, 248–52, 276
antelope. *See* pronghorn
Antiquities Act, 61, 63, 70n57, 71n64
Antonio, John, 25–26, 27–28, 32
Apache Tribe, 5, 18; Havasupai, 101; Hualapai, 101; Mescalero, 5, 25, 27, 31, 32–33, 143n7; San Carlos, 130n31, 360;
White Mountain, 5, 14, 18, 25, 27–28, 31, 32–33, 101, 142, 143n3, 144n14, 360
Arapaho Tribe, 7, 81–82; Northern, 100
assimilation, of Indigenous people, 68n14, 124, 143n5, 146, 233, 275
Assimilation and Allotment Period, 13
Assiniboine (Nakota) Tribe, 8, 76–77; Fort Peck, 112
Association of Village Council Presidents, 216
Athabascan Tribe, 208, 321
Aurapakan, Chief (Sinixt Tribe), 152
avocets, 340

Babbitt, Bruce, 59, 112
badgers, 153, 263
Badlands National Park, 323–24, 327
balance, as Indigenous cultural value, 252, 309, 366–67, 382
Bald and Golden Eagle Protection Act, 139–40
Banks, John, 27–28
Barr, William, 153, 160n41
basket weaving, 25, 309–11
bats, 18, 47, 289, 294
Baumflek, Michelle, 310–11
bears (*Ursus*), 46, 254, 292; black (*U. americanus*), 108, 147, 149, 254, 272, 292, 305; brown (*U. arctos*), 147, 149, 256–60, 329; grizzly (*U. arctos horribilis*), 147, 149, 152, 155, 193, 329; hunting, 146, 147, 149, 152, 155, 290, 291–92
Bears Ears Commission, 62, 63
Bears Ears Inter-Tribal Coalition, 60, 61–62, 63, 68n9, 70n44, 101
Bears Ears National Monument, 54, 58, 60–63, 66, 70n52, 70n57, 101; consultation process regarding, 54, 58, 60–63, 66, 101
Beattie, Mollie, 26, 28
beavers (*Castor canadensis*), 46, 114, 147, 148, 152, 153, 155, 277–78, 297, 300, 303, 305; dams, 147, 281, 297, 357–57; as nuisance species, 279, 280, 281
Belloni, Robert, 58–59, 69n28
Belt, Tom, "Tōhi: The Cherokee Concept of Well-Being," 309
Bering Sea, 16, 175–76, 221. *See also* Pribilof Islands
BeringWatch, 170–71, 177
Bich, Joel, 262
Biden, Joseph, 40, 42, 43, 60, 63, 65, 66, 71n64, 101, 349

387

big game hunting and management, 19, 27, 29, 146, 156–57, 253–54. *See also names of individual game species*
biodiversity, 46, 47, 135, 139, 165, 229, 316, 332, 333
Biodiversity Research Institute, 292
birds, 316; in Alaska Native cultures, 210–12; O'odham names for, 338–45; population decreases, 208, 213, 221, 278; as subsistence food, 209–10; threatened or endangered, 348. *See also* avian species richness; Quitobaquito Springs ('A'al Vaupia); shorebirds; waterfowl; *names of individual bird species*
Bishop, Rob, 60–61
bison *(Bison bison)*, 110, 113, 273, 327; cultural significance, 18, 110, 111, 112, 113, 316, 325, 327; restoration and management programs, 17, 18, 110–13, 117, 193, 254, 316, 318, 325
Bison Stewardship Strategy, 112
bitterns, 341
Black, Samuel, 150
Black Elk, 313
Blackfeet Community College, 200, 201
Blackfeet Reservation, 200, 201
Blackfeet Tribe, 27, 200, 201
Blaeser, Kimberly, "Glyph," 190–93
Blazer, Arthur "Butch," 31–33, 35
blueberries *(Vaccinium* spp.), 45, 183, 276
blue jays *(Cyanocitta cristata)*, 303
boarding schools, 23, 208, 233, 259, 317
bobcats *(Lynx rufus)*, 108, 279, 282, 300, 315, 328
Boldt decision *(United States v. Washington)*, 16–17, 102–3, 105, 114
Bonneville Power Administration, 114
Bouchard, Randy, 146, 148
Boundary Waters Canoe Area, 278
Bradley, Bill, 27–28
Bragg-Gabriel, Rosalie, 153, 154
Bristol Bay Native Association, 216
British Columbia, Indigenous Guardians programs, 186
Brown, Paul, 347
Bryan, Kirk, 336
Buffalo, Henry, Jr., 106, 109, 115, 117
buffalo. *See* bison
Bureau of Indian Affairs (BIA), 15, 26, 27, 29, 31, 64, 76, 104, 255, 261, 275, 294; Northeast Region, 293; Relocation Program, 14
Bureau of Land Management (BLM), 64–65, 66, 71n67
Bush, George W., 68n17

Cahuilla Indians, Agua Caliente Band, 83–85
California Coastal Commission, 220
Campbell, Ben, 26
Campbell, Carmilita Theresa, 153, 160n48
Campbell, Noel, 154
Canada, 324; co-production of knowledge projects, 240–41; Intertribal Timber Council, 24; migratory bird treaty, 206, 213, 214, 218; Wildlife Act, 159
cane, 335–36; river *(Arundinaria gigantea)*, 309–12
caracaras, 342
cardinals, northern *(Cardinalis cardinalis)*, 337, 345, 348
caribou/reindeer *(Rangifer tarandus)*, 146, 147, 148, 149, 153, 155, 173, 357
Carlisle, Ashley, 35–38
Carrasco, Diego, 335
Carroll, Clint, 311
Carson, Rachel, 6–7
Carufel-Williams, Sally, 34, 35
ceremonial and religious activities: brush dance, 379–84; ceremonial sweats, 154; government prohibition, 154; of O'odham peoples, 332, 346; of Sinixt Tribe, 154, 156; Spirit Quests, 154; Sun Dance, 317; treaty rights related to, 121
ceremonial and sacred objects, 44, 139–40, 212, 215, 233–34, 272. *See also* sacred places
ceremonial significance: bison, 325; cottonwood trees, 317; eagles, 139–40; elk, 367, 373, 374, 375, 376; plants, 274; squirrels, 271. *See also* cultural significance; sacred places
cetaceans, 175. *See also* whales
Chaco Culture National Historical Park, 239, 244n1; Chaco Canyon oil and gas lease sales, 58, 63–66, 67, 71n64, 71n67
Chaffetz, Jason, 60–61
char, brook *(Salvelinus fantinalis)*, 357
Cherokee identity, 235
Cherokee language, 309–10, 311, 312
Cherokee Nation, 7, 229, 309–12, 355; Eastern Band, 18, 239–40; Environmental Resources Group, 310; Medicine Keepers, 311; principle of balance (tōhi and osi), 309, 312; river cane basket weaving tradition, 309–11
Chief Seattle, 25
children and adolescents. *See* Native youth
Chino, Wendell, 32–33
Chinoo (Winter) Dances, 154
Chippewa Ottawa Resource Authority (CORA), 17, 100–101, 104, 117

INDEX **389**

Chippewa Tribe: Eastern Band of Cherokee Indians, 239–40; Fond du Lac Band, 5, 9–10, 275–78; Lac Courte Oreilles Band, 106, 107, 108–9, 127, 130n32, 131n44; Lake Superior, 106, 275; Lake Superior, Bad River Band, 34; Lake Superior, Lac du Flambeau, 34, 109; Minnesota Mille Lacs, 34; Red Cliff Band, 106; Sault Ste. Marie Band, 117
Chischilly, Samuel, 35–38
Choctaw Tribe, Mississippi Band, 19, 279–81
Choctaw Wildlife and Parks Department, 279–81
Christian, Alexander, 155
Christian, Mary, 155
Christianity, 23, 248, 252
chronic wasting disease, 19, 264
Chugach Regional Resources Commission, 216
Chugach tribes, 220
civil rights, 14, 109, 131n40
clans, 181–82, 191, 272–73, 275, 296, 309
Clark, William, 194–95
climate change, 6, 39, 40, 42, 158, 232, 278, 347, 349, 366; impact on Arctic and subarctic ecosystems, 170, 175, 176–77, 184, 209, 361; tribal adaptation plans, 18–19, 105, 255
Clinton, Bill, 57, 58–59, 61, 62, 65, 68n17
Coastal Stewardship Network (British Columbia), 186
Cobell, Gerald "Buzz," 24, 25, 27–28, 35
Cobell settlement, 41
Coeur d'Alene Tribe and Reservation, 85–86, 92–93, 97n129
Cohen's Handbook of Federal Indian Law, 54, 67n2
collaboration, 6, 10, 328; differentiated from co-production of knowledge, 231; Indigenous Stewardship Model approach, 319, 321, 322, 323, 328; in research, 230, 231–34, 263–64, 358
collaboration, federal-tribal, 19, 30, 262; Bears Ears Inter-Tribal Coalition, 60, 61–62; bison restoration, 112, 113; elk management and conservation, 374, 375–76; Endangered Species Act issues, 141–42; Indigenous Guardians Network, 181–89; of intertribal wildlife organizations, 112, 113–14, 116–17; as NAFWS priority, 29; salmon restoration, 113–14; US Fish and Wildlife Service partnerships, 142. *See also* co-management
collaboration, state-tribal, 19, 230, 262
Colorado Parks and Wildlife, 254
Colorado State University, 35, 325
Columbia Basin Fish Accords, 113–14
Columbia River Inter-Tribal Fish Commission (CRITFC), 17, 27–28, 100–101, 104, 113–14
Columbia River Inter-Tribal Police Force, 104
Columbia River treaty tribes, 58, 69nn28–29
Columbus, Christopher, 12–13, 234–35
Colville Reservation, Confederated Tribes of, 27–28, 87–88, 89, 100, 145, 156–58. *See also* Sinixt Tribe
co-management, 15–16, 138–39, 229; across international border, 145–46, 156–59; Alaska Migratory Bird Co-Management Council, 206, 214, 215–22; Bears Ears National Monument, 60, 61–62; definition, 215; elk, 376; legal basis, 16–17; marine mammals, 172; salmon, 58, 101; sicklefin redhorse, 239–40; in Southeast Alaska, 188; trumpeter swans, 196–98
Comcáac people, biosystematic lexicon, 356–57, 358
Confederated Salish and Kootenai Tribes, 18, 27–28, 193–205
Conroy, Pat, 152–53
conservation easements, 129n3, 142
conservation ethic, 6–7, 9–10
conservation movement, 6–7, 11n3
consultation, federal-tribal, 8, 53–72; case studies, 58–66, 68n27; collaborative management framework, 54, 61–62, 67n8; executive orders-based, 8, 53, 55, 57, 58–59, 61–62, 65, 66, 68n17, 138, 231–32; government-to-government framework, 53–54, 55, 57, 62, 66, 68n14, 68n17, 137–39; under National Historic Preservation Act, 54, 55–57, 64–65, 68n23
consultation, in co-production of knowledge, 230, 235, 236
Convention for the Protection of Migratory Birds, 213, 214
coots, 337–38, 340
co-production of knowledge, 6, 229–47; authorship equity and inclusion in, 236–37, 242–44; tribal consultation in, 230, 235, 236
cormorants, 207, 341
cottonwood trees (*Populus* spp.), 317, 338
Court of Appeals decisions: *Agua Caliente Band of Cahuilla Indians v. Coachella Valley Water District*, 83–84, 83–85; *Colville Confederated Tribes v. Walton*, 87–88, 89, 90; *Kittitas Reclamation District v. Sunnyside Valley Irrigation District*, 90–91, 93; *Lac Courte Oreilles Band of Lake Superior*

Court of Appeals decisions (*cont.*)
Chippewa Indians v. Voigt, 107, 127; *Michigan v. United States Army Corps of Engineers*, 128; *Navajo Nation v. US Forest Service*, 101; *US v. Adair*, 88–90, 92; *US v. Dion*, 140; *US v. Washington* (Boldt decision), 16–17, 102–3, 105, 114
Courtois, Valerie, 186
Coushatta Tribe, 269
COVID-19 pandemic, 65, 71n67, 71n72, 72n85, 113, 274, 291, 347
coyotes (*Canis latrans*), 153, 263, 279, 293, 294, 301, 315, 329, 336, 384
Crabb, Barbara, 108–9
crabs, Dungeness (*Metacarcinus magister*), 183
cranes, 207, 208, 211, 296
creation stories, 22–23, 248, 296
Creek Tribe (Muscogee Nation), 267–74, 296; Ocmulgee Mounds ancestral homeland, 267–70, 272–74; Trail of Tears, 268, 273
crissal thrashers (*Toxostoma crissale*), 337, 343
Crow Tribe, 14, 100
cui-ui (*Chasmistes cujus*), 142
cultural appropriation, 40
cultural review boards, 242
cultural significance: baskets, 25, 309–11; bears, 256–60; Bears Ears National Monument, 54, 58, 60–63, 101; birds, 211, 212; bison, 18, 110, 111, 112, 113, 316, 325, 327; Chaco Canyon, NM, 58, 63–66, 239, 241n1; drums, 25; elk, 367, 370, 373, 374, 375, 376, 379–83; fish and fisheries, 88; intertribal areas, 101; migratory bird harvesting, 210–12, 217–18; old-growth timber, 188; small mammals, 238; subsistence activities, 210, 212
cultural sustainability, 275
cultures, Indigenous, suppression, 23, 233, 367–68
currants (*Ribes* spp.), 183
Cushing, Frank Hamilton, 233
cypress, 274, 296

dams, 9, 23–24, 73, 87, 90, 94n1, 114, 151, 239, 262; of beavers, 147, 281, 297, 357–58; removal, 18, 68n27, 114
Dawes Act, 13, 136, 143n5
Death Valley National Monument, Devils Hole reserved water rights, 78–79, 80
DeConcini, Dennis, 26, 136, 143n5
deer, 276, 292–93, 305, 323–24; mule, 253, 254, 323–24; Sitka black-tailed (*Odocoileus hemionus sitkensis*), 183; white-tailed (*Odocoileus virginianus*), 108, 279, 282, 292, 323–24

deer hunting, 108, 145, 146, 147, 149, 150–51, 153, 156, 158–59, 271, 272, 279, 290, 293, 327; with dogs, 147, 150–51; night hunting, 109; with snares, 277
Desautel, Richard Lee, 145, 154, 158–59
desert oasis. *See* Quitobaquito Springs
De Smet, Pierre-Jean, 194–95
Devils Hole pupfish (*Cyprinodon diabolis*), 78–79
Dingell Act, 267
Dingell-Johnson Federal Aid in Sport Fish Restoration Act, 10n1, 15
District Court, OR: *Sohappy v. Smith*, 58, 69nn28–29; *US v. Oregon*, 58, 69n28
District Court, SD, *US v. Turtle*, 140
Dodge, Natt N., 338
dogs, 147, 150–51, 210–11, 278, 283–84, 348
doves, 337, 339, 348
Dompier, Doug "The Old Dog," 24, 25, 27–29
DuBray, Fred, 110
ducks: blue-winged teals (*Anas discors*), 337; bufflehead (*Bucephala albeola*), 176, 337; of desert oasis (Quitobaquito Springs), 335–36; eiders (*Somateria* spp.), 172, 175, 176, 358; goldeneye (*Bucephala* spp.), 175, 176; harlequin (*Histrionicus histrionicus*), 172, 211, 357; harvest in rural Alaska, 211, 217; hunting regulations, 207, 213–14; long-tailed (*Clangula hyemalis*), 172, 176; mergansers, red-breasted (*Mergus serrator*), 172; northern pintail (*Anas acuta*), 337, 338–39; northern shovelers (*Spatula clypeata*), 337; scoters, white-winged (*Melanitta deglandi*), 172, 175; sea, 171, 172–73, 175–76, 217; wood (*Aix sponsa*), 297, 338
duck stamps, 220

eagle feathers, ceremonial use, 44, 139–40
eagles, 16, 305; bald (*Haliaeetus leucocephalus*), 43, 44, 139–40, 360; golden, 329; spotted, 329
ecology, 353–54, 356
economy, local community-based, 314–16
Ecosystem Conservation Office (ECO), 171–72, 173–75, 177
ecosystems: Arctic and subarctic, climate change impact, 170, 175, 176–77, 184, 209, 361; benefits of bison to, 113, 316; geopolitical boundaries, 99–100; Indigenous understanding of, 9, 314–15, 316, 317; of tribal lands, 229
Eden, Jim, 338
Edward, Chief (Sinixt Tribe), 153
egrets, 337, 341, 348
Eiler, Lorraine Marquez, 333

INDEX **391**

elders, tribal, 10; in co-production of knowledge, 116, 229, 232, 237, 238, 242; participation in NAFWS youth practicums, 34; respect for, 43; in transmission of TEK, 326–27, 333
elk, 273; ceremonial and cultural significance, 367, 370, 373, 374, 375, 376, 379–93; hunting, 5, 148, 149, 151, 153, 156, 158; Roosevelt (*Cervus canadensis roosevelti*), 366–78
elk management and conservation, 5, 41, 157, 193, 254, 261, 367, 374; TEK-based, 366–78
Emerson, Ramona, "Shash," 256–60
endangered or threatened species, 15–16, 32, 156, 236, 254, 255, 263, 271, 272, 291, 300; cultural significance, 18; freshwater aquatic, 46, 78–79, 334; marine species, 169, 170, 171–75, 359; of Quitobaquito Springs desert oasis, 334, 346–47, 348; TEK-based approach, 354–57, 359–60. *See also* recovery and restoration programs
Endangered Species Act: consultation process, 54, 58–60, 139, 140–41; executive branch directives, 58–60, 62, 66, 68n11, 137–38, 141–42; tribal perspectives, 135–44. *See also* endangered and threatened species; recovery and restoration projects
environmental impact statements, 47
environmental justice movement, 184
Environmental Protection Agency (EPA), 115, 201, 289
Environmental Systems Research Institute (ESRI), 42, 44n1
equal footing doctrine, 123, 124, 231
ethnobotany, 229, 319, 322, 326–27
European colonization, of Indigenous lands, 13, 121–22, 146, 147, 150, 151, 157, 169–70, 233, 373, 374
Everglades, 282–86
executive branch administrative directives, 137–38, 142
executive orders: for establishment of reservations, 77, 78, 84; for tribal consultation, 8, 53, 55, 57, 58–59, 61–62, 65, 66, 68n17, 138, 231–32
Eyak Tribe, 208

Fairbanks, Ed, 27–28
fair share doctrine, 58, 99–100, 125
Faith, Mike, 113
falcons, 342
Family Planning Services and Population Research Act, 233
Father Kino's tryonia springsnail (*Tryonia quitobaquitae*), 334

Federal Code of Regulations, 24
Federal Subsistence Management Program, 183
federal-tribal relationship: historical overview, 12–14. *See also* consultation, federal-tribal; federal trust responsibility; *names of individual federal departments and agencies*
federal trust responsibility, 8–9, 16, 53–54, 123, 130n19, 135–36, 137; conflicts with tribal sovereignty, 137; Endangered Species Act and, 138; government-to-government relationship, 53–54, 55, 57, 62, 66, 68n17, 137–39; reserved water rights, 74, 83, 84, 85–86; tribal lands removed from, 14
ferns, fiddlehead (*Matteuccia* spp.), 183
ferrets, black-footed (*Mustela nigripes*), 18, 142, 262–63
figs (*Ficus* spp.), 346
finches, 344
fires and fire management, 7, 19, 157–58, 276, 278, 281, 373, 374–75, 382
First Nations: co-production of knowledge projects, 240–41. *See also* Canada
fish, 23; barriers to passage, 23–24, 87, 90, 114, 239; community-based monitoring projects, 177; counts, 114; cultural significance, 115; hatcheries, 16, 17, 27, 105; invasive, 282; mercury tissue sampling, 294; water quality issues, 115, 116. *See also individual fish species*
fishers (*Martes pennant*), 108
fishing and fisheries, 318; Alaska Native peoples, 169, 176–77, 185, 209–10; commercial, 322; competition with northern fur seals, 170; electro-fishing, 294; Indian Interim Fisheries Program, 31–32; National Fish Hatchery closures, 27; permits and regulations, 104, 105, 117, 123, 271; recreational, 32, 264; restoration projects, 185–86; seasonal practices, 148, 149; spearfishing, 109; sustainable, 7
fishing rights: Boldt decision, 16–17, 102–3, 105, 114; Confederated Salish and Kootenia Tribes, 193; fair share doctrine, 58, 99–100, 125; intertribal, 100–101; Lake Superior Chippewa, 278; off-reservation, 14, 17, 90–94, 100–101, 102–3, 104, 106, 108–9; reserved water rights for, 87–94; salmon fisheries, 58, 69nn28–29, 90–92, 100, 140; as tribal rights, 30; Yakama Nation, 69n29, 90–92, 102–3
Flathead Indian Reservation. *See* Salish (Flathead) Tribe
Floyd, James R., Chief, 267–74

flycatchers (bird species), 141, 342, 345, 346–47
Fond du Lac Reservation, MN, 275–78
food plots, 262, 276
food security, 169–80, 185, 206, 210
forestry and logging, 32–33, 108, 157–58, 185, 188, 276, 278, 290, 321, 354
Fort Belknap Indian Community, 201
Fort Belknap Reservation, 76–77, 200
Fort Belknap Tribal Council, 200
foxes, 262–63, 300, 304
Frank, Jr., Billy, 105
frogs, 46; bullfrogs, 203; northern leopard (*Lithobates pipiens*), 198–203; Sierran treefrogs (*Pseudacris sierra*), 202
funding: collaborative research, 232; co-production of knowledge projects, 240–41, 242, 243; federal trust responsibility, 26, 294
funding, of tribal fish and wildlife management programs, 6, 137–38, 253, 254, 262, 264, 289–90, 292, 293; inequity and inadequacy, 15–16, 26, 29, 33; species restoration projects, 263
fur trade, 148, 149, 150, 151, 169–70, 208

Gadsden Purchase, 333
gallinules, 340
Garrett, Jim, 34
gathering activities, 146, 148–49, 150, 151, 152, 169, 209–10, 326–27
gathering rights, 17, 87–90, 106, 108–9, 123, 124, 193, 278
geese, 211, 215, 217, 219, 220, 261
General Allotment Act (Dawes) Act, 13, 136, 143n5
geographic information system (GIS), 37, 42, 44n1
Geography of Memory: Recovering Stories of a Landscape's First People (Pearkes), 149, 155
George, Levi, 24, 27–28
Gila National Forest, 79, 80
ginseng, American (*Panax quinquefolius*), 108
godwits, bar-tailed, 207
Gorman, Amanda, 43
Gorsuch, Neil, 122–23
Gover, Kevin, 32
government-to-government framework, 53–54, 55, 57, 62, 66, 68n14, 68n17, 137–39
grass and prairie management, 113, 374, 375
Grassel, Shaun, 261–64
Greater Chaco Coalition, 64
Great Lakes Indian Fish and Wildlife Commission (GLIFWC), 17, 18, 34, 100–101, 104, 105–9, 110, 114, 115, 116, 117, 118n2
grebes, 207, 337, 339, 346

grosbeaks, 338, 345, 346
Gros Ventre (Aaniiih) Tribe, 8, 76–77
grouse, 153, 318
Grunlose, Philip, 153
Guardian Watchmen (Canada), 186–87
gulls, 211; glaucous-winged (*Larus glaucescens*), 172; Heermann's (*Larus heermanni*), 341; ivory (*Pagophila eburnean*), 358; monitoring program, 171, 172–73, 176; ring-billed, 297
Gwich'in people, 40
Gwin, Pat, 310

Haaland, Deb, 20, 63, 71n62, 190
habitat: connectivity, 18; geopolitical boundaries, 99–100; protection and restoration, 7, 29, 39, 125, 254, 262, 264, 316, 317, 332, 357; trust funds, 262
Haida Tribe, 181–89, 208
Hall-Cleveland, Sheilah, 154
Harmon, Alexandra, 103–4
Harvard University, Academic Research Guidelines, 236–37
Havasupai Tribe, 101, 234
hawks, 328, 341–42, 348
Heaton, Timothy, 244
herons, 45, 296, 305, 337, 338, 341
Herrera, Clayvin, 100
herrings, Pacific (*Clupea pallasii*), roe, 183
Heye, George Gustav, 234
Hoagland, Serra J., "Power Parade in Pablo, Montana," 384–85
hogs, feral, 19, 271, 279, 280
holism, 20, 157, 232, 315, 316, 320, 353–54, 376
Hollowed, John, 105
Hoonah community, 185
Hoonah Indian Association, 185, 187
Hoonah Native Forest Partnership, 185, 188
Hopi Cultural Preservation Office, 234, 238–39
Hopi language, 238–39
Hopi Tribe, 8, 9, 26, 101, 238–39
Houten, Beverly, 27–28
Hudson's Bay Company, 148, 149, 150, 151, 155, 195
Huey, Laurence M., 337, 338–45
human remains, 24, 234, 244, 332
hummingbirds, 339, 346–47
Huna Totem, 185
hunting: ceremonial, 368; codes, 253; commercial operations, 261; cross-border, 100; illegal, 198; lodges, 261; at night, 109; permits, 5, 14, 15, 142, 207, 254, 271, 290, 291, 311, 314; reciprocity in, 315–16; recreational, 368; seasonal patterns, 147, 151;

sustainable, 7; traditional moral codes of, 156, 376; use of dogs in, 147, 150–51, 160n31; young people's training in, 212, 255. *See also names of specific bird and animal species*
hunting regulations and codes, 123–24, 198, 255; on Creek Nation lands, 271–72; enforcement, 29, 32; historic tribal hunting rights *versus,* 156, 158–59; lead ammunition prohibitions, 198; migratory birds, 108, 206–8; tribal opposition, 30
hunting rights: across international borders, 145–63; under Alaska Native Claims Settlement Act, 208–9; Confederated Salish and Kootenia Tribes, 193; intertribal, 100–101; Lake Superior Chippewa, 278; for non-Indians, 14; off-reservation, 14, 17, 90–94, 100–101, 104; protection, 124–29; reserved water rights for, 89–90; as treaty rights, 14, 16–17, 100–101, 106, 193; as tribal rights, 30–31, 156, 158–59. *See also* treaty rights, for fishing, hunting, and gathering, off-reservation (usufructuary)
hunting seasons, 171, 206–7, 211, 213, 214, 215, 216, 217–18, 262, 281
Hydaburg community, 185

ibis, white-faced *(Plegadis chichi),* 337, 341
Indian Civil Rights Act, 14
Indian Claims Commission, 93, 151
Indian Fish and Wildlife Resource Management Act, 26
Indian Health Service, 233
Indian Interim Fisheries Program, 31–32
Indian peaches, 274
Indian Removal Act, 13
Indian Reorganization Act (IRA), 13, 14, 15, 111
Indian Self-Determination and Education Assistance Act, 14
Indian Tribal Land Acquisition Program, 326
indicator species, 177
Indigenous Guardians (Canada), 186–87
Indigenous Guardians Network, 181–89
Indigenous identity, 235
Indigenous knowledge, 231. *See also* Traditional Ecological Knowledge (TEK)
Indigenous Leadership Initiative, 186
Indigenous peoples, as percentage of world's population, 46
Indigenous Sentinels Network, 169–80
Indigenous Stewardship Model, 313–31; case examples, 321–29; obstacles to, 323–27
Inouye, Daniel, 26
insects, 23, 315, 316, 334
institutional review boards (IRBs), 230, 237, 241–42

intellectual property, 41–42, 230, 233–35, 242–43
International Association of Fish and Wildlife Agencies, 33
International Sonoran Desert Alliance, 349
International Whaling Commission, 101
InterTribal Buffalo Council (Intertribal Buffalo Cooperative), 17, 31, 110–13, 117, 261
intertribal wildlife management organizations, 40–41, 99–119
Inuit, 241, 321, 357, 358
Inupiaq people, 208, 357, 358
Inupiat people, 116
invasive and nuisance species, 19, 100, 108, 128, 271, 279, 280, 281, 282–86

Jackson, Andrew, 13
Jim, Nathan "Eight Ball," 23–25, 27–28
Joint Chiefs' Restoration Initiative project, 188
Joint Secretarial Order on American Indian Tribal Rights, Federal-Tribal Trust Responsibility, and the Endangered Species Act, 54, 58–60, 66
Jojola, Jovon, 35–38
Jones, Tamra, 35–38
Judd, Steve, 27–28

Kake, AL, Organized Village of, 185, 187
Kake Tribal Corporation, 185
Kane, Paul, 150–51
Karuk Tribe, 18, 144n21
Kasaan community, 185
Kawerak Inc., 216
Keex' Kwaan Community Forest Partnership, 185, 188
kelp, bull *(Nereocystis luetkeana),* 183
Kennedy, Dorothy, 146, 148
keystone species, 375
Kimmerer, Robin, 40
kingfishers, 337, 342
kinglets, 343
Kino, Eusebio, 335
Kittson, William, 150
Klamath Tribe, 88–90, 124
Klawock community, 185
Kluane First Nation, 324
Knife Chief Buffalo Nation Society, 325
Kootenai Tribe, 193. *See also* Confederated Salish and Kootenai Tribes (CSKT)
Kuwanwisiwma, Leigh, 234; *Footprints of Hopi History: Hopihiniwtiput Kukveni'at,* 238–39

Lacey Act, 140, 213
Laforet, Andrea, 154

Lakȟóta language, 328–29
Lakȟóta Tribe, 100, 313–19; agriculture-based land leasing, 324–26; Oglala, 317, 318, 324, 326; worldview, 313, 314–16, 327
Lakin, Ruth, 152
lampreys, 18, 100, 108, 114
land management, 7, 241, 309, 311, 312, 316–18; agriculture-based land leasing, 324–26; tribal consultation and, 53, 54, 55, 58, 59, 60–62, 64–67
languages, Indigenous: cultural significance, 275; loss or suppression, 23, 35, 36, 233; traditional knowledge base, 358; tribal elders' use, 242. *See also specific languages*
Lathrop, Rob, 27
law enforcement, 25, 29, 32, 104, 207, 213–14, 253, 254
Lee, David A., 34
Leech Lake, 27–28
Lemery, Lewis Butch, 154
Lemery, Peter, 151
Lenape (Delaware) Tribe, 13
Leopold, Aldo, 6–7
Lewis, Meriwether, 194–95
Littlemoon, Walter, 317, 318
Livingston, Chip, *Swamp Boy's Pet and Field Guide...* (memoir), 295–305
Loomis, Lorraine, 114
loons, 207, 211, 292, 294
Louie, Martin, Sr., 153
Lovell, John, 112
Lower Brule Sioux Tribe Wildlife Enterprises, 261, 262
Lumholtz, Carl, 336
Lummi Tribe, 103
Lynch, Karen, 36
lynx, 153, 289, 315

Makah Tribe, 103
Makoce Agriculture Development, 325
Maniilaq Association, 216
Marchand, Edward, 153
Marchand, Joseph, 153
Marchand, Mary, 152–53
Marine Mammal Protection Act, 101
marine mammals: monitoring, 169–80; subsistence hunting, 169–80, 209–12; tissue biosampling, 174–75. *See also* sea lions; seals; whales
Marshall, John, 8, 53–54
Martel, Wes, 27–28
martens, 149, 152, 153, 294
Mashantucket Pequot Tribal Nation, 22
masks, 212, 215
Mays, Herschel, 24, 27–28
McCain, John, 26

McCarran Amendment, 74–75, 81, 85, 86, 91
Menominee Tribe, 14, 34
Mescalero Apache Reservation, 5
mesquite (*Prosopis* spp.), 346
Mexican Commission for Protected Natural Areas, 332
Mexico, migratory bird treaty, 206, 213, 214
Miccosukee Tribe, Burmese python management program, 282–86
mice (*Mus musculus*), 300
migratory birds: Alaska Migratory Bird Co-Management Council role, 206, 214, 215–22; cultural significance, 210–12, 217–18, 220; egg harvests, 206–7, 210–11, 214, 217; flyways, 208, 215, 217, 218, 219, 222; impact of hunting/harvesting regulations on, 108, 206–8, 213–15; Western harvest management policies, 213–15. *See also* ducks; geese; seabirds; waterfowl
Migratory Bird Treaty Act (MBTA), 206, 213, 214, 218
Miller, Adrian "Dusty," 34
mining, 47, 63, 108, 109, 116, 125, 144n21, 154, 182, 239
Minnesota Department of Natural Resources, 45, 46
Minnesota Forestry Department, 278
minnows, Rio Grande silvery (*Hypognathus amarus*), 140–41
Mission Mountains Tribal Wilderness Area, 193
Mississippi Department of Wildlife, Fisheries, and Parks, 280, 281
Mohs, Gordon W., 148
Montana, Fish, Wildlife and Parks, 194, 195
Montana Waterfowl Foundation, 197
Montgomery-Anderson Brad, *Cherokee Reference Grammar*, 309–10
moose, 148, 157, 211, 278, 290, 291, 292, 293, 305
moths, catalpa sphinx (*Ceratomia catalpae*), 297
mountain goats and sheep, 146, 147, 149. *See also* bighorn sheep
Muckleshoot Tribe, 41, 103
Muir, John, 6–7
Multiple-Use Sustained-Yield Act, 80
Murkowski, Lisa, 39
murrelets, marbled, 141–42
muskrat food; sweet flag (*Acorus calamus*), 315
muskrats, 152, 305
mussels, 46, 100

Naseyowma, Gilbert, 239
National Congress of American Indians, 27, 241

National Fish Hatchery, 26, 27
national forests, 79–80, 157–58, 182
National Guard, 276
National Historic Preservation Act, 54, 232; section 106 consultation process, 55–57, 64–65, 68n23
National Marine Fisheries Service, 116
National Oceanic and Atmospheric Administration (NOAA), 172
national parks, bison herds, 111–12
National Park Service, 12, 112, 135, 237, 267, 272, 332–33, 334, 337–38, 349
National Register of Historic Places, 56, 346
National Snow and Ice Data Center, 175–76
National Wildlife Refuges: Ninepipe, 194; Pablo, 194, 197; Red Rock Lakes, 195
Native American Educational Services College, 34
Native American Fish and Wildlife Society (NAFWS), 17, 39–40, 41; annual conference (1999), 22; interviews with founders, 22–38; regional organization and conferences, 24, 29, 31, 33, 36; student interns, 35–38; youth practicums, 30–31, 32, 34–35
Native American Rights Fund, 60
Native Migratory Bird Working Group, 215
Native youth: attitudes toward tribal hunting rights, 30–31; boarding schools, 23, 208, 233, 259, 317; NAFWS programs, 25, 30–31, 32, 34–38; science and research initiatives, 238, 241; stewardship program participation, 185, 186; training in subsistence skills, 210, 212; in tribal wildlife management, 42–43; urban, 43
natural resource management, Western scientific model, 9, 41, 42, 252, 313, 323; comparison with Indigenous stewardship concept, 9–10, 313, 316, 319, 320
natural resources, tribal authority over, 5–6, 232
Natural Resources Conservation Service, 185, 262; Joint Chiefs' Restoration Initiative project, 188
nature, Indigenous views, 41, 44, 165, 275–76, 314–16, 317–18, 327; comparison with non-Indigenous views, 9–10, 248–52, 315
Nature Conservancy, 185, 186, 187
Navajo Nation, 15, 25, 101, 135, 142, 256–60; Bears Ears National Monument and, 60–63; Chaco Canyon oil and gas leases and, 63–66; climate change adaptation plan, 255; cooperative agreements with Arizona, 253; Fish and Wildlife Department, 253–55
Navajo Reservation, 36

New Mexico Department of Game and Fish, 5
Nez Perce Tribe, 142
Nishnawbe Tribe, 41
Nisqually Tribe, 105
Nixon, Richard, 55, 57, 62, 136
North American Model, of wildlife management, 7
Northern Arizona University, 6
Northrup, Vern, 5, 9–10, 275–78; *Iskode* (Fire), 277; *Wisdom of the Trees,* 276
North Slope Borough, 216
Northsun, Nila, "We Always Knew"; "Wetlands," 165
Northwest Indian Fisheries Commission (NWIFC), 16–17, 101–5, 106, 107, 110, 114, 116, 117, 118n1
nuisance law, 125–29

Obama, Barack, 54, 60, 61, 62, 67n7, 68n17
Ocmulgee National Park, GA, 267–70, 272–74
Oglala Sioux Parks and Recreation Authority, 318
Oglala Sioux Tribal Council, 324
oil and gas lease sales, 58, 63–66
oil and gas pipelines, 8, 45–46, 47, 73, 208–9
oil spills, 45–46, 47
Ojibwe language, 248, 276
Ojibwe Tribe, 17, 107, 109, 117; Great Lakes, 116; Leech Lake Band, 27–28; White Earth, 47
Old Person, Earl, 27
Oneida Tribe of Wisconsin, 34
O'odham language, 335, 336, 356–57
O'odham people, Hia c-ed and Tohono, 332–51; historic and cultural prehistoric resources, 332–33, 334; Quitobaquito Springs ('A'al Vaupia) management, 332–51; Tohono, 332
Organic Administration Act, 79–80
Organ Pipe National Monument, 332, 334, 337, 346; Cactus Monument, 347
orioles, 344
Orr, Veronica M., 154
Osiyo TV, 312
otters, 46, 300
Outah, Terrance, 239
owls, 276, 305; spotted, Mexican (*Strix occidentalis lucida*), 32–33, 110; spotted, Northern, 141–42

Pacific Flyway Council, 217
Paiute-Shoshone Tribe, 128
Paiute Tribe, Pyramid Lake, 142
panthers, 273, 302

partnerships, for conservation, 279–81
Passamaquoddy Tribe, 33, 289–94
Pearkes, Eileen Delehanty, *Geography of Memory: Recovering Stories of a Landscape's First People*, 149, 155, 160n31
pelicans, 341
Pend d'Oreille Tribe, 193. *See also* Confederated Salish and Kootenai Tribes (CSKT)
Penobscot Tribe, 27–28
pentastomes, reptilian invasive *(Raillietiella orientalis)*, 284
persistent organic pollutants (POPs), 175
pets, wildlife as, 295–305
phalarpopes, 340
pheasants, 318
Picard, Elsie M., 153, 154
Piel, Vince, Jr., 27–28
pike, northern, 19
Pinchot, Gifford, 6–7
Pine Ridge Reservation, 314–15, 323–25
Pinkham, Joe Jay, 27–28
Pittman-Robertson Federal Aid in Wildlife Restoration, 10n1, 15
place, Indigenous people's relationship to, 9–10, 278
Plains tribes, 110
plants: of desert oasis (Quitobaquito Springs), 335–36, 337, 346; ecosystem interconnectivity, 9; impact of climate change on, 40; Indigenous interconnectivity with, 315, 316; invasive, 100, 108, 282; pollen analysis, 40
plants, culturally important, 101, 107–8; of the Cherokee Nation, 273–74, 309–12; ethnobotany, 229, 319, 322, 326–27; of the Great Lakes Objiwe Tribe, 116; Indigenous names, 248, 328–29, 371; of the Sinixt Tribe, 148–49, 151; TEK about, 229, 356–57; of the White Earth Ojibwe, 47
plants, edible: of Alaska Native peoples, 183, 209–10; of Creek Nation, 274; depletion, 324; of the Lakȟóta Tribe, 318, 326–27; of the O'odham people, 332, 346; of the Sinixt Tribe, 150, 151, 152, 153–54
plants, medicinal, 148–49, 151, 273, 311, 313, 316, 324, 326–27
plovers, 141, 340
poaching, 32, 111, 254, 271
pollution, 45–47, 71n67, 125, 128, 170, 174–75
polybrominated diphenyl ethers, 175
pomegranates *(Punica* spp.), 346
population tracking, 156
Potawatomi Tribe, 41
Powell, Lewis F., Jr., 122
Poynter, Ken, 33
prairie dogs, black-tailed, 263

Pratt, Richard Henry, 13, 233
predator control programs, 253
Pribilof Islands, Indigenous Sentinels Network, 169–80
private lands, 14, 41, 124, 136, 139, 230–31, 232, 235, 271, 291, 333, 367
pronghorn *(Antilocapra americana)*, 146, 147, 154, 254, 323, 327, 328; endangered *(A. americana sonoriensis)*, 357
Pryce, Paula, 154–55
Public Land Initiative, 60–61
Public Law 116-260, 18
public nuisance law, 127–28
public trust doctrine, 124, 126, 128, 131n37
Pueblo land, 140–41
Pueblo of Laguna, 27–28
Pueblo of Zuni, 233
Pueblo tribes, 63–66, 244n1
Puget Sound Indigenous peoples, 103–4. *See also* Northwest Indian Fisheries Commission (NWIFC)
Pyramid Lake Fisheries, 27–28
pyrrhuloxia *(Cardinalis sinuatus)*, 337

quails, Gambel's *(Callipepla gambelii)*, 337
Quinault Nation, 141–42
Quinault Tribe, 102–3
Quintana, Mathis, 35–38
Quintasket, Charles, 153
Quitobaquito desert pupfish *(Cyprinodon eremus)*, 334
Quitobaquito Springs ('A'al Vaupia), avian species richness, 332–51; border wall construction impact, 335, 337, 338, 346, 347, 348, 349; under federal management, 332–33, 334, 337–38, 337–49, 346–49; during O'odham occupation, 332–33, 337–45, 348; water resources and, 333, 335–36, 337–38, 346, 347, 349

rabbits, 147, 148, 152, 271, 279; cottontail *(Syvilagus* spp.), 318; jackrabbits *(Lepus californicus)*, 318
raccoons *(Procyon lotor)*, 297, 298, 302, 303
rails (bird species), 340
Rankel, Gary, 31
ravens, 211
Ray, Verne F., 147, 148
recovery and restoration programs, 142–43; bighorn sheep, 157; bison, 17, 18, 110–13, 117, 193, 254, 261, 316, 318, 325; black-footed ferrets, 18, 142, 262–63; elk, 5, 41, 261; geese, 219; mule deer, 253; northern leopard frog, 198–203; pronghorns, 157; Recovery Champion Award, 142; sicklefin redhorse, 239–40; Steller sea lions, 170;

swift foxes, 262–63; trumpeter swans, 193–98; wolves, 18, 141–42, 373, 375–76
Redwood National Park, 367, 373, 375, 376
Reed, Mitzi, 279–81
Removal Era, 13
Rendon, Marcie, 248–52; "Cholla Cacti," 250, 251; "Desert," 251–52; "Mahnomen Waboose Moose Wawashkeshi," 249; "Prickly Pear Cacti," 251; "Scorpion," 250; "The Red River of the North," 249–50
research, with tribes, 229–47; collaborative approach, 263–64; incorporation of TEK, 39–44, 229; institutional review boards, 230; Lower Brule Sioux wildlife department, 263–64; observation-based, 40; racist, 233–34; tribal distrust of, 233–34. *See also* co-production of knowledge
research review boards, 237
Reservation Era, 13
reservations: agricultural purpose, 76–78, 80–82, 83, 85, 86, 87, 88, 89–90; created by executive order, 77, 78, 84; establishment, 13; intertribal, 100; jurisdictional diversity, 16; land allotments, 13, 82; land purchases within, 41; permanent homeland purpose, 81, 82–87, 88, 92; US Army Corps of Engineers land within, 262. *See also* tribal lands; *names of individual reservations*
reserved water rights, 45–46, 73–98, 141; determination of scope, 80–87; establishment, 75–78; extension to traditional subsistence activities, 81, 85, 87–94; federal violations, 73–74; of non-Indian reservations, 78–80, 81, 82–83, 85; off-reservation, 90–94; primary/secondary purposes test, 79–80, 82–84, 85–86, 89; species recovery efforts and, 141; state supreme court decisions, 81–83, 85–87; state water law and, 75, 76, 77–78, 80; traditional ways of life and, 87–94; treaties and, 77, 78, 81–82, 83, 85–87, 89–90, 91, 93; in trust status, 74, 83; US Supreme Court decisions, 73, 74, 75–80, 81, 82–84, 85, 86, 87, 88, 94, 125; Winters doctrine, 75–77, 78, 80, 81, 82, 84, 87, 88, 89, 125, 141
resource inventories, 185
Restatement (Second) of Torts, 127
Reyes, Lawney L., 155
Richardson, Dill, 26
Rights of Nature movement, 47
rivers: environmental damage, 9; protection, 45–48; subsistence species monitoring programs, 177
Roberts, John, 63, 71n64
Roosevelt, Franklin, 332
Roosevelt, Theodore, 6–7
Ross, et al, *Indigenous Peoples and the Collaborative Stewardship of Nature...*, 320
Rural Alaska Resources Association, 27–28

sacred places, 8, 39, 101, 111, 236, 332, 337. *See also* Bears Ears National Monument; Chaco Culture National Historical Park
St. George Traditional Council Kayumixtax Eco-Office, 177
salamanders, 18, 304
Salazar, Ken, 60
Salish (Flathead) Tribe, 193–205. *See also* Confederated Salish and Kootenai Tribes (CSKT)
salmon (*Salmo* spp.), 357–58; Chinook, 114, 149; coho, 114, 144n21, 149; cultural significance, 18, 110; migrations, 149–50, 151, 357–58; Pacific, 177; sockeye, 114, 149; steelhead, 114; as traditional food, 183
salmon fisheries, 322; co-management, 16–17, 58; destruction, 23–24; off-reservation, 90–92, 100, 140; reserved water rights and, 87, 88, 90–92; restoration and recovery projects, 18, 113–14; of Sinixt Tribe, 149–50, 151; threats to, 321–22; Tlingit stewardship management program, 321–22; as treaty rights, 58, 69nn28–29
Salmon People Partnership (SSP), 185
Sand Creek Massacre, 7
San Diego Natural History Museum, 337
Sandier, Don, 27–28
sandpipers, 340, 348
Schwalenberg, Dewey, 31, 34
Schwalenberg, Patty, 34
science, Indigenous, 114–16
seabirds, 171, 172–73, 175–76, 211, 221, 358
Seacoast Indigenous Guardians Network, 188
Sealaska, 183, 185, 187, 188
sea lions, Steller (*Eumetopias jubatus*), 169, 170, 172, 173–75
seals, northern fur (*Callorhinus ursinus*), 169, 171–75
Seattle, Chief, 25
self-determination, 14, 54, 55, 57, 67n5, 86, 120–21, 184, 241
Self-Determination Era, 14
Seminole Tribe, 140, 269
Sewell, John, 289–94
sheep, 146; bighorn (*Ovis canadensis*), 157, 193, 254, 323–24, 328, 357; Dall, 324; mountain, 146
shellfish, 186, 209–10, 220
Shell River, MN, 45–48
Shelly, Ben, 60

shorebirds, 211, 220, 338–45, 348
Shoshone Tribe, Eastern, 81–82, 100
Sicangu Lakota Oyate Tribe, 8
sicklefin redhorse (*Moxostoma* sp.), 239–40
Sierra Club, 7
Sinixt Tribe (Lakes Tribe), 145–63; cross-border wildlife management activities, 145–46, 156–59; gathering activities, 146, 148–49, 150, 151, 152, 153–54; hunting methods and traditions, 147–52, 153; impact of white settlement on, 150–51, 153, 155, 157
Sioux Tribe, 112; Cheyenne River, 16, 29, 142, 238, 262; Dakota, 100; Lower Brule, 262–64; Nakota, 100; Standing Rock, 73; Yankton, 140. *See also* Lakȟóta Tribe
Sitka community, 185
Skates, Ron, 29–31
Skokomish Tribe, 103
skunks, striped (*Mephitis mephitis*), 302–3
Smith, Feather, 309, 312
Smith, John, 24, 27–28
Smithsonian Institution, National Museum of the American Indian, 234
snakes: Burmese pythons (*Python bivattatus*), 282–86; coachwhip (*Masticophis flagellum*), 299; eastern hognose (*Heterodon platirhinos*), 298, 299; eastern indigo (*Drymarchon couperi*), 300; eastern ribbon (*Thamnophis suarita*), 302, 303; garter (*Thamnophis sirtalis*), 302, 304, 305; green, rough (*Opheodrys aestivus*), 298, 304; green, smooth (*Opheodrys vernalis*), 296–97; kingsnake, ringed (*Lampropeltis getula*), 300, 304; as pets, 296–97, 298–300, 301, 302, 304–5; rat, gray (*Pantherophis spiloides*), 300; rat, red; corn snake (*Pantherophis guttatus*), 299–300, 302; rattlesnake, diamondback (*Crotalus adamanteus*), 300; rattlesnake, pygmy (*Sistrusus miliarius*), 300; ring-necked (*Diadophis punctatus*), 305; southern black racer; blacksnake (*Constrictor priapus*), 298–99; water moccasin; cottonmouth; swamp moccasin (*Agkistrodon piscivorus*), 301, 304; water snakes, banded; common (*Nerodia sipedon; N. fasciata*), 304
South Dakota Game and Fish, 263
Southeast Alaska Sustainability Strategy, 188
Southeast Alaska Tribal Ocean Research (SEATOR), 186, 187
Southeast Conference, 188
Southwestern Indian Polytechnic Institute, 36
Southwest Tribal Fisheries Commission (SWTFC), 17
sovereignty, tribal, 6, 8, 13, 135; Alaska Natives, 217; constitutional basis, 8, 136, 230; co-production of knowledge and, 230, 232, 235, 241; federal violations, 136; in fish and wildlife management, 14, 20, 291; government-to-government relationship, 53–54, 55, 57, 62, 66, 68n14, 68n17, 137–39; Indigenous Guardians Network, 181–89; intertribal wildlife management organizations and, 116–17; in off-reservation usufructuary activities, 121–22, 128, 129; overlap with federal and state sovereignty, 121, 122, 291; relation to federal trust responsibility, 54, 55, 135–37; termination, 13–14
Spirit Quests, 154
spirituality, 22–23, 42–44, 137, 138, 317, 356
Spokane Tribe, 41
Spotted Elk, Clara, 34
spruce budworms, 292, 293
Spruce Root, 187
Squaxin Island Tribe, 103
squirrels, 18, 271, 279, 297, 303
Stago, Phil, 27–28
Standing Bear, Luther, 313
state lands, adjoining reservations/tribal lands, 232
states: criminal jurisdiction, 14; failure to regulate corporations, 45–46, 47; trust doctrine, 128. *See also* supreme courts (states) decisions
state-tribal relations, 19; in Alaska, 184, 321–22; conflicts, 29; cultural appropriation issue, 40; gathering rights, 123, 124; hunting regulations, 271–72; lack of authority over tribes, 122–23; in Maine, 291; off-reservation usufructuary activities, 122–24, 128–29; tribal jurisdiction issue, 27
State Wildlife Grant program, 15
Stay, Alan, 27
Stevens, Isaac Ingalls, 104
Stevenson, Matilda Coxe, 233
stewardship, Indigenous-led, 5, 181, 221, 361; in bison management, 112; criticisms, 7; of desert oases, 332–51; of endangered or threatened species, 139; local community-based economy and, 314–15; Lower Brule Sioux Tribal Wildlife Department model, 261–64; resource management approach *versus*, 316–18; shared, 181–89; in Southeast Alaska, 181–89; as spiritual responsibility, 22–23. *See also* Indigenous Stewardship Model
Stewart, Alice M., 153
stilts (bird species), 340

storks, 341
subsistence activities, 19, 368; Alaska Native peoples, 209–10; cultural significance, 210, 212; definition, 183; Lakȟóta Tribe, 313, 314–16, 318; maritime and saltwater, 169, 290; Passamaquoddy Tribe, 290; reciprocity in, 315–16; seasonal, 148–50. *See also* fishing and fisheries; gathering activities; hunting
subsistence-cash mixed economy, 209–10
subsistence resource monitoring: Bering-Watch, 170–71; Indigenous Sentinels Network, 169–80
subsistence rights, 81, 85, 87–94
Sun'aq Tribe of Kodiak, 216
Supreme Court (Canada), *Regina v. Richard Lee Desautel*, 145, 154, 158–59
supreme courts (states) decisions: *Aquavella*, 91, 93; *Bighorn I*, 81–82, 94; *Gila V*, 82–84, 86; *Hopi Tribe v. Arizona Snowbowl Resort Ltd. Partnership*, 127; *In re CSRBA*, 94; reserved water rights, 81–83, 85–87, 91–92, 94. *See also* US Supreme Court decisions
Sustainable Southeast Partnership (SSP), 182, 185, 186, 188
sustained yield model, 264
Swan, Lucy, 152
swans, 211; trumpeter *(Cygnus buccinator)*, 193, 194–98

Tanana Chiefs Conference, 177, 216
Tanka Fund, 325
technology, in natural resource management, 36, 37–38
Teit, James, 146, 155
TEK. *See* Traditional Ecological Knowledge (TEK)
Termination Act, 14
Termination Period, 13–14, 55, 68n14, 143n5
terns, 341
Thayer, Gordon, 107
Thompson, David, 194–95
Thorstenson, Julie, 35, 36, 38, 238
Thunder Valley Community Development Corporation, 324–25
Tlingit people, 181–89, 208, 244; Eagle/Wolf (Ch'aak'/Gooch) moieties, 181–82; fishery hook-offs, 321–22; forestry project, 321; Raven (Yéil), 181–82
Tohono O'odham Nation Legislative Council, 338
Tom, Gloria, 253–55
Traditional Ecological Knowledge (TEK), 18, 20, 36, 156, 229, 307, 313, 368; in climate change research, 39, 40; in conservation biology, 353–65; in co-production of knowledge, 240–41, 243; definition, 231, 336, 353, 354–55; in elk management and conservation, 336–78; foundations, 355; incorporated into scientific research, 39–44, 54, 235, 240–41, 243; Indigenous Guardians Network, 181–89; Indigenous science and, 114, 115–16; intergenerational transfer, 10, 313, 321–22, 326–27, 355–56; racist attitudes toward, 7, 11N3, 233–34; relationship to Western science, 6, 314, 323–24, 353–54, 355–60; researchers' unauthorized appropriation, 235; in subsistence resource monitoring, 170. *See also* stewardship, Indigenous-led
trapping, 16, 89, 153, 155, 277–78, 280, 281
treaties, 13; 1855 treaty, 90, 91, 124, 193; federal trust responsibility and, 53–54; Hellgate (1855), 193; for migratory bird protection, 206, 213–15; of Oregon, 146, 150; Treaty of No Point (1855), 124; Wolf River (1854), 14
Treaty Authority, 17
treaty rights, for fishing, hunting, and gathering: defense of, 28–29. *See also* fishing rights; gathering rights; hunting rights
treaty rights, for fishing, hunting, and gathering, off-reservation (usufructuary), 14, 17, 104, 120–32; abrogation or limitations, 122, 124, 136, 140; across jurisdictional boundaries, 100–101; Environmental Protection Act and, 140; history, 120–21; intertribal, 100–101; legal basis and interpretations, 14, 121–29; nuisance claims against, 125–29; principles, 121–24; protection, 124–29. *See also* reserved water rights
tree fungus, "chicken of the woods" *(Laetiporus conifericola)*, 183
trees, 23; balsam fir *(Abies balsamea)*, 108; birch *(Betula* spp.), 108; interconnectivity, 9; oak, 295–96; Objiwe word for, 276; as path markers, 273–74; planting programs, 262; sugar maple *(Acer saccharum)*, 108
tribal commissions, 16–17, 18
Tribal Conservation Districts (TCDs), 185–86, 188
tribal councils, 35, 323–24
tribal fish and wildlife management: across international border, 156–58; collaboration in, 19, 366; conflict with state and federal agencies, 323–24; employees, 28–29; future challenges, 39–44; future directions in, 20; independence from state oversight, 255; integrated resource management plans, 156; intertribal organizations, 101–13; landmark court cases, 14;

tribal fish and wildlife management (*cont.*) Lower Brule Sioux, 261–64; Navajo Nation, 253–55; opposition, 99; overview, 5–11; regionally-oriented organizations, 16–17; regulations enforcement, 104, 117; size and diversity, 16; Southeastern US, 279–80; staffing, 42–43; tribal councils' role, 323–24; tribal lands diversity and, 16; tribal organizations for, 16–17

tribal governments: legal status, 54; legitimacy challenges, 117

tribal identity, 41, 42; false claims of, 235

tribal lands: buyback programs, 41; ceded, 9, 106, 124, 142, 193; dispossession, 99, 136, 143n5; diversity and status, 16; economic and resource development, 139, 142; on international borders, 16; land area, 135, 229; research on, 229–47; trust status, 14, 16, 41, 271

tribal law, intersection with federal and state law, 13

Tribal Wildlife Grant (TWG) program, 15

tribes: diversity, 6; federally-recognized, 100, 135, 143n5, 208

trout: Apache, 142; bull (*Salvelinus confluentus*), 92, 141–42, 145; cutthroat, Lahontan (*Oncorhynchus clarkii henshawi*), 88, 142; cutthroat, west slope (*Oncorhynchus clarkii*), 92; eastern brook, 294; steelhead, 28

Truman, Harry, 79

Trump, Donald, 60, 63, 101, 267

trust obligation. *See* federal trust responsibility

Tsimshian people, 208

Tulalip Tribes, 140

turkeys, 254, 271, 279

turtles and tortoises, 271, 272, 297, 300–301, 304, 305, 334, 356–57

Unangax̂-Aleut people, 169–70, 208

UNESCO, biosphere reserve, 332, 346

US Army Corps of Engineers, 23, 114, 128, 262, 264, 347

US Bureau of Reclamation, 114

US Code of Federal Regulations, 237

US Constitution, 53, 230

US Customs and Border Patrol, 332, 347

US Department of Agriculture, 63, 255, 294, 310–11, 317; Environmental Quality Incentives Program, 317; Farm Service Agency, 326; Mississippi Wildlife Services, 280, 281; Southeast Alaska Sustainability Strategy, 188

US Department of the Interior, 64, 140, 218; secretarial orders, 58–60, 62, 66, 67n5, 137, 138, 141–42; Secretary, 20, 60, 63, 71n62, 73–74

US Fish and Wildlife Service, 28, 29, 158, 194, 215, 294; Alaska Migratory Bird Co-Management Council and, 215–16; black-footed ferret recovery programs, 263; Bureau of Sport Fisheries and Wildlife (BSFW), 213–14; cooperative agreements with the Navajo Nation, 253; Endangered Species Act and, 135, 140–41; Native American Policies, 137–39, 241; Office of Native American Liaison, 39–44; relation with NAFWS, 29, 31–32; seabird monitoring, 172; Tribal Wildlife Grant program, 15, 289; trumpeter swan restoration program, 195

US Forest Service, 18–19, 30, 181, 182, 185, 187–89, 218

US Geological Survey, 263; Climate Adaptation Science Centers, 18–19

US Navy, 318

US Supreme Court decisions: *Antoine v. Washington*, 158; *Arizona v. California*, 77–78, 80, 81, 82, 87, 94; *Cappaert v. US*, 78–79, 80, 82, 83; *Cherokee Nation v. Georgia*, 53–54; *Confederated Tribes of the Colville Indian Reservation v. State of Washington*, 14; *Fletcher v. Peck*, 121–22; *Herrera v. Wyoming*, 14, 100, 124, 144n10; *Inyo County*, 128; *McGirt v. Oklahoma*, 122–23; *Menominee Tribe of Indians v. United States*, 14; *Mille Lacs v. Minnesota*, 122; *Montana v. Blackfeet Tribe*, 122; *Native Village of Kivalina v Exxon-Mobil Corp*, 128; *Nevada v. United States*, 128; *New Mexico v. Mescalero Apache Tribe*, 14, 31; *In re Exxon Valdez*, 128; *US v. New Mexico*, 78, 79–80, 81, 82–84, 85, 86; *US v. Washington*, 114; *US v. Winans*, 99–100, 136–37; *Washington v. Wash.State Commercial Passenger Fishing Vessel Assoc.*, 99–100, 125; *White Mountain Apache v. the State of Arizona Department of Game and Fish*, 14; *Winters v. US*, 75–77, 78, 80, 81, 82, 84, 87, 88, 89, 125, 141

University of Arizona, 347

urban Indigenous populations, 36, 43, 208, 219

Utah Diné Bikéyah, 60

Utah State University, 318

Ute Tribe, 101; Mountain Ute, 61; Ouray, 61; Uintah, 61

Voight Intertribal Task Force, 107, 108

Wabanaki Exposure Study, 115, 116

walrus, Pacific (*Odobenus rosmarus divergens*), 357

walrus teeth dice, 212
warblers, 344–45
Warm Springs Tribe of Oregon, 27–28
waterfowl: of desert oasis (Quitobaquito Springs), 335, 338–45, 348; harvest regulation enforcement, 213–14; harvests, 217; hunting seasons, 262; monitoring programs, 171, 172–73, 175–76; protection areas, 194; subsistence hunting, 171, 172–73, 175–76. *See also* ducks; geese; *names of individual species*
waterfowl stamps, 207
Water Resources Development Act, 262
watersheds, 80, 103, 105, 114, 116, 185–86, 188, 292
wetlands, 18, 85, 165, 195, 279, 349; as bird habitat, 195, 196, 198, 334, 338–45, 348, 349; as frog habitat, 201, 202–3; lead ammunition contamination, 198
whales, 23; beluga (*Delphinapterus leucas*), 357–58, 359; bowhead (*Balaena mysticetus*), 113, 116, 358; killer, 170
whaling, 101, 113, 116, 358
whitefish, 177
whole systems management approach, 156
widgeons, 172, 337
wildlife: definition, 248; as pets, 295–305; spiritual significance, 275–76
wildlife biologists, tribal, 29–31, 34, 253–55, 261–64, 289–94
Wildlife Conservation International, 200
wild rice (*Zizania palustris*), 47, 108
Wilkinson, Charles, 59–60

Wilson, John, 267, 269, 270, 273–74
Wind Caves National Park, 327
Wind River Reservation, 27–28, 81–82
Winters, Harry Winters, Jr., 335, 336
Wisconsin Natural Resources Board, 142
wolfberry (*Lycium* spp.), 346
wolves, 46, 142, 153, 190–93, 277–78, 373, 375; Mexican, 18, 142; recovery and restoration projects, 18, 142, 373, 375–76; Rocky Mountain (*Canis lupus*), 142
woodpeckers, 342
Work, John, 148, 149, 150
Wyoming Wetlands Society, 196–97

Yakama Nation, 27–28, 32, 69n29, 90–92, 102–3
Yakutat Tribe, 185
Yallup, Wilfred, 27–28
Yellowstone Interagency Bison Management Plan, 112
Yellowstone National Park, 111–12, 195, 198, 373, 375–76
Yuchi Tribe, 269
Yukon-Kuskokwim Delta Goose Management Plan, 219
Yup'ik language, 358
Yup'ik Tribe, 116, 212, 215, 357; Central, 208; St. Lawrence Island, 208
Yurok language, 368, 370–71, 376, 379–83
Yurok Tribe, TEK-based elk management and conservation, 366–83

Zuni Tribe, 25, 61, 101